Nelson MindTap + **You** = Learning amplified

"I love that everything is interconnected, relevant and that there is a clear learning sequence. I have the tools to create a learning experience that meets the needs of all my students and can easily see how they're progressing."

— **Sarah,** Secondary School Teacher

Nelson Science Year 10 Queensland Student Book
1st Edition
Brett Steeples
Anne Disney
Charlotte Donovan
Florence Coghlan
Stephen Zander
Sarah Langley
Ariell Rose
Celia McNeilly
ISBN 9780170463041

Series publisher: Catherine Healy
Project editor: Alan Stewart
Editor: Bradley Smith
Series text design: Leigh Ashforth
Series cover design: Leigh Ashforth
Series designer: Linda Davidson
Cover image: Shutterstock.com/Mihai_Andtritoiu
Permissions researcher: Debbie Gallagher, Brendan Gallagher
Production controller: Bradley Smith
Typeset by: MPS Limited

Any URLs contained in this publication were checked for currency during the production process. Note, however, that the publisher cannot vouch for the ongoing currency of URLs.

For product information and technology assistance,
in Australia call **1300 790 853**;
in New Zealand call **0800 449 725**

For permission to use material from this text or product, please email
aust.permissions@cengage.com

National Library of Australia Cataloguing-in-Publication Data
A catalogue record for this book is available from the National Library of Australia.

Cengage Learning Australia
Level 5, 80 Dorcas Street
Southbank VIC 3006 Australia

Cengage Learning New Zealand
Unit 4B Rosedale Office Park
331 Rosedale Road, Albany, North Shore 0632, NZ

For learning solutions, visit **cengage.com.au**

Printed in China by 1010 Printing International Limited
1 2 3 4 5 6 7 27 26 25 24 23

nelson science. 10

Brett Steeples
Anne Disney
Charlotte Donovan
Florence Coghlan
Stephen Zander
Sarah Langley
Ariell Rose
Celia McNeilly

LEARNING DISCOVERY
VITAMAN C

This cover image shows Vitamin C crystals, which have been magnified 10 times and illuminated with polarised light.

Vitamin C, also known as ascorbic acid, is vital to human health. Unlike many animals, our bodies can't make vitamin C, which is why we have to obtain it through our diet. It is naturally found in many foods, such as citrus fruits and some vegetables. The Kakadu plum, a fruit native to Australia, has 100 times more vitamin C than an orange. Vitamin C is an organic compound and has the formula $C_6H_8O_6$.

QLD
Australian Curriculum

FIRST NATIONS AUSTRALIANS GLOSSARY

Country/Place

Spaces mapped out that individuals or groups of First Nations Peoples of Australia occupy and regard as their own and that have varying degrees of spirituality. These spaces include lands, waters and sky.

Cultural narrative

A broad term that encompasses any cultural expression that includes (but is not limited to) knowledge and community values that are central to the identity of a particular group of First Nations Peoples.

Cultural narratives can hold information about almost anything, such as the origins of life, or can teach people about acceptable behaviour and rules, such as caring for Country. They can take the form of songs, stories, visual arts or performances. 'Cultural narrative' is a more accurate and respectful term than 'myth', 'story' or 'fable'; terms that often diminish their importance.

First Nations Australians

'First' refers to the many nations/cultures who were in Australia before British colonisation. This a collective term that refers to all Aboriginal Peoples and Torres Strait Islander Peoples. The term 'Indigenous Australians' is also used to refer to First Nations Australians.

Nation

A self-governed community of people based on a common language, culture and territory.

Peoples and Nations

We use the plural for these terms because First Nations Australians do not belong to one nation/culture. There are many distinct Peoples and Nations. Also, some Nations consist of distinct clans or groups, so are referred to as Peoples.

ACKNOWLEDGEMENT OF COUNTRY

Nelson acknowledges the Traditional Owners and Custodians of the lands of all First Nations Peoples of Australia. We pay respect to their Elders past and present.

We recognise the continuing connection of First Nations Peoples to the land, air and waters, and thank them for protecting these lands, waters and ecosystems since time immemorial.

Warning – First Nations Australians are advised that this book and associated learning materials may contain images, videos or voices of deceased persons.

Contents

Authors and contributors

Lead author

Brett Steeples

Contributing authors

Anne Disney

Charlotte Donovan

Florence Coghlan

Stephen Zander

Celia McNeilly

Sarah Langley

Ariell Rose

Consultants

Joe Sambono
First Nations curriculum consultant

Dr Silvia Rudmann
Digital learning consultant

Judy Douglas
Literacy consultant

Science communication consultants

Additional science investigations

Reviewers

Xenia Pappas, Scott Adamson, Megan Mackay, Faye Paioff, Pete Byrne, Karen Cantwell, Noelle Finnerty
Teacher reviewers

Aunty Gail Barrow, Nicole Brown, Christopher Evers, Associate Professor Melitta Hogarth, Carly Jia, Jesse King, Dr Jessa Rogers, Theresa Sainty
Reviewers of First Nations Science Contexts pages

nelson science. Learning Ecosystem

Nelson Science 10 caters to all learners

Nelson Cengage has developed a **Science Learning Progression Framework**, which is the foundation for Nelson's Science 7–10 series. An editable version is available on Nelson MindTap.

LEARN
Reinforce
Nelson MindTap provides a wealth of differentiated activities and resources to meet the needs of all students.
LEARN
Evaluate prior knowledge
Students complete a quiz to test their prior knowledge.
nelson science.
Nelson MindTap
Engage
Each chapter showcases fascinating, real-world science in action, while our hands-on activities, short videos and fun interactives keep students engaged.
LEARN
Practise
Our differentiated, scaffolded activities and investigations allow all learners to build essential skills and knowledge.
LEARN
Assess
Allocate and grade assessments using our differentiated end-of-topic tests and summative portfolio assessment tasks in Nelson MindTap.
LEARN

Nelson MindTap

A flexible and easy-to-use online learning space that provides students with engaging, tailored learning experiences.

- Includes an eText with integrated activities and online assessments.
- Margin links in the student book signpost multimedia student resources found on Nelson MindTap.

Video activity
Cells

For students:

- Short, engaging videos with fun quizzes that bring science to life.
- Interactive activities, simulations and animations that help you develop your science skills and knowledge.
- Content, feedback and support that you can access as you need it, which allows you to take control of your own learning.

For teachers:

- 100% modular, flexible courses let you adapt the content to your students' needs.
- Differentiated activities and assessments can be assigned directly to the student, or the whole class.
- You can monitor progress using assessment tools like Gradebook and Reports.
- Integrate content and assessments directly within your school's LMS.

How to use this book

Nelson Science 10 has three digital-only chapters available on Nelson MindTap. These chapters are designed to extend students and help prepare them for Year 11 Biology, Chemistry and Physics.

Big science, real context: The opening page begins the chapter by placing the science topic into a real-life context that is both interesting and relevant to students' lives.

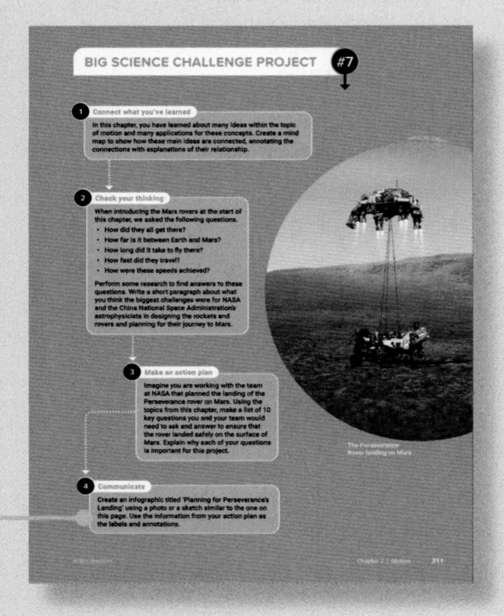

Think, do, communicate: You are encouraged to reflect on and apply your learning to a set of activities, which allows you to make meaningful connections with the content and skills you have just learned.

Learning modules: Content is chunked into key concepts for effective teaching and learning.

Learning objectives: Clear, concise objectives give you oversight of what you are learning and set you up for success.

Key words: These are defined the first time they appear.

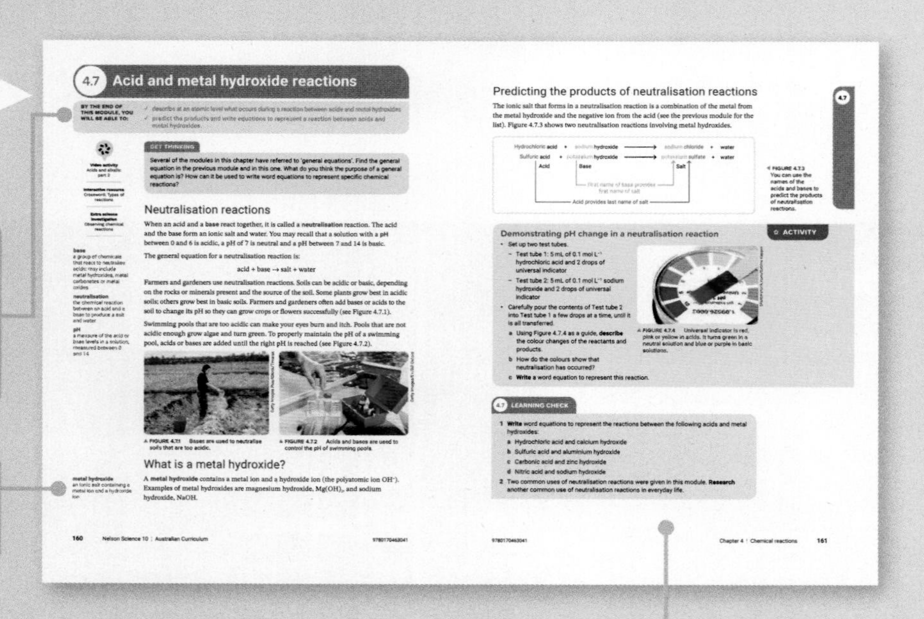

Learning check: These are engaging activities to check your understanding. Activities are presented in order of increasing complexity to help you confidently achieve the module's learning objectives. **Bolded** cognitive verbs help you clearly identify what is required of you. Activities are presented in order of increasing complexity.

Science as a Human Endeavour: Elaborations are explicitly addressed with interesting, contemporary content and activities.

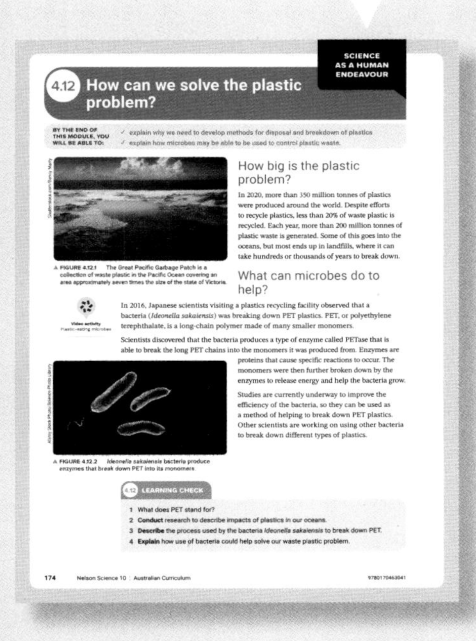

First Nations science contexts: This content was developed in consultation with a First Nations Australian curriculum specialist. It showcases the key Aboriginal and Torres Strait Islander History and Cultural Elaborations, with authentic, engaging and culturally appropriate science content.

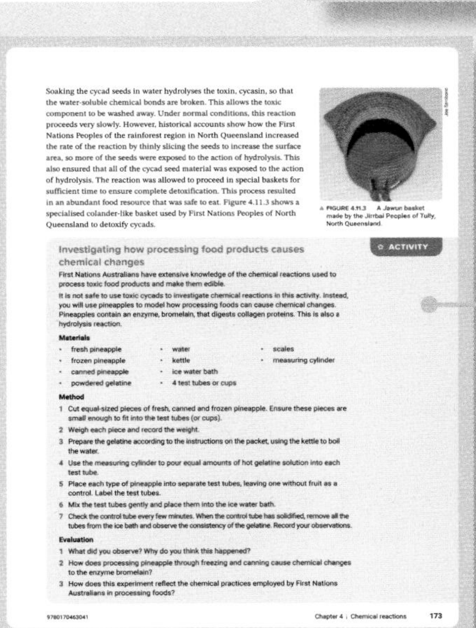

Activities: Activities are open-ended and often hands on, helping you understand the connections between First Nations cultures and histories and science.

Science skills in focus: Each chapter focuses on a specific science investigation skill. This is explained and modelled with our Science skills in a minute animation, before you put it into practice in a science investigation. The science skill is reinforced with our Science skills in practice digital activities.

Investigations: Practise and reinforce good scientific method through fit-for-purpose, hands-on science investigations.

1 Genetics

1.1 The structure of DNA (p. 4)
DNA is the genetic make-up of organisms and has the same basic structure in all living things.

1.2 Genetic material (p. 7)
Genetic material controls all cell activities and characteristics.

1.3 Mitosis (p. 10)
Mitosis is the process of cell division for growth and repair.

1.4 Meiosis (p. 14)
Meiosis is the process of cell division for reproduction.

1.5 Mendelian inheritance (p. 18)
Mendelian inheritance explains the patterns of inheritance seen in sexually reproducing organisms.

1.6 Sex determination (p. 22)
Particular chromosomes determine the sex of organisms.

1.7 Punnett squares (p. 26)
Punnett squares are used to predict the traits of offspring.

1.8 Pedigree charts (p. 30)
Drawing and interpreting pedigree charts can determine inheritance patterns in families.

1.9 Cancer and genetic disorders (p. 34)
Cancer is uncontrolled cell division; genetic disorders can be attributed to errors in the DNA.

1.10 FIRST NATIONS SCIENCE CONTEXTS: First Nations Australians' kinship structures (p. 38)
First Nations Australians have long observed patterns of inheritance and have developed and use complex societal structures to prevent relationships between closely related people.

1.11 SCIENCE AS A HUMAN ENDEAVOUR: The race to discover DNA (p. 40)
James Watson and Francis Crick discovered the structure of DNA in 1953.

1.12 SCIENCE INVESTIGATIONS: Modelling DNA (p. 42)
1 Building a DNA molecule
2 Extracting DNA
3 Mitosis and meiosis under the microscope
4 Phenylthiocarbamide analysis

BIG SCIENCE CHALLENGE 

Shutterstock.com/Lopolo

▲ FIGURE 1.0.1 Identical twins

Have you ever wondered why you look more like one parent than the other?

Perhaps you have already heard of genetic screening and genetic counselling that prospective parents may go to before they start having children.

You may already know the terms 'cloning' and 'genetically modified'.

All these questions and issues link back to the field of genetics: a branch of biological science concerned with inheritance and variety in living things.

▶ **What do you think are the advantages of understanding more about your genes?**

▶ **What is your opinion on changing genes to produce a particular characteristic? Or to make a product that can help treat a disease?**

▶ **If you could 'cure' cancer through gene editing, do you think this would be a medical breakthrough?**

#1 SCIENCE CHALLENGE ACCEPTED!

At the end of this chapter, you can complete the Big Science Challenge Project #1. You can use the information you learn in this chapter to complete the project.

Assessments
- Prior knowledge quiz
- Chapter review questions
- End-of-chapter test
- Portfolio assessment task: Research project

Videos
- Science skills in a minute: Modelling DNA **(1.12)**
- Video activities: What is DNA? **(1.1)**; Inheritance **(1.4)**; Mendel and inheritance **(1.5)**; Sex determination **(1.6)**; Cancer and genetics **(1.9)**; Rosalind Franklin **(1.11)**

Science skills resources
- Science skills in practice: Modelling DNA **(1.12)**
- Extra science investigations: Karyograms **(1.2)**; Modelling mitosis **(1.3)**; Inheritance and chance **(1.5)**

Interactive resources
- Label: DNA molecule **(1.1)**; Chromosomes **(1.2)**; Phases of mitosis **(1.3)**; Meiosis I and II **(1.4)**; Pedigree charts **(1.8)**
- Drag and drop: Punnett squares **(1.7)**

Nelson MindTap

To access these resources and many more, visit: **cengage.com.au/nelsonmindtap**

1.1 The structure of DNA

BY THE END OF THIS MODULE, YOU WILL BE ABLE TO:

- ✓ describe the structure of DNA
- ✓ label the key components of DNA (sugar, phosphate, nitrogenous base, nucleotide and hydrogen bond)
- ✓ define 'nucleotide', 'nitrogenous base' and 'hydrogen bond'
- ✓ list the nitrogenous bases in DNA and their complementary base pairs.

Video activity
What is DNA?

Interactive resource
Label: DNA molecule

GET THINKING

Did you know you have two metres of DNA in each of your cells? Could you calculate the total length of DNA found in your entire body?

What is DNA?

deoxyribonucleic acid
the molecule that determines the characteristics of most living things

molecule
a group of atoms bonded together

double helix
the shape of DNA, similar to a twisted ladder

DNA stands for **deoxyribonucleic acid.** It is a complex **molecule** made up of two chains coiled together, like a twisted ladder. It was successfully described in 1953 by two scientists, James Watson and Francis Crick, who determined its shape using an X-ray diffraction image of DNA taken by Raymond Gosling, working under the supervision of Rosalind Franklin. The image, named *Photo 51*, showed a clear cross in the centre (see Figure 1.1.1). This indicated to Watson and Crick that a DNA molecule has a **double helix** shape (see Figure 1.1.2).

Despite the great diversity shown in all living things, the basic structure of DNA is the same in the cells of living things and can be shown as a simple model.

Getty Images/Donaldson Collection

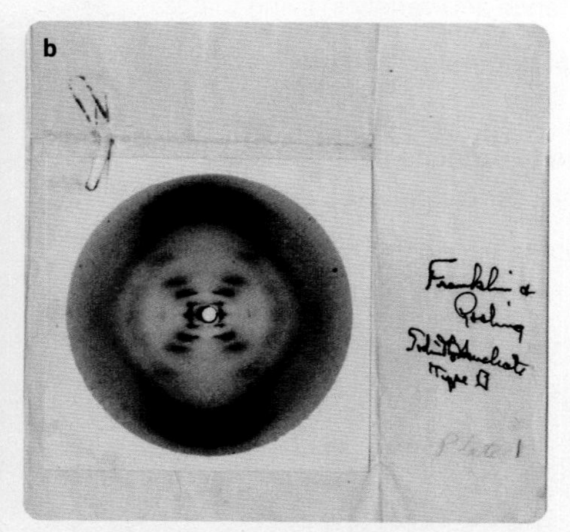

▲ **FIGURE 1.1.1** (a) Rosalind Franklin; (b) *Photo 51* displaying the centre cross, indicating a helix shape

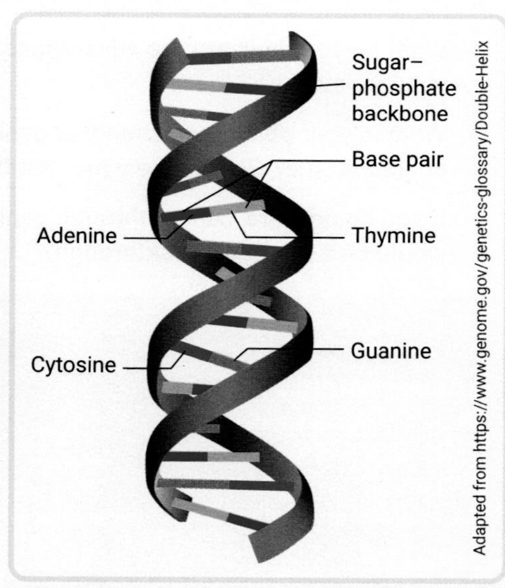

Adapted from https://www.genome.gov/genetics-glossary/Double-Helix

▲ **FIGURE 1.1.2** The double helix shape of DNA

The structure of DNA

DNA's double helix consists of two strands that twist around each other. Each strand is made from alternating sugars (**deoxyribose sugars**) and **phosphate groups**, with a **nitrogenous base** attached to each sugar molecule.

A **nucleotide** is a repeating unit in the DNA molecule. We often refer to nucleotides as the building blocks of DNA. Each nucleotide consists of a deoxyribose sugar, a phosphate group and a nitrogenous base, as shown in Figure 1.1.3.

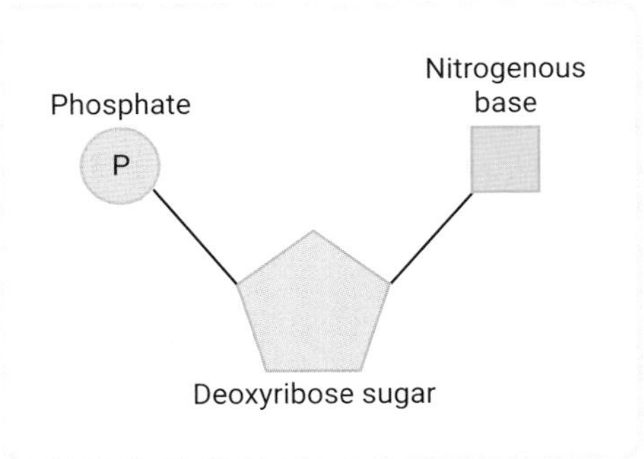

▲ **FIGURE 1.1.3** The three components of nucleotides

You can think of a DNA molecule as a ladder; the alternating sugar–phosphate molecules form the side rails, and the nitrogenous bases form the rungs. Each nucleotide is attached to the one before and the one after it, creating a long chain that is the DNA molecule. This is shown in Figure 1.1.4.

deoxyribose sugar
one of the components of a nucleotide in DNA

phosphate group
one of the components of a nucleotide in DNA

nitrogenous base
a base that contains nitrogen: adenine (A); thymine (T); cytosine (C) and guanine (G)

nucleotide
the building block of DNA

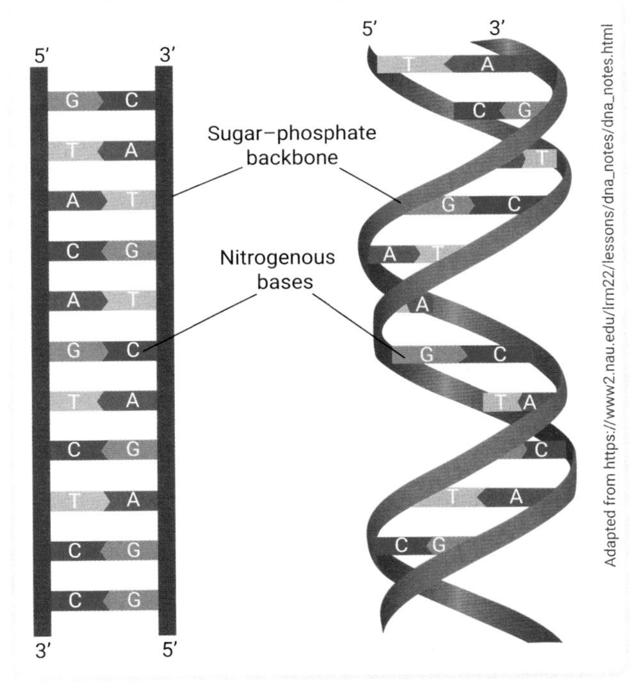

▲ **FIGURE 1.1.4** DNA as a ladder and twisted into the double helix shape

The four bases of DNA

There are four nitrogenous bases, adenine (A), thymine (T), guanine (G) and cytosine (C). The sequence of these bases forms the **genetic code**, as shown in Figure 1.1.4.

genetic code
the sequence of nitrogen-rich bases in an organism's DNA

In the late 1940s, the scientist Erwin Chargaff observed that DNA contained equal amounts of adenine and thymine and equal amounts of cytosine and guanine.

Chargaff's rule

Chargaff's observation provided the clue behind how the nitrogenous base pairs form and led to 'Chargaff's rule'. This states that in DNA, there is always an equal quantity of the bases A and T and the bases G and C.

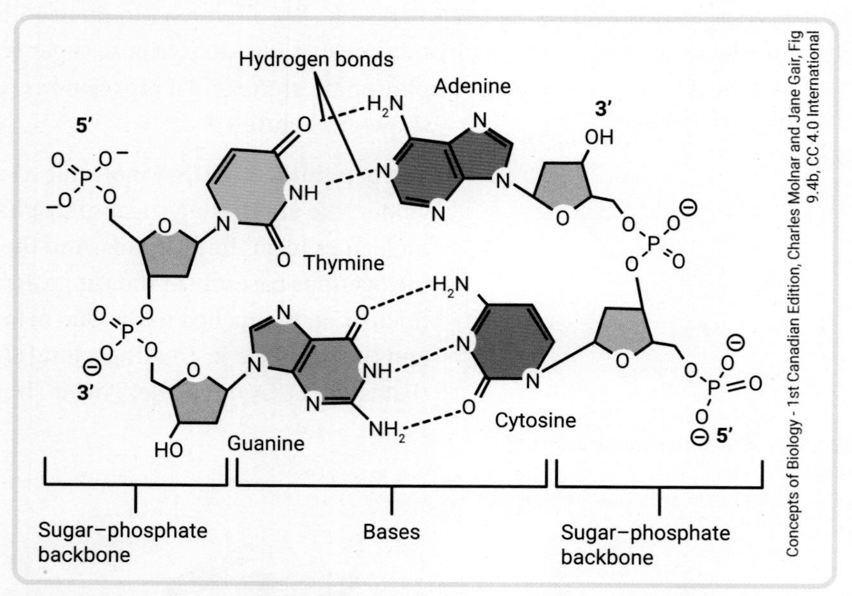

▲ **FIGURE 1.1.5** DNA is made up of complementary bases: A and T, and G and C.

hydrogen bond
a type of attraction between molecules

For the two DNA strands to join, **hydrogen bonds** form between complementary nitrogenous bases. These bonds effectively 'stick' the strands together to form the double helix. Adenine bonds with thymine and uses two hydrogen bonds to join. Cytosine bonds with guanine and uses three hydrogen bonds to join.

1.1 LEARNING CHECK

1 What does DNA stand for?
2 **State** the building blocks of DNA.
3 **Name** the shape of DNA and explain how DNA is like a ladder.
4 **Draw** a labelled diagram of DNA, including the key components of a deoxyribose sugar, phosphates, the four nitrogenous bases, a nucleotide and the hydrogen bonds.

Extension

5 If a molecule of DNA is found to have 35% of its bases as guanine, what would the percentage of thymine be? **Explain** your answer using Chargaff's rule.

1.2 Genetic material

BY THE END OF THIS MODULE, YOU WILL BE ABLE TO:

- ✓ define 'chromosome', 'gene' and 'karyotype'
- ✓ describe the relationship between chromosomes, genes and DNA.

GET THINKING

The organism with the most chromosomes is the Adder's tongue fern. This plant has an amazing 1440 chromosomes per cell! Do you know how many chromosomes humans have per cell?

Interactive resource
Label: Chromosomes

Extra science investigation
Karyograms

What is DNA?

DNA is a molecule found in all living things. Often called the blueprint of life, DNA holds the instructions to build proteins, provides the instructions for the cell's activities and contributes to the characteristics of an organism.

How does DNA fit into a cell?

DNA is found in the nuclei of most cells. The exception is red blood cells, which do not contain DNA, and many single-celled organisms such as bacteria, where DNA exists simply within the cell walls of a central area called a nucleoid. In this chapter, we'll be looking at DNA in the nuclei of cells in multi-celled organisms.

When a cell is not dividing and multiplying, DNA can be found as **chromatin** and looks like spaghetti in a bowl. During cell division, DNA condenses and takes the form of **chromosomes**. Chromosomes are thin, thread-like structures of tightly coiled chromatin found in the nucleus of a cell. They are made of two **chromatids**, as shown in Figure 1.2.1.

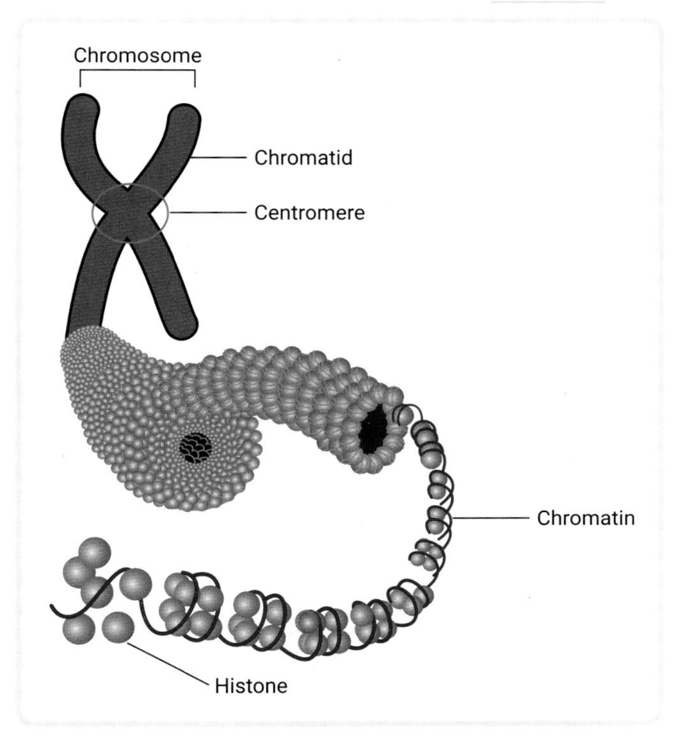

▲ **FIGURE 1.2.1** DNA coils to form chromosomes made from two chromatids attached at the centromere.

Each chromosome is one long DNA molecule coiled around proteins called histones. One way to think about chromosomes is like cotton wrapped around a spool, as shown in Figure 1.2.2. The cotton is the DNA, and the spool acts as the histone proteins. This allows the long molecule to take up much less space in the nucleus and the cell.

▲ **FIGURE 1.2.2** Chromosomes consist of DNA wrapped around histones, a bit like thread wrapped around a spool.

chromatin
unpackaged DNA found within the nucleus of a non-dividing cell

chromosome
a thread-like structure found in the cell, composed of DNA

chromatid
one half of a duplicated chromosome

The number of chromosomes in a cell varies greatly between different types of organisms. A higher number of chromosomes doesn't relate to an organism's size or complexity. For example, elephants have 56 chromosomes in their cells, while the Atlas blue butterfly has between 448 and 452 chromosomes.

How many chromosomes do humans have?

meiosis
cell division producing cells that will specialise into gametes with half the number of chromosomes of the parent cell

karyotype
a picture of an organism's complete set of chromosomes

autosomes
all of the chromosomes in a cell, except for the sex chromosomes

sex chromosomes
a pair of chromosomes that determine the sex of an individual

Your cells have 46 chromosomes; half came from your mother and the other half from your father. This combining of chromosomes occurs during a process called **meiosis**. We will look at this in more detail later in Module 1.4.

Chromosomes form X-shapes and are visible under a light microscope just before cell division. They can be arranged and numbered from longest to shortest and then photographed, forming a **karyotype**. The karyotype allows doctors to check for serious abnormalities in the chromosomes.

Of the 46 chromosomes, or 23 pairs, 22 pairs are called **autosomes**. The final pair are the **sex chromosomes**. These chromosomes determine the sex of the body you are born with and whether you are biologically female or male. Figure 1.2.3 shows a female karyotype with two XX chromosomes. Figure 1.2.4 shows a male karyotype, arranged with one X and one Y chromosome in the same location.

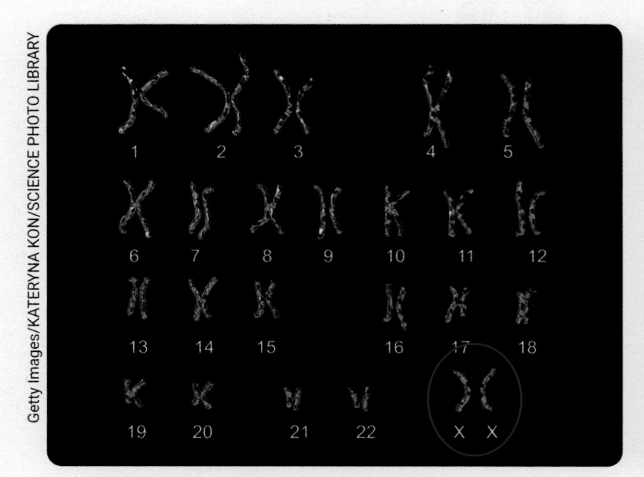

▲ **FIGURE 1.2.3 A female karyotype with the XX chromosomes circled**

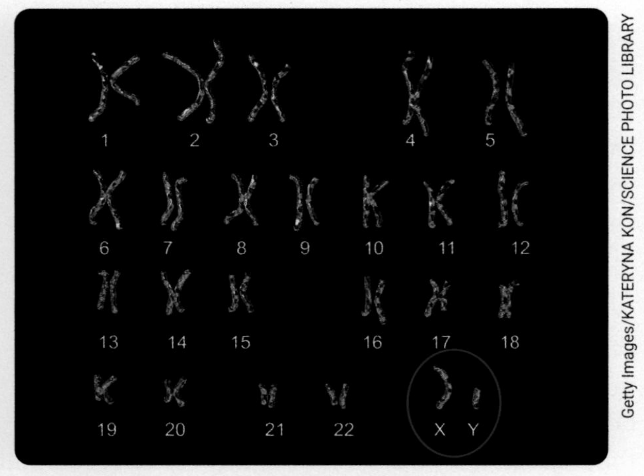

▲ **FIGURE 1.2.4 A male karyotype with the XY chromosomes circled**

homologous pairs
maternal and paternal chromosomes with genes found at the same location

gene
a section of DNA that codes for a protein or a certain trait

centromere
the point on a chromosome where the two chromatids are joined

In the karyotype, the chromosomes are arranged in **homologous pairs**. The chromosomes in each homologous pair have the same **genes** at the same location. They are the same length and have their **centromere** located at the same position. Consider Figure 1.2.4 showing a male karyotype. The chromosomes are in numbered, homologous pairs. Note the difference between the X and Y chromosome.

If you were to compare two people's karyotypes, they are likely to look very similar. So, how are our differences expressed in our chromosomes?

Genes and the chromosomes

Genes are the instructions to build a protein. They are sequences of bases within the DNA and found along the length of the chromosomes, as shown in Figure 1.2.5. Each chromosome can hold hundreds to thousands of genes.

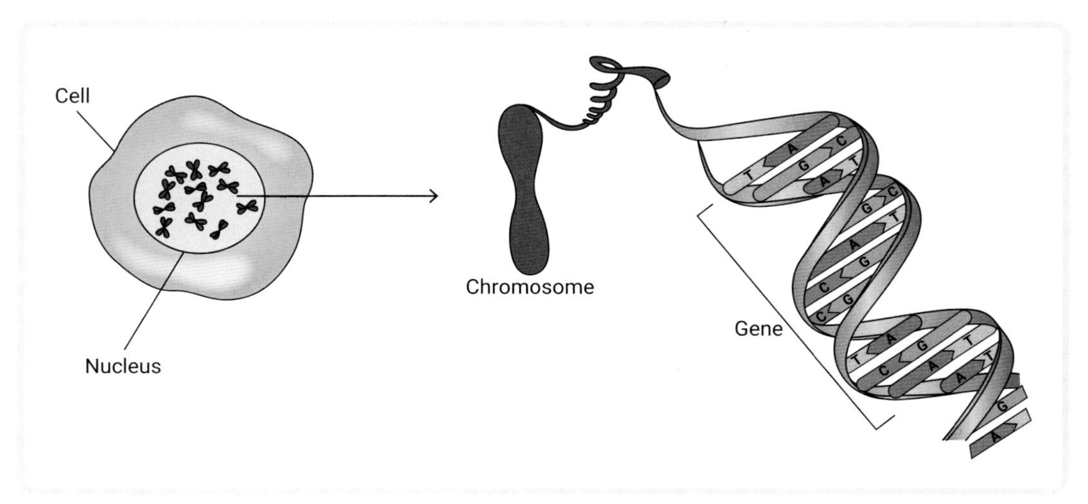

▲ **FIGURE 1.2.5** Each chromosome can have many thousands of genes, which consist of sequences of bases within DNA.

Genes differ in the sequence and number of nitrogenous bases they contain. The sequence of bases is the genetic code. The proteins created from the instructions in the genes are needed to complete cell activities and produce the cell and organism's characteristics.

1.2 LEARNING CHECK

1 **Define** 'genetic code'.
2 **Define** 'chromosome' and 'chromatin'.
3 **Describe** how DNA is packaged into a cell's nucleus.
4 **Explain** why DNA is sometimes called the 'blueprint of life'.
5 **Explain** how a karyotype can identify chromosome abnormalities.
6 **Compare** the structure and function of genes and chromosomes.

1.3 Mitosis

BY THE END OF THIS MODULE, YOU WILL BE ABLE TO:

- ✓ describe the cell cycle
- ✓ list the stages of mitosis
- ✓ recognise the stages of mitosis
- ✓ recall the importance of mitosis for growth and repair.

Interactive resource
Label: Phases of mitosis

Extra science investigation
Modelling mitosis

GET THINKING

The cells in your stomach lining have an average life span of 3 days. However, the cells in your brain have an average life span of more than 80 years! Do you know the name of the process organisms undergo to repair and replace cells?

Mitosis

Imagine if every time you cut your finger or broke a bone, your body could not repair itself. **Mitosis** is an important process that takes place in the nucleus and drives the growth, repair and replacement of cells. Mitosis occurs in all **somatic** cells of organisms.

mitosis
cell division for growth, replacement and repair of somatic cells

somatic
relating to the cells that make up the body other than the reproductive cells

cell cycle
the series of events that takes place in a cell as it grows and divides

interphase
the resting phase of the cell cycle

The cell cycle

Each cell in your body is at some point along its **cell cycle** (see Figure 1.3.1). This is a series of events that allows your cells to perform normal cellular functions and prepare for cell division.

The cell cycle can be broken down into two major phases: **interphase** and mitosis. During interphase, the cell grows and functions normally (G1 phase). Cells spend most of their time in interphase. In this stage, the DNA can be found as chromatin, unwound in the nucleus. Throughout the S phase, the DNA is duplicated and then condensed or folded into chromosomes containing two identical chromatids.

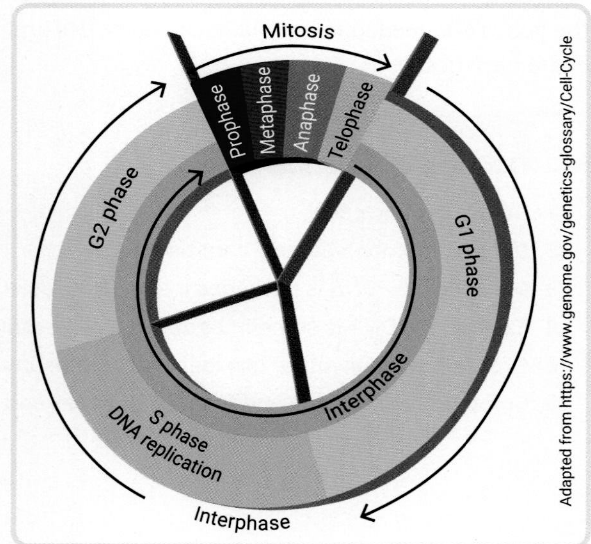

▲ **FIGURE 1.3.1** The cell cycle

Mitosis phases

Mitosis is the process of nuclear division. This process is further broken down into four successive phases, as shown in Figure 1.3.2.

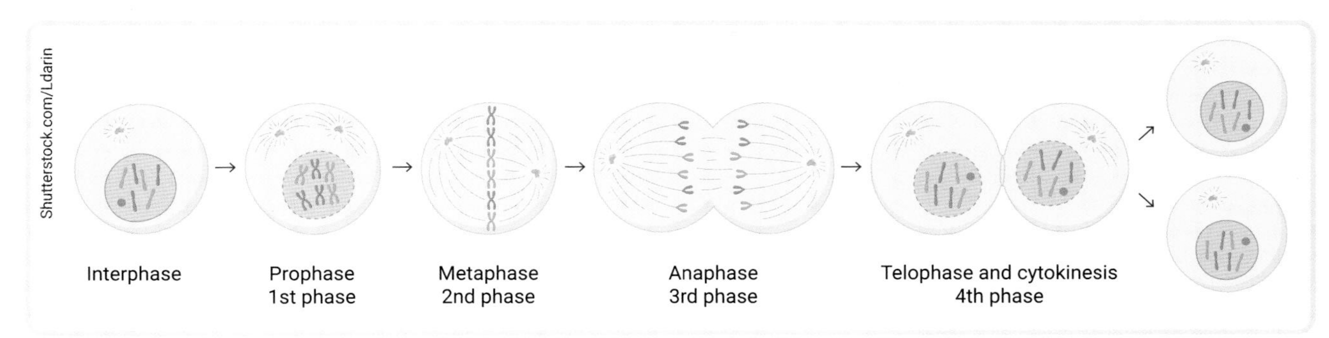

▲ **FIGURE 1.3.2** The process of mitosis consists of four phases and produces two genetically identical daughter cells.

Prophase is the first phase of mitosis. During this phase, the chromosomes duplicate and condense, forming visible 'X' shapes made from identical chromatids, as seen in Figure 1.3.3. Other important organelles in the cell, the **centrioles**, also duplicate and start to move to the opposite poles of the cell. The role of the centrioles is to produce **spindle fibres**. The nuclear membrane breaks down during prophase.

prophase
the first phase of cell division, when chromosomes duplicate and condense

centrioles
organelles in the cell that produce spindle fibres

spindle fibres
protein structures that separate the chromosomes during cell division

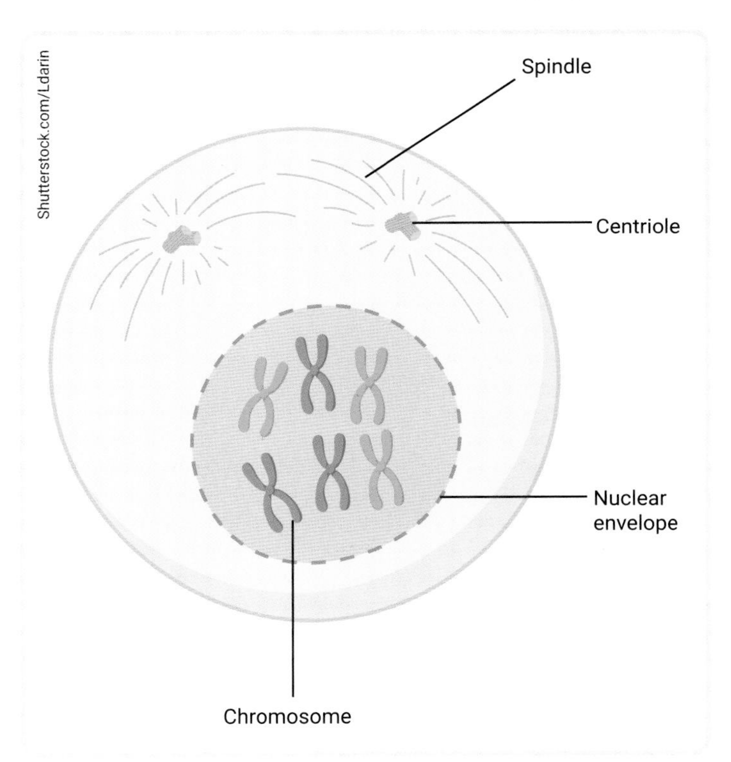

▲ **FIGURE 1.3.3** During prophase, the chromosomes duplicate and condense.

metaphase
the second phase of cell division, when chromosomes line up in the centre of the cell

anaphase
the third phase of cell division, when the chromosomes are pulled apart

telophase
the fourth and final phase of cell division, when the nucleus re-forms and the chromosomes unravel

Metaphase is the second phase of mitosis. During metaphase, the chromosomes line up on the equator of the cell along an imaginary line called the metaphase plate, as seen in Figure 1.3.4. The spindle fibres, extending from the centrioles, attach to the centromeres of the chromosomes.

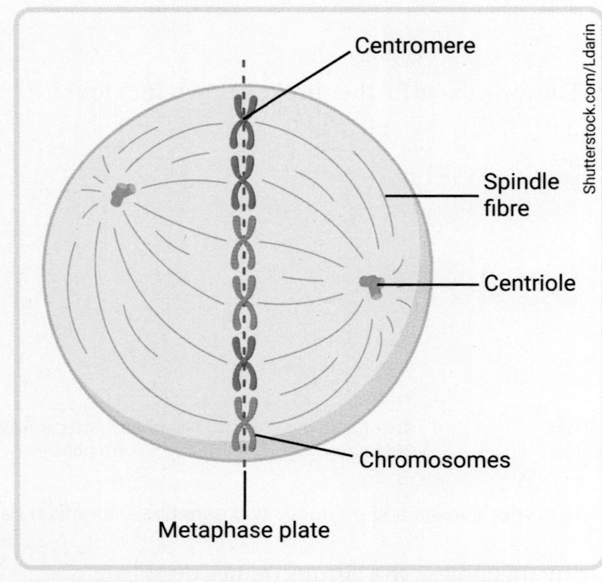

▲ **FIGURE 1.3.4** During metaphase, chromosomes line up on the metaphase plate.

Anaphase is the third phase of mitosis. During anaphase, the spindle fibres retract, pulling the chromosomes apart at the centromeres. This drags the chromosomes to opposite poles of the cell, as seen in Figure 1.3.5.

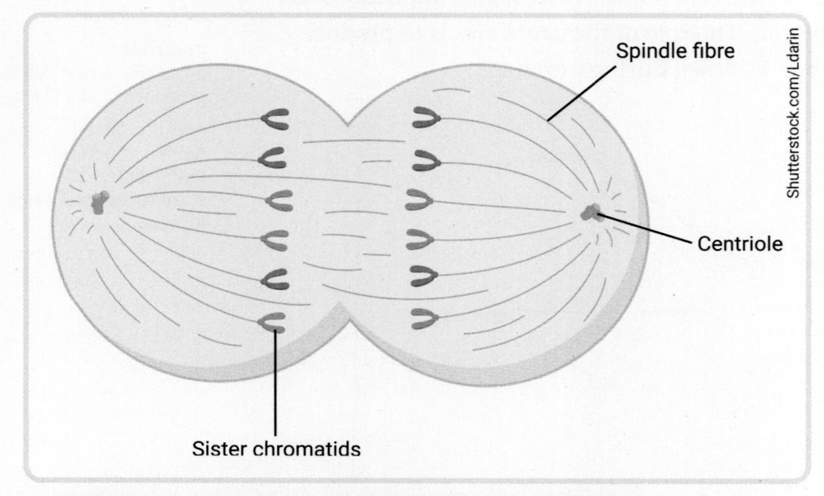

▲ **FIGURE 1.3.5** During anaphase, the chromosomes separate.

Telophase is the fourth and final phase of mitosis. During telophase, the spindle fibres detach, the nuclear membrane re-forms around each set of genetic material, and the cell membrane starts to pinch inwards, as shown in Figure 1.3.6. The chromosomes unravel, forming chromatin that will be encased in the nucleus of each daughter cell.

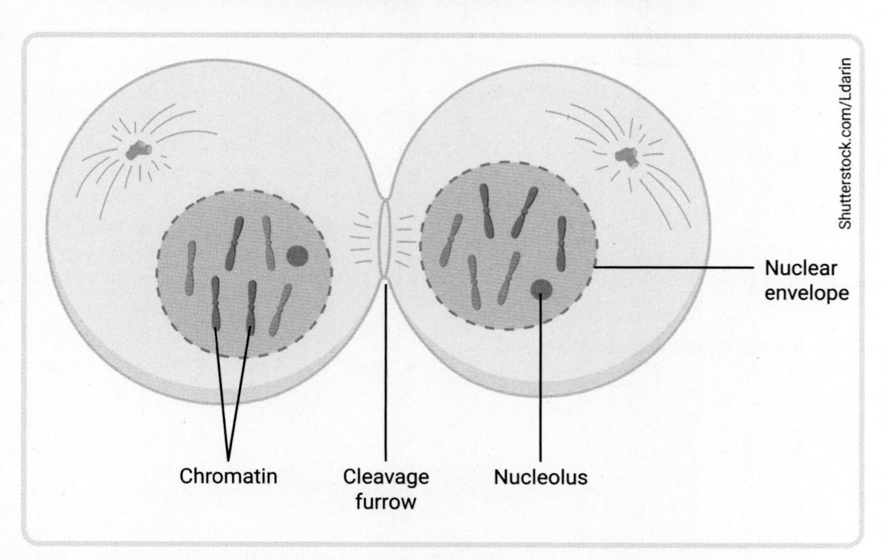

◀ **FIGURE 1.3.6** Telophase is the last phase of mitosis. The division of the cytoplasm, known as cytokinesis, is the final step of mitosis.

At the completion of telophase, **cytokinesis** begins. Cytokinesis is the division of the cytoplasm, resulting in two **daughter cells**. Each daughter cell has an identical copy of the parent DNA and a full complement of paired chromosomes with only one chromatid each. This condition is called **diploid** ($2n$). The cells are identical due to the DNA replication during interphase.

cytokinesis
the division of the cytoplasm after mitosis

daughter cells
the two cells that are produced as a result of cell division

diploid
the full complement of DNA, represented as $2n$

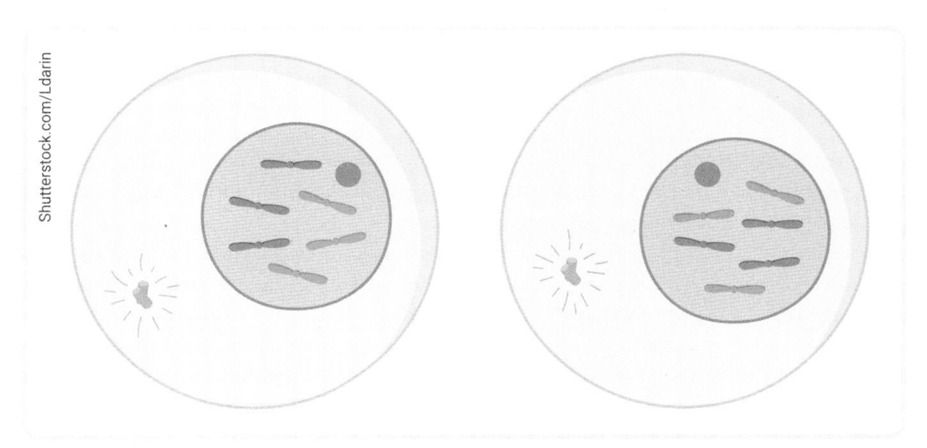

▲ **FIGURE 1.3.7** Two identical daughter cells are produced by the process of mitosis.

Mitosis is essential in the growth of organisms, the repair of damaged tissue and the replacement of dead cells in the body.

1.3 LEARNING CHECK

1. **Recall** the five phases of the cell cycle in order.
2. **Explain** the role of the centrioles in mitosis.
3. **Define** 'diploid'.
4. **Explain** why it is important that all multicellular organisms undergo the process of mitosis.
5. **Design** and make a flip book showing the phases of the cell cycle. Each page should have the name of the phase and a labelled diagram representing the condition of the cell during that phase.

1.4 Meiosis

BY THE END OF THIS MODULE, YOU WILL BE ABLE TO:

- ✓ list the stages of meiosis
- ✓ recognise the stages of meiosis
- ✓ recall the importance of meiosis to produce cells needed for sexual reproduction
- ✓ compare and contrast the processes of mitosis and meiosis.

Video activity
Inheritance

Interactive resource
Label: Meiosis I and II

GET THINKING

Human sex cells – sperm and eggs – contain 23 chromosomes. When a sperm and an egg unite, the resulting cell has 46 chromosomes. What would happen if the sex cells each had 46 chromosomes to start with?

Cell division for reproduction

Mitosis is an essential process in the cell cycle that results in genetically identical daughter cells needed for growth, replacement and repair. However, many organisms also need to make cells that can be used in sexual reproduction. As you learned in Module 1.2, humans have 46 chromosomes in each cell, 23 from the mother and 23 from the father. Meiosis explains how this is possible and how chromosome numbers are maintained in a species from one **generation** to the next.

generation
the time taken for one individual to produce offspring

gonads
the sex organs of an organism; where meiosis occurs

germline cells
the cells that form the ovum and the sperm

gametes
the sex cells of a sexually reproducing organism

zygote
a diploid ($2n$) cell resulting from the joining of two haploid gametes

Where does meiosis occur?

Meiosis occurs in the **gonads** of sexually reproducing organisms. The gonads contain **germline cells** that are responsible for the creation of **gametes**. In humans, the gonads are the ovaries (female, which produce ova) and testes (male, which produce sperm). Gametes are also called sex cells. When the sex cells of a male and a female combine at fertilisation, the cell they form has the potential to develop into a new individual.

Meiosis is a cell division process that reduces the chromosome number by half so that when the gametes combine to form a **zygote**, the correct number of chromosomes are present. Meiosis also results in variation in the species because it produces sex cells that are genetically different from each other. The stages of meiosis are depicted in Figure 1.4.1.

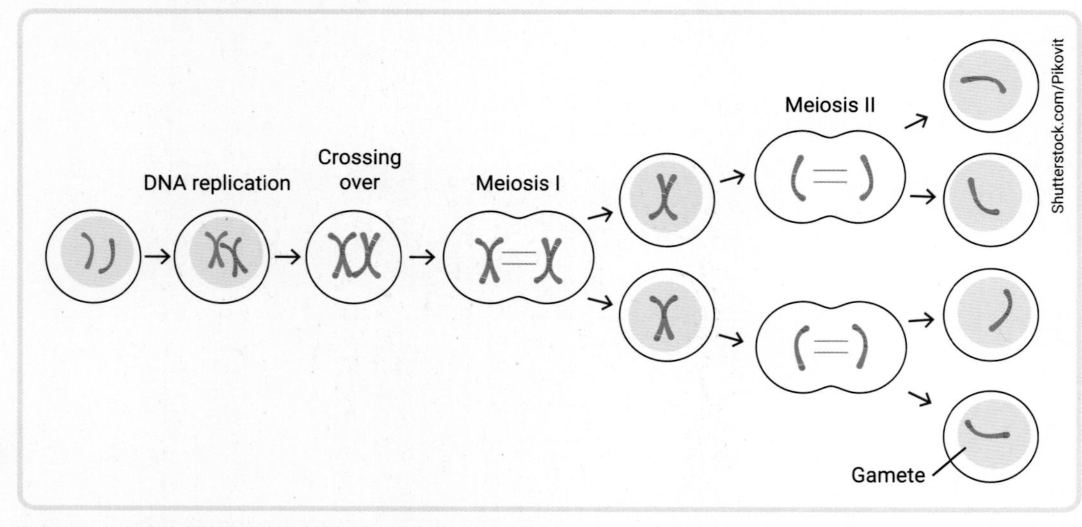

▶ **FIGURE 1.4.1** Meiosis

The phases of meiosis

Meiosis has more phases than mitosis does. In meiosis, the cell replicates its DNA once but divides twice. Meiosis is divided into two stages: meiosis I and meiosis II.

The parent cell grows during interphase, replicates its DNA and prepares the chromosomes for cell division. This is similar to mitosis. After interphase, the chromosomes are in their X-shape, consisting of two sister chromatids. At this stage, the chromatids are genetically identical.

Meiosis I

The cell progresses into meiosis I, which has four phases – prophase I, metaphase I, anaphase I and telophase I – as shown in Figure 1.4.2. Cytokinesis occurs at the completion of telophase I. There is a short interphase before the cell begins meiosis II.

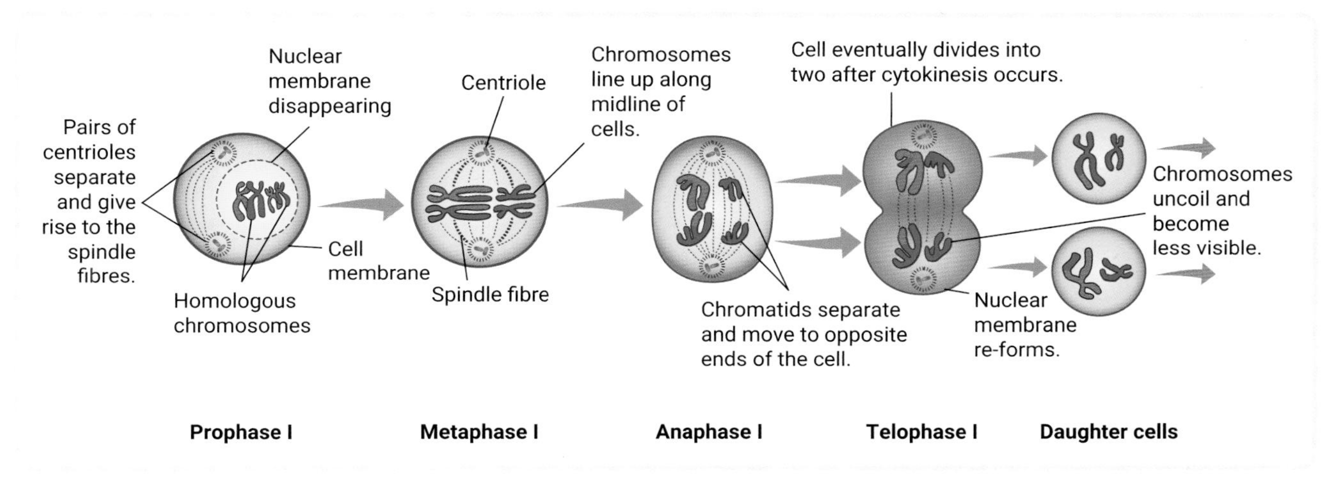

▲ **FIGURE 1.4.2** The phases of meiosis I

During prophase I, the nuclear membrane disappears, the centrioles duplicate and migrate to the opposite sides of the cell, and the chromosomes are visible in their X-shape. The chromosomes move into their homologous pairs and **crossing over** occurs, as shown in Figure 1.4.3.

crossing over
the exchange of genetic material between non-sister chromatids on homologous chromosomes

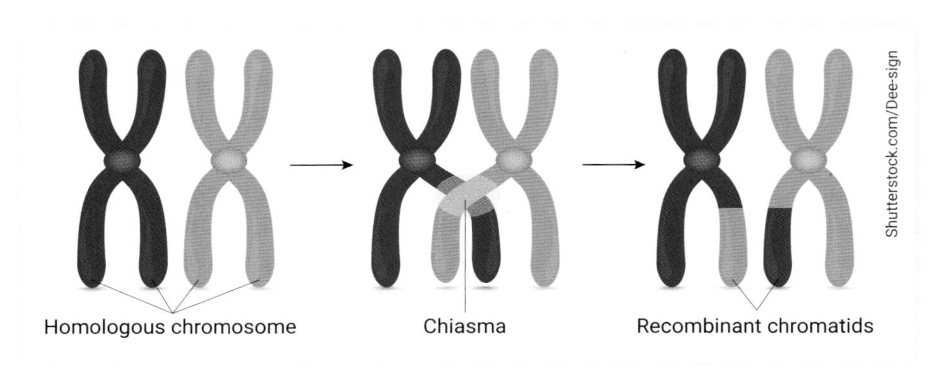

▲ **FIGURE 1.4.3** The crossing over of non-sister chromatids

chiasma
the point on the chromatids where crossing over occurs

random assortment
the way chromosomes line up during metaphase I, resulting in random combinations of genes

homologous
carrying the same genes for characteristics at the same locations on a chromosome

haploid
having one copy of each chromosome, represented as *n*

In crossing over, the non-sister chromatids exchange genetic material with each other at a point known as the **chiasma** (plural: chiasmata). This results in a recombination of genetic material and is a major source of genetic variation.

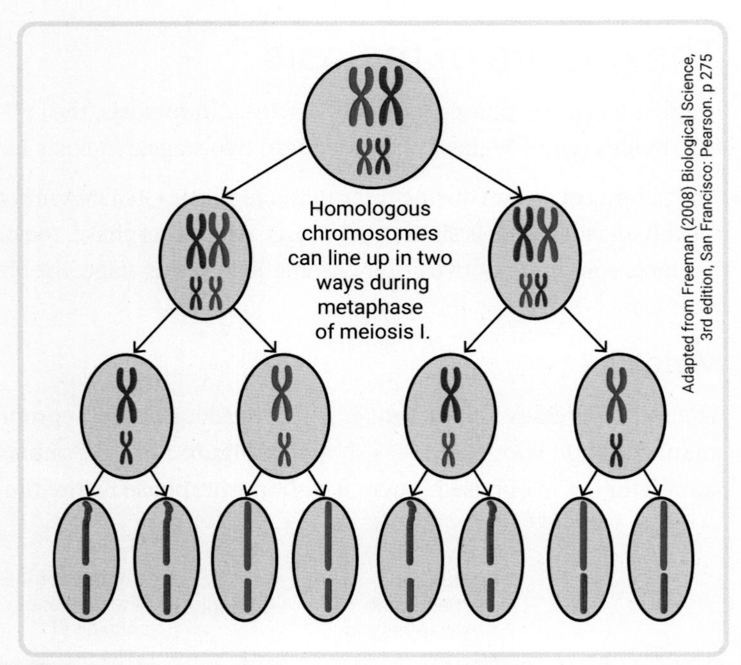

▲ **FIGURE 1.4.4** **Chromosomes line up randomly in metaphase I, resulting in the recombination of genes in the daughter cells.**

During metaphase I, the chromosomes line up at the equator of the cell, and the spindle fibres attach to the centromeres. The chromosomes line up randomly in a process called **random assortment**, as shown in Figure 1.4.4. This is another source of genetic variation, as the resulting cells receive different combinations of genes.

During anaphase I, the **homologous** chromosomes are pulled to the opposite poles of the cells. The chromatids are not separated at this point, and the chromosomes are not in homologous pairs. Each chromosome consists of two chromatids.

Telophase I and cytokinesis result in the production of two daughter cells that are **haploid** and genetically different. The cell will enter a short interphase; however, DNA replication does not occur.

Meiosis II

Meiosis II commences in both daughter cells, starting with prophase II. During this phase, the centrioles in each cell duplicate and migrate to opposite poles of the cell. The chromosomes line up at the equator of the cell and the spindle fibres attach to the centromeres.

Anaphase II occurs when the spindle fibres contract, and the chromatids are separated and pulled to the opposite poles of the cells. Telophase II and cytokinesis follow, resulting in four haploid cells that are genetically different. Each chromosome now consists of one chromatid. The phases of meiosis II are depicted in Figure 1.4.5.

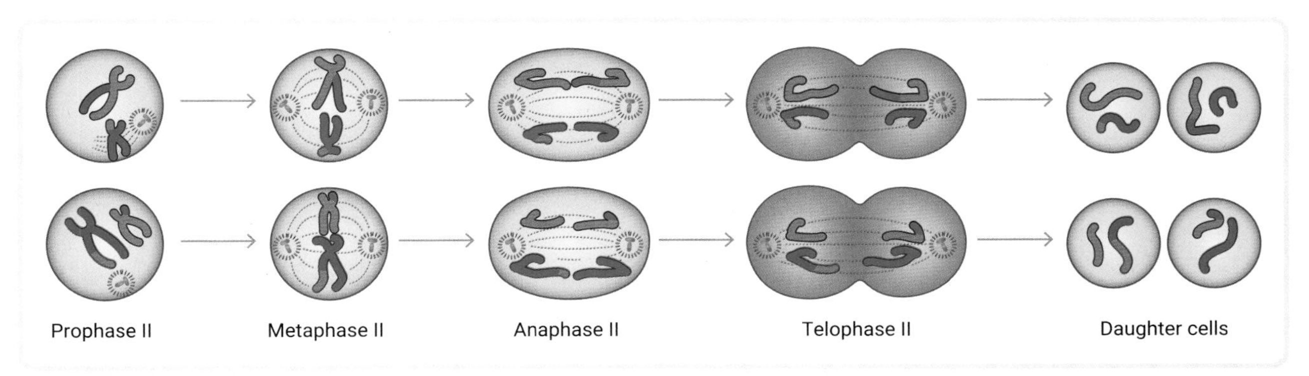

▲ FIGURE 1.4.5 The phases of meiosis II

Mitosis v. meiosis

Mitosis and meiosis are similar, but key differences in some of the steps and the number of phases result in very different outcomes. Table 1.4.1 provides a comparison of mitosis and meiosis.

▼ TABLE 1.4.1 A comparison of mitosis and meiosis

Characteristic	Mitosis	Meiosis
Type of cells that use this process	Body cell	Germline cell
Number of nuclear divisions	1	2
Number of daughter cells produced	2	4
Chromosome number of the parent cell	Diploid (2*n*)	Diploid (2*n*)
Chromosome number of the daughter cells	Diploid (2*n*)	Haploid (*n*)
Involves the separation of homologous chromosomes?	No	Yes, during anaphase I
Involves the separation of sister chromatids?	Yes, during anaphase	Yes, during anaphase II
Purpose of the process	Growth, replacement and repair of cells	Production of haploid gametes for reproduction

1.4 LEARNING CHECK

1 **Recall** the location where meiosis occurs.
2 **Define** 'haploid'.
3 **Identify** two sources of genetic variation due to meiosis and the phases they occur in.
4 **Draw** a labelled diagram showing the process of crossing over.
5 **Distinguish** between the terms 'chromosome' and 'chromatid'.
6 **Draw** a Venn diagram to compare and contrast mitosis and meiosis.

1.5 Mendelian inheritance

BY THE END OF THIS MODULE, YOU WILL BE ABLE TO:

- ✓ define the common terms used in genetics
- ✓ recognise dominant and recessive traits
- ✓ classify genotypes as homozygous dominant, heterozygous or homozygous recessive.

GET THINKING

Why do you look more like one parent than the other? Why are some characteristics seen in all your family members, whereas others occur randomly? The answer is genetics!

wikimedia/Professor William Bateson

▲ FIGURE 1.5.1 Gregor Mendel

Video activity Mendel and inheritance

Extra science investigation Inheritance and chance

Mendel: the father of genetics

Gregor Mendel was an Austrian monk who was interested in meteorology, mathematics and biology (see Figure 1.5.1). His breeding experiments with pea plants (*Pisum sativum*) in his monastery in the 1860s resulted in the development of the principles of inheritance. Mendel cross-pollinated his pea plants and meticulously recorded the statistics of the resulting offspring. This data and his observations were published but largely ignored until the 1900s when other scientists discovered his work and linked it to their own studies in genetics.

Important discoveries that explain the mechanisms of inheritance

The later discovery of chromosomes by Walther Flemming in 1879, and the link between chromosomes and heredity described by Theodor Boveri in 1902, helped scientists further understand the mechanisms behind inheritance. In addition, Walter Sutton's observations in 1903 of chromosome behaviour during cell division and gamete formation was consistent with Mendel's work.

Mendel's laws of inheritance

Mendel proposed three principles of inheritance, which are now known as Mendel's laws of inheritance.

- The law of segregation: each inherited trait is defined by a single gene pair. The parent genes are randomly separated into the gametes. Offspring inherit one **allele** from each parent when the gametes join at fertilisation.
- The law of independent assortment: traits are inherited separately from one another. The inheritance of one trait is not dependent on the inheritance of another.
- The law of dominance: where an organism has two alternative forms of a gene, the **dominant allele** will be expressed.

allele an alternative form of a gene

dominant allele the allele that will be expressed in the phenotype

Mendel's pea plant experiments

Mendel studied seven of the characteristics of pea plants, including plant height, pea colour, flower colour, pod shape and seed shape (see Figure 1.5.2). He worked with plants that self-pollinated and consistently produced the same characteristics from one generation to the next. Mendel used these observations to conclude that the parental lineage of these pea plants was **purebred**.

purebred the same alleles for a given gene; see homozygous dominant/recessive

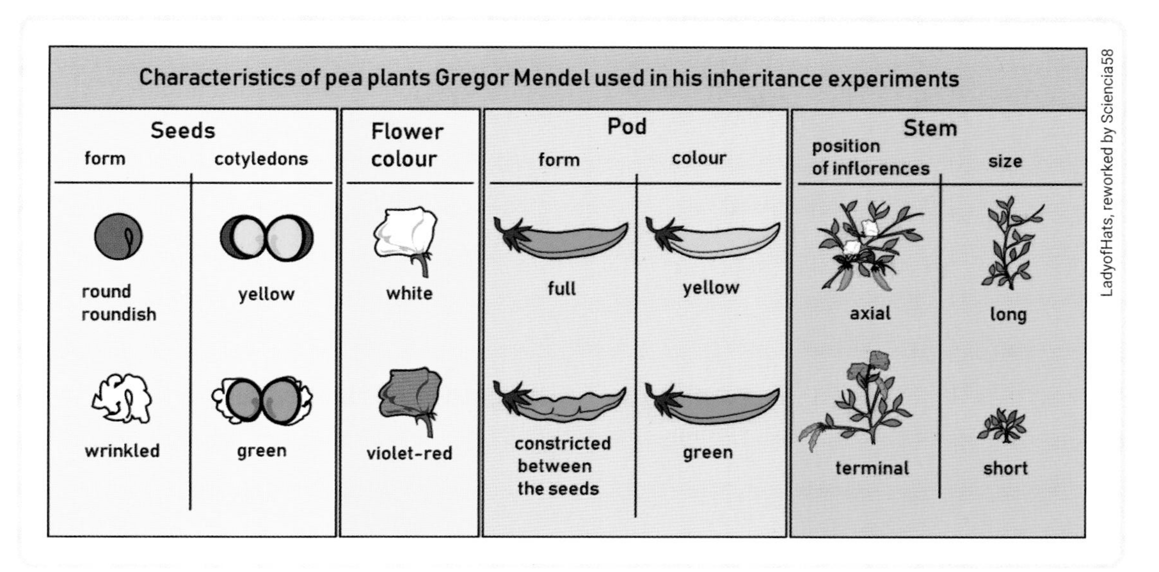

▲ **FIGURE 1.5.2** Mendel studied seven of the characteristics of pea plants.

Mendel cross-pollinated one variety of purebred plant with another and discovered that the offspring looked like either one of the parent plants, not a blend of the two. For example, crossing a long stem plant with a short stem plant resulted in offspring having long stems, not stems somewhere between long and short. In general, if the offspring (or **progeny**) of crosses between purebred plants looked like only one of the parents, Mendel called the expressed trait the **dominant** trait. The trait that was not seen in the progeny, and was masked by the dominant trait, was called the **recessive** trait.

progeny
offspring

dominant
an inheritance that identifies the dominant allele in a genotype

recessive
an inheritance that identifies the recessive allele in a genotype

filial
first set of offspring from a cross between parents

Mendel designated the two pure-breeding parental generations as P_1 and P_2 and identified the progeny as the **filial** or F_1 generation. Although the F_1 generation looked uniformly like one parent of the P generation, they had inherited a different allele from each parent plant. This type of breeding is known as a **monohybrid cross**, as only one gene is being investigated (see Figure 1.5.3).

Mendel then crossed the F_1 generation by allowing the F_1 generation to self-pollinate, creating the F_2 generation. The F_2 generation had offspring that were either long or short in stem length, showing that the recessive trait had been carried down in the F_1 generation. The ratio of long to short stems was roughly three long stems to one short stem.

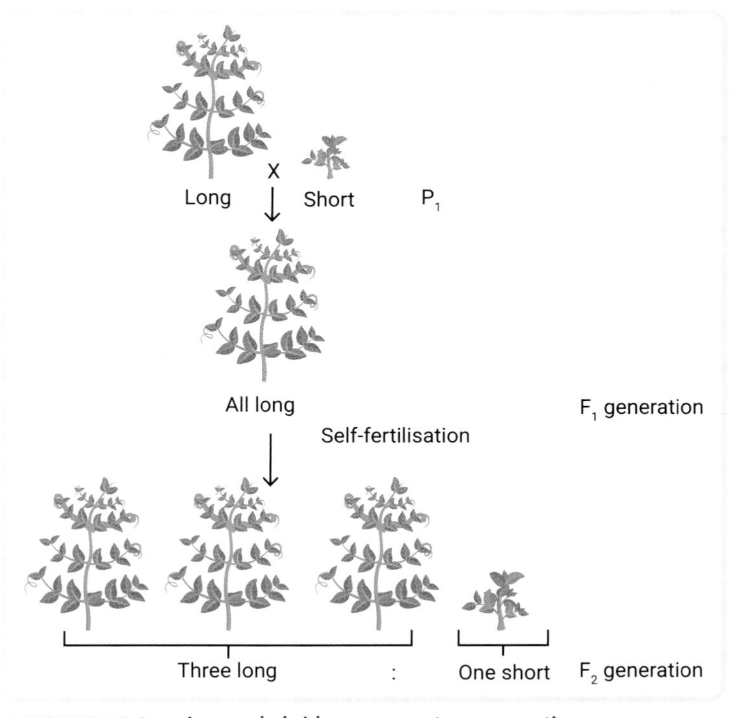

▲ **FIGURE 1.5.3** A monohybrid cross over two generations

monohybrid cross
a cross between two organisms with two alleles at one gene location

DNA, genes and alleles

Mendel's data suggested that each parent contributes some particulate matter to their offspring, and he called this hereditary material 'elementen'. Today we know this as DNA and genes. We also know that there are forms of genes called alleles. For example, when considering stem height, there is an allele for long and an allele for short stem height.

Mendel's genetic notation

recessive allele
the allele that is masked by a dominant allele and is only expressed in the homozygote

homozygous dominant
having two of the same dominant alleles on homologous chromosomes

homozygous recessive
having two of the same recessive alleles on homologous chromosomes

heterozygous
having two different alleles on homologous chromosomes

genotype
the combination of alleles for a specific gene

phenotype
the observable characteristics of the genotype

Mendel used a specific notation to represent his data. For a dominant allele he used a capital letter, and he used a lowercase letter to represent the **recessive allele**. Purebred long stem plants, which are **homozygous dominant**, have a notation of LL. Pure-breeding short stem plants, which are **homozygous recessive**, have a notation of ll. The hybrids in the F_1 generation, which are **heterozygous**, have a notation of Ll. Heterozygous plants carry the recessive allele, despite physically showing the dominant allele.

This notation is known as the **genotype** of the individual, as it is a representation of the combination of alleles found in a specific gene. The observable expression of the genotype is called the **phenotype**. For example, Figure 1.5.4 shows a pea plant with the genotype Ll displaying the phenotype of a long stem.

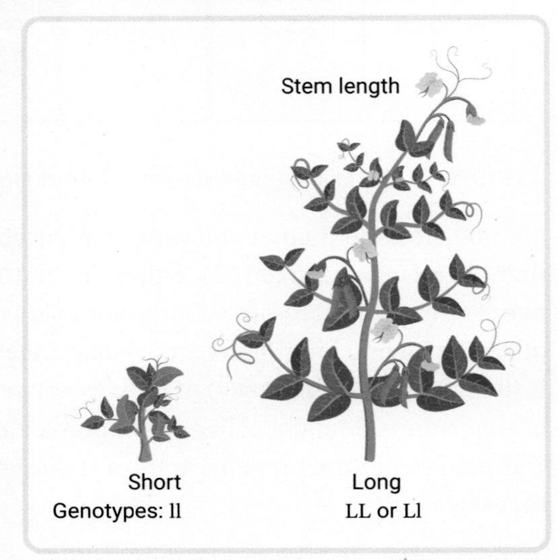

▲ **FIGURE 1.5.4** A pea plant with the genotype Ll displaying the phenotype of a long stem

1.5 LEARNING CHECK

1 Figure 1.5.5 shows the characteristics of pea plants and the classification of dominant and recessive alleles.
   a **Assign** a key to each characteristic of a pea; for example, seed shape R = round and r = wrinkled.
   b **Draw** this as a table in your workbook.
2 **Write** the genotypes for the following pure-breeding plant characteristics:
   a wrinkled seeds.
   b inflated pod shape.
   c white flower.
   d tall stem height.

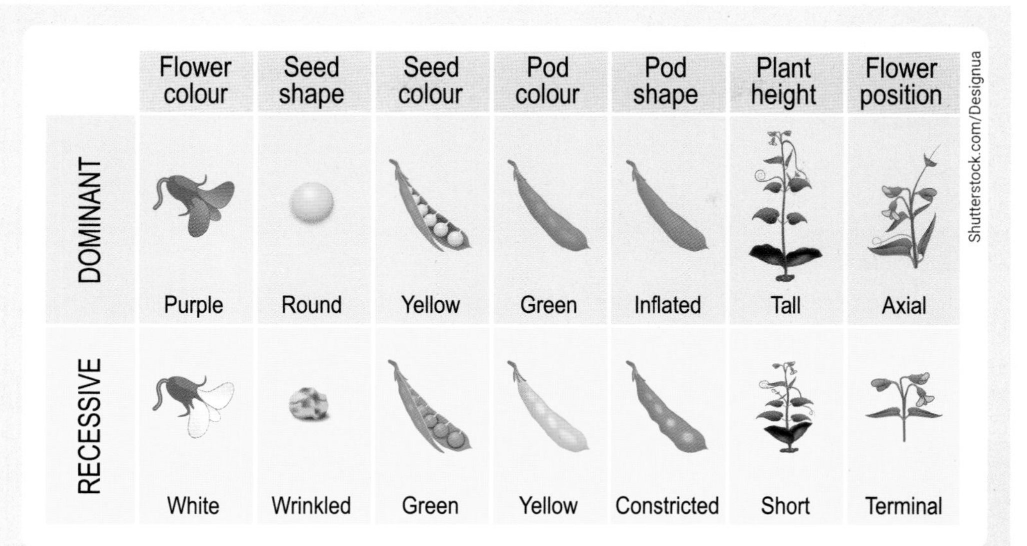

▲ **FIGURE 1.5.5** The dominant and recessive traits of pea plants

3 **Predict** the phenotypes of the offspring in the following crosses. Use Figure 1.5.5 to help you.

Example: A purebred inflated seed pod crossed with a purebred constricted seed pod will produce offspring that all have inflated seed pods because inflated seed pod is the dominant trait.

- **a** A purebred purple flower crossed with a purebred white flower
- **b** A purebred yellow seed crossed with a purebred green seed
- **c** A purebred tall stem crossed with a purebred short stem

4 Write the genotypes for the following phenotypes into your workbook:

- **a** homozygous green pod.
- **b** homozygous terminal flower position.
- **c** homozygous wrinkled seed.
- **d** heterozygous purple flower.
- **e** heterozygous inflated pod shape.

5 **List** Mendel's three laws.

6 **Define:**

- **a** purebred.
- **b** monohybrid cross.
- **c** genotype.
- **d** phenotype.

7 **Conduct** some research on the Internet to determine if the following human characteristics are dominant or recessive.

- **a** The ability to roll your tongue into a U shape
- **b** Free earlobes
- **c** Interlocking fingers with the left thumb on top
- **d** A widow's peak hairline on the forehead

1.6 Sex determination

BY THE END OF THIS MODULE, YOU WILL BE ABLE TO:
- ✓ identify the chromosomes responsible for sex determination in different species
- ✓ predict the frequency of male and female offspring
- ✓ describe some conditions caused by variation in sex chromosome inheritance.

GET THINKING

In crocodiles, male hatchlings are born if the eggs incubate at 34°C and female hatchlings are born if the eggs incubate at 30°C. The average body temperature of humans is 37°C. So what, if not temperature, determines our sex?

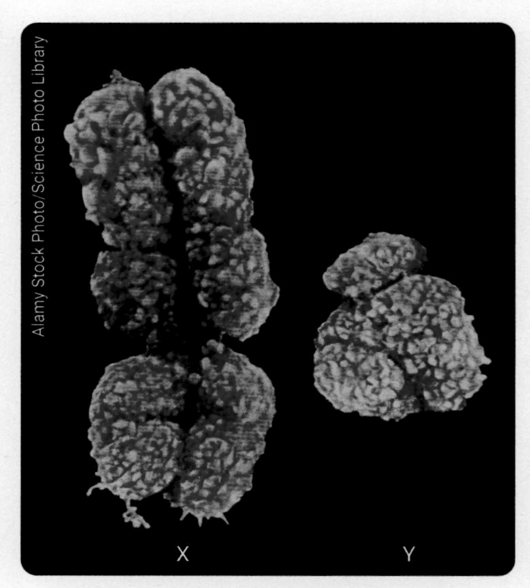

▲ **FIGURE 1.6.1** A scanning electron micrograph showing the X chromosome (left) and Y chromosome (right)

Video activity
Sex determination

XX and XY chromosomes

You have previously learned that humans have 23 pairs of chromosomes, of which 22 are the autosome chromosomes and the 23rd pair are the sex chromosomes. In humans, and most mammals, these sex chromosomes are labelled X and Y (see Figure 1.6.1). The combination of the sex chromosomes determines the sex of the offspring: XX for female and XY for male.

How is sex determined?

The sex of offspring is determined by what happens during the separation of the chromosomes during meiosis (see Figure 1.6.2). In meiosis, the chromosome number halves, with each daughter cell receiving a haploid number of chromosomes from the parent cell. In humans, all the eggs cells (ova) produced by the female carry one X chromosome. Half the sperm cells produced by the male carry one X chromosome and the remaining half carry the Y chromosome.

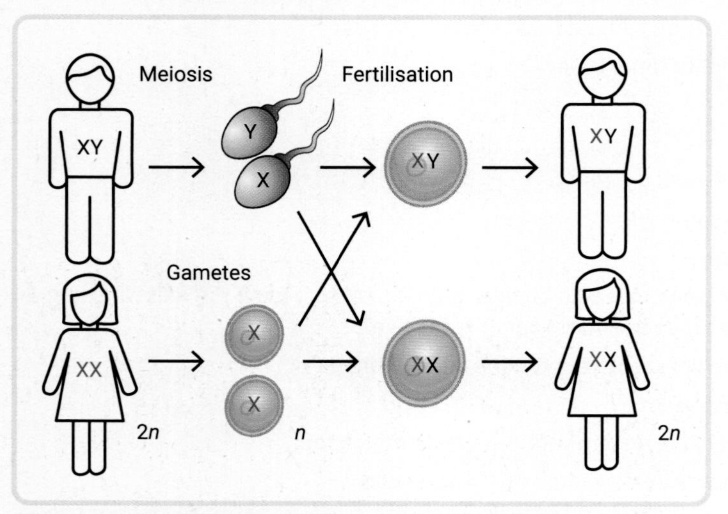

▲ **FIGURE 1.6.2** How sex chromosomes are determined in humans

At fertilisation, there is a 50 per cent probability that the female egg cell will be fertilised by an X-carrying sperm and a 50 per cent probability that the female egg cell will be fertilised by a Y-carrying sperm. The resulting zygote will be diploid ($2n$) and will have a full complement of sex chromosomes.

Sex chromosomes in other animals

Other animals have different sex-determining systems from most mammals. For example, birds have a ZW system, where males are ZZ and females are ZW (see Figure 1.6.3). Reptiles and amphibians have a mixture of XX/XY and ZZ/ZW depending on the species. These animals, in particular turtles and crocodiles, are influenced by the environment, usually temperature, which can influence the sex of the offspring. For example, turtle eggs that are incubated below 27.7°C will result in male hatchlings and eggs incubated at temperatures above 31°C will result in female hatchlings.

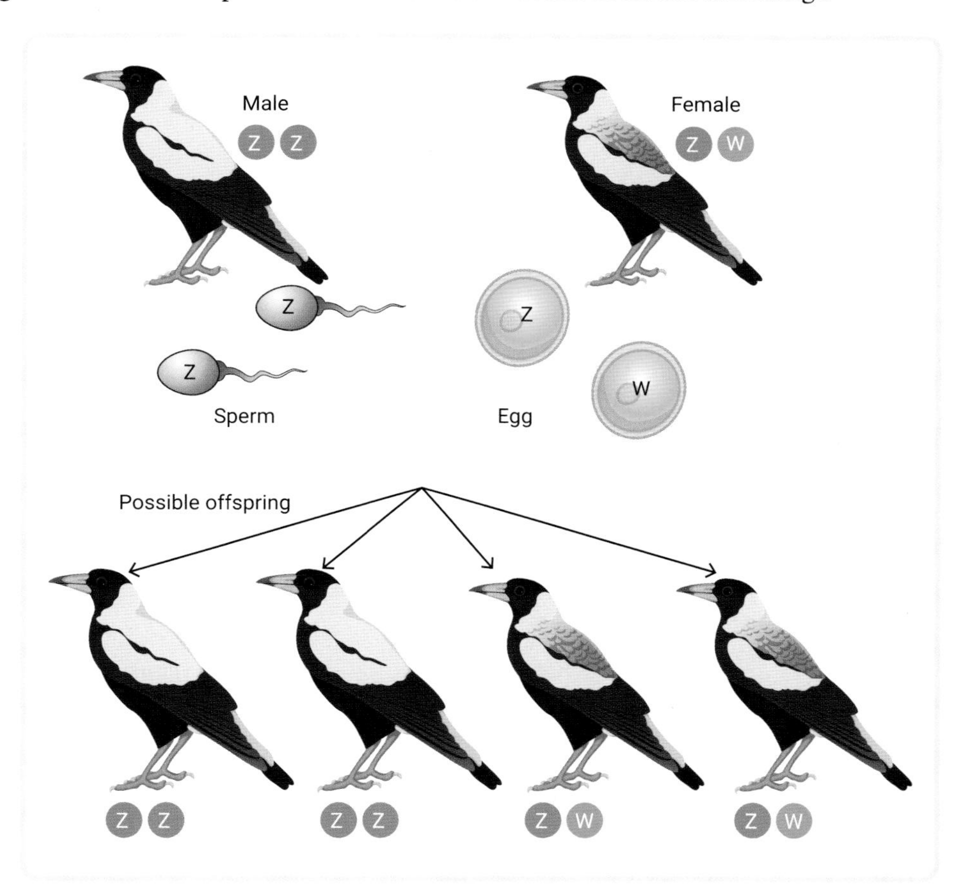

▲ **FIGURE 1.6.3** The inheritance of sex chromosomes in birds

Insects also inherit sex chromosomes, although they do so differently from vertebrate animals. Honeybees are an interesting example where all drones (males) are formed from unfertilised eggs and are, therefore, haploid. Female worker bees develop from fertilised eggs and are diploid. The queen bee is a female (diploid) that was fed royal jelly in the hive. She lays all the eggs in the hive. The inheritance of sex chromosomes in honeybees is shown in Figure 1.6.4.

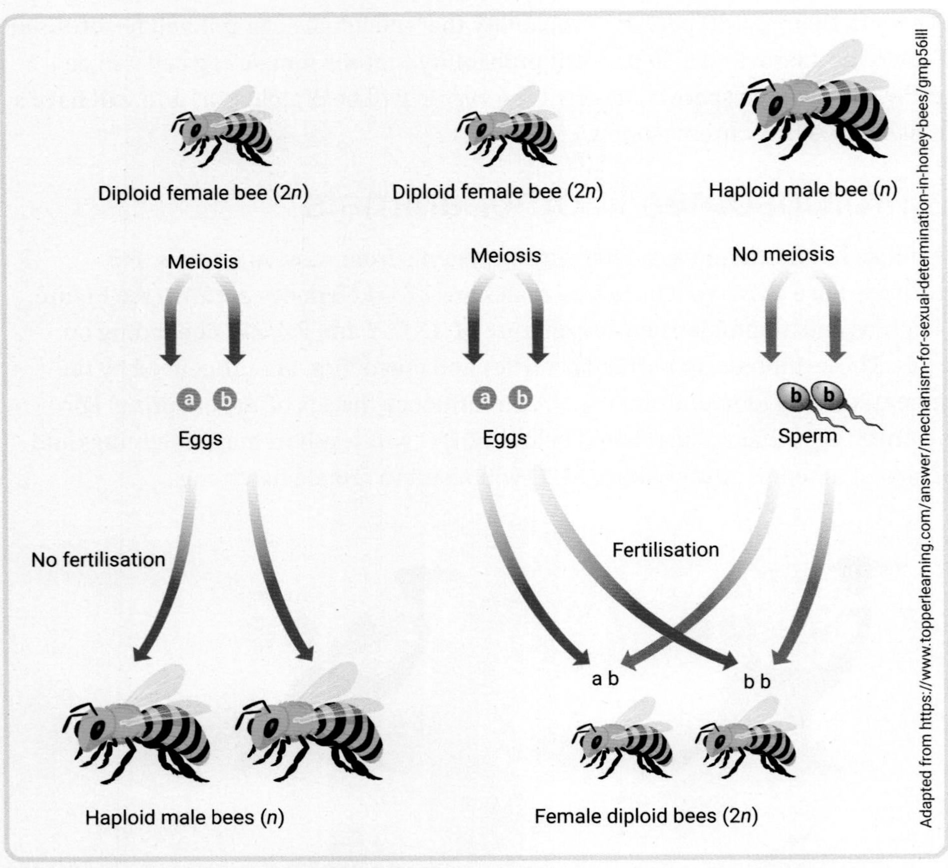

▲ **FIGURE 1.6.4** The inheritance of sex chromosomes in honeybees

non-disjunction
the incorrect separation of chromosomes or chromatids at the centromere, resulting in gametes with an unusual chromosome number

Extra or missing sex chromosomes

Sometimes the chromosomes do not separate normally during meiosis, resulting in gametes that have one extra or one fewer sex chromosome. This is called **non-disjunction** and can be seen in Figure 1.6.5.

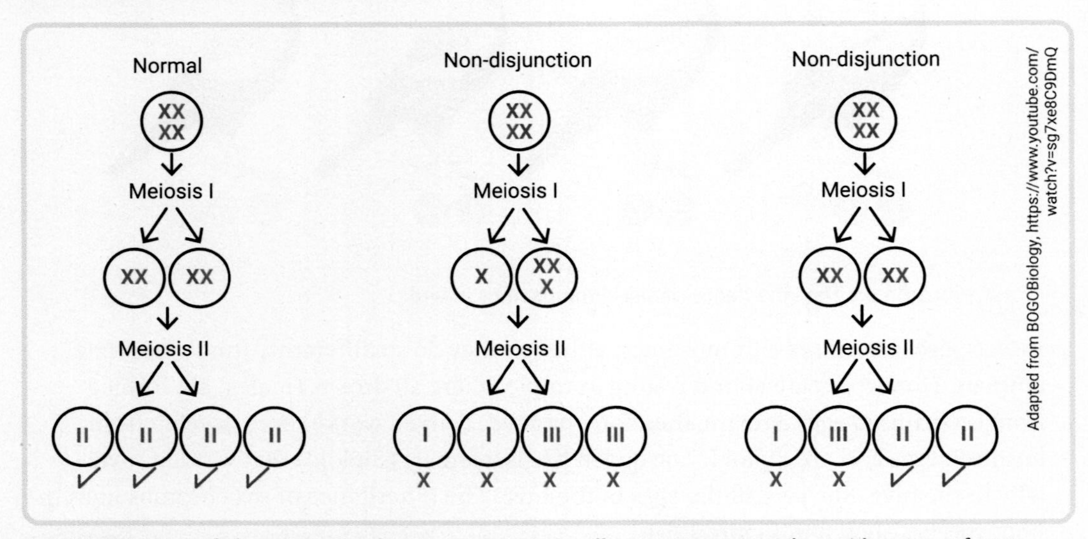

▲ **FIGURE 1.6.5** If chromosomes do not separate normally, gametes can end up with more or fewer chromosomes than normal.

Klinefelter syndrome occurs in individuals who have XXY as their sex chromosomes. They are male, taller than average with less facial and body hair. Their testes are smaller and produce less testosterone and sperm. A karyotype of a person with Klinefelter syndrome is shown in Figure 1.6.6.

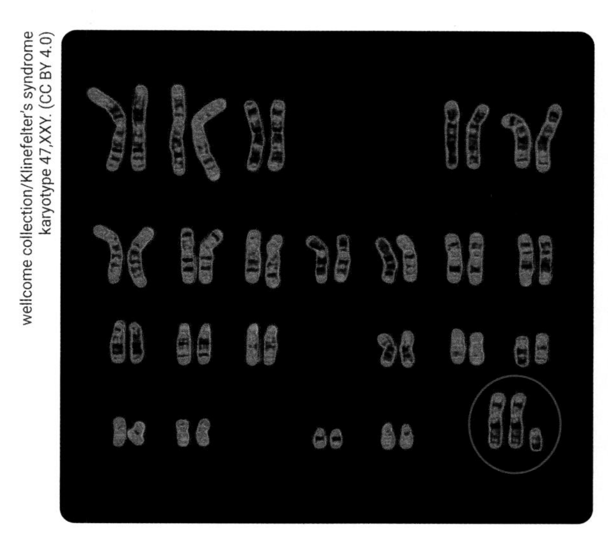

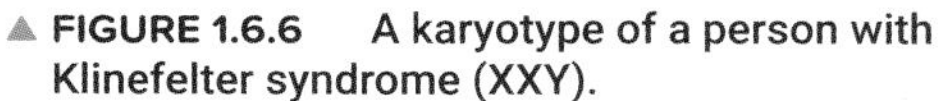

▲ **FIGURE 1.6.6** A karyotype of a person with Klinefelter syndrome (XXY).

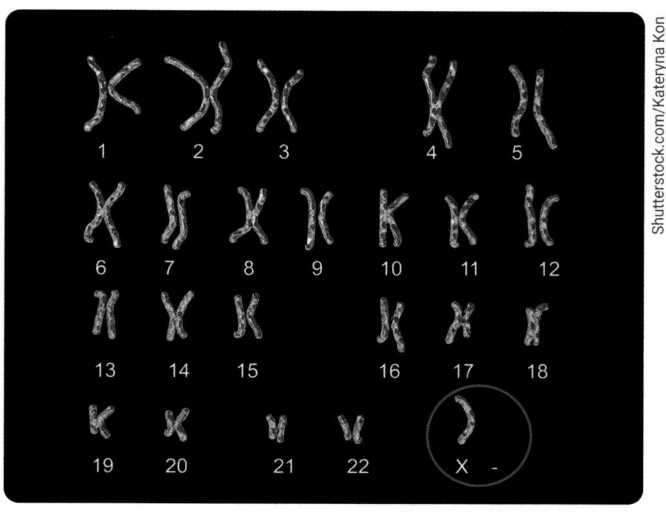

▲ **FIGURE 1.6.7** A karyotype of a person with Turner syndrome (XO)

Turner syndrome occurs in females who inherit only one X chromosome. Their genotype is XO. Characteristics of the condition can include infertility, a webbed neck, short stature and irregularities in the development of their reproductive organs. A karyotype of a person with Turner syndrome is shown in Figure 1.6.7.

1.6 LEARNING CHECK

1 **Recall** the sex chromosomes for male and female in humans.

2 Use a diagram with correct notation to **predict** the probability of male offspring in:

  a mammals.

  b birds.

3 Using the Internet, **research** the SRY gene, located on the Y chromosome in male mammals. Identify what role the SRY gene plays in determining the maleness of a mammal.

4 **Research** and **describe** the full characteristics of Klinefelter syndrome and Turner syndrome. In your description include the frequency of occurrence of both syndromes in Australia and the phase(s) of meiosis where incorrect separation of chromosomes can occur.

1.7 Punnett squares

BY THE END OF THIS MODULE, YOU WILL BE ABLE TO:

- ✓ define Punnett squares and explain how to use them
- ✓ use Punnett squares to predict the genotypes and phenotypes of offspring from monohybrid crosses.

Interactive resource
Drag and drop: Punnett squares

GET THINKING

Studying your ancestry can reveal some amazing stories about your family history. What if we could examine the genetic history of your family too? Are there interesting stories hidden in your genes?

What are Punnett squares?

Punnett squares are used to show the possible genotypes and phenotypes that can be produced from monohybrid crosses. As you learned in Module 1.5, Mendelian inheritance shows that alleles (versions of genes) occur in pairs, and these allele pairs separate independently during meiosis so that each allele for the trait appears in a gamete. At fertilisation, gametes unite randomly to produce potential offspring.

Punnett squares are a simple way to show this process. Capital letters are used for dominant traits, and lowercase letters are used for recessive traits. It is conventional to keep the letters the same for the trait being examined; for example, AA, Aa, or aa.

Punnett squares in action

As Mendel observed in peas, round seed shape (R) is dominant to wrinkled seed shape (r). Let's cross a homozygous round seed (RR) with a homozygous recessive wrinkled seed shape (rr) to determine the offspring produced in the **first filial generation** (F_1) generation.

first filial generation the first set of offspring from a parent cross

1 Show the parent genotype cross:

$P_1 \times P_2$ cross = RR × rr

2 Complete the Punnett square by placing each parent allele into the top and side boxes. You can then fill in the boxes, showing the probability of the offspring. A completed Punnett square showing P_1 and P_2 genotypes and the F_1 genotypes is shown in Figure 1.7.1.

3 State the genotypes of the first filial generation (F_1) generation.

100 per cent of the genotypes for the F_1 is Rr, or heterozygous.

4 State the phenotypes of the F_1 generation.

100 per cent of the progeny will display the dominant phenotype trait of round, as the round dominant allele (R) will mask the wrinkled recessive allele (r).

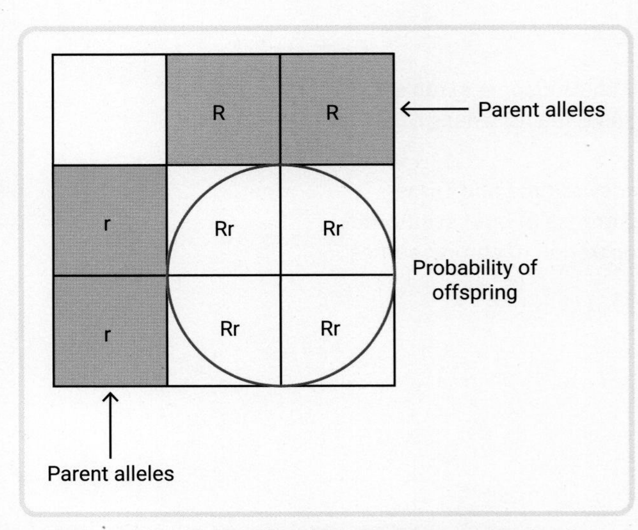

▲ **FIGURE 1.7.1** A completed Punnett square showing P_1 and P_2 genotypes and the F_1 genotypes

By separating the parent genotypes into the boxes at the top and side of the Punnett square, you are representing the independent assortment of chromosomes and production of gametes from meiosis.

Now, if you cross the F_1 generation you will see the Mendelian ratio appear in the **second filial generation** (F_2 generation). Follow the same four steps from the first cross. Table 1.7.1 shows a completed Punnett square demonstrating a cross between two F_1 genotypes and the resulting F_2 genotypes.

$$F_1 \times F_1 = Rr \times Rr$$

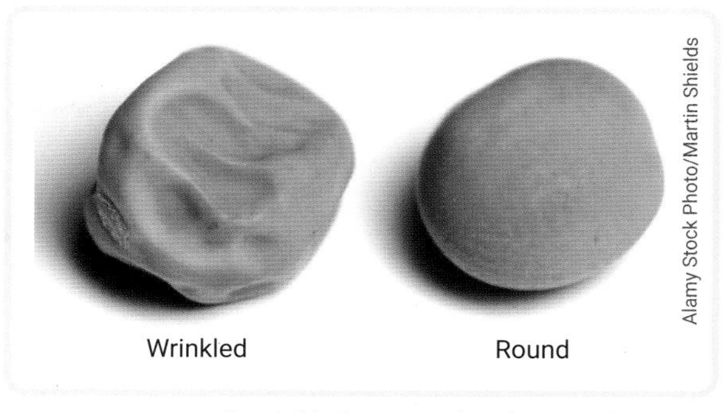

▲ **FIGURE 1.7.2** A wrinkled pea seed and a round pea seed

second filial generation
the set of offspring from the first filial parent cross

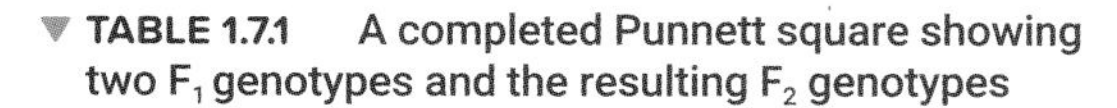

▼ **TABLE 1.7.1** A completed Punnett square showing two F_1 genotypes and the resulting F_2 genotypes

	R	**r**
R	RR	Rr
r	Rr	rr

Twenty-five per cent of the genotypes are RR, homozygous dominant, 50 per cent are Rr, heterozygous and 25 per cent are rr, homozygous recessive. Phenotypically, 75 per cent of the offspring will display round seeds and 25 per cent will show wrinkled seeds. As a ratio, the genotype can be expressed as 1 : 2 : 1 and the phenotype as 3 : 1. Figure 1.7.3 demonstrates the potential seed shapes produced in the F_2 generation.

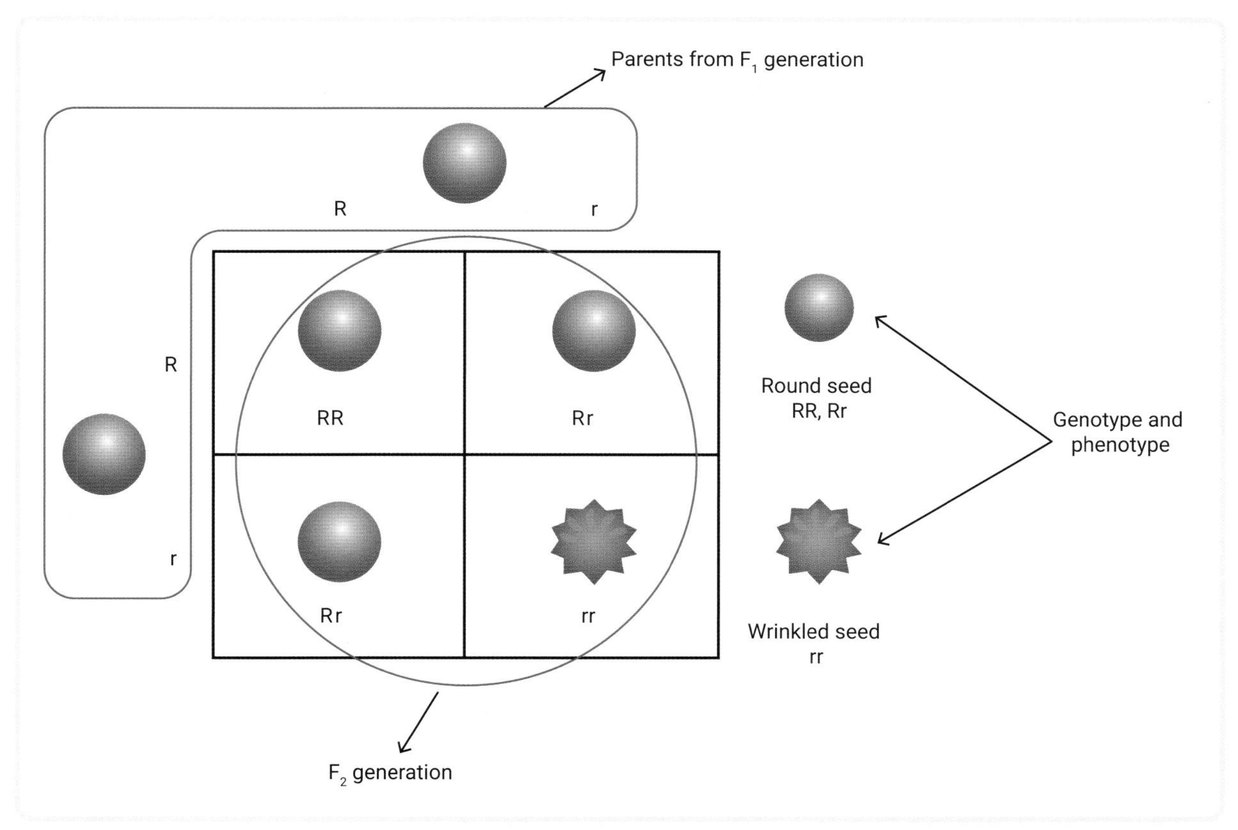

▲ **FIGURE 1.7.3** A Punnett square showing the potential seed shapes produced in the F_2 generation

1.7 LEARNING CHECK

1 In your workbook, copy and **match** the terms to the correct definitions.

Term	Definition
Homozygous	Another term for offspring
Heterozygous	The trait masked in the heterozygote
Dominant	Having two different alleles for a gene
Recessive	Having two of the same alleles for a gene
Allele	An alternative form of gene
Progeny	The trait expressed in the heterozygote

2 For each genotype listed below, **determine** whether it is homozygous dominant, homozygous recessive or heterozygous.

a AA
b Tt
c tt
d Gg
e Mm
f ss

3 For each of the genotypes listed below, **determine** the phenotype given the following information.

a Purple flowers are dominant to white flowers.
  A PP
  B Pp
  C pp
b Brown eyes are dominant to blue eyes.
  A Bb
  B bb
  C BB
c Grey fur is dominant to white fur.
  A GG
  B gg
  C Gg

4 **Complete** the following monohybrid crosses. Remember to set a key, show the parent cross, complete the Punnett square and calculate the probabilities of both genotype and phenotype of the offspring.

a In seals, the gene for the length of whiskers has two alleles. Long whiskers are dominant to short whiskers.

i **Determine** the genotypes and phenotypes of the offspring from a cross between a homozygous dominant seal and a heterozygous seal.

ii If one parent seal is heterozygous and the other is short whiskered, what is the probability that their offspring will have short whiskers?

b Curly hair is dominant in humans and straight hair is recessive.

i A woman with curly hair has children with a man who is homozygous for straight hair. Predict the genotypes and phenotypes of their children.

ii A man with straight hair, whose mother was curly haired, has children with a woman with curly hair. **Predict** the genotypes and phenotypes of their children.

iStock.com/spfoto

c In humans, right-handedness is dominant and left-handedness is recessive.

i Two parents, both heterozygotes, have children. What are the predicted genotypes and phenotypes of their children?

ii If those parents have already had one left-handed child, what is the probability of a second left-handed child?

1.8 Pedigree charts

BY THE END OF THIS MODULE, YOU WILL BE ABLE TO:

- ✓ define, interpret and draw pedigree charts
- ✓ predict genotypes and phenotypes of individuals based on a pedigree chart
- ✓ justify the possible genotypes of individuals in a pedigree chart
- ✓ model the occurrence of a certain trait over several family generations.

Interactive resource
Label: Pedigree charts

GET THINKING

When you hear the word pedigree, what do you think of? For many, it means dogs or cats that have papers showing the parents of the offspring to prove a 'pure line'. Pedigrees can also be applied to humans, as they display the lineage of inheritance over many generations. How many generations do you know of in your family tree?

Inheritance in families

pedigree chart
a diagram showing patterns of inheritance over generations; also called a family tree

In Module 1.7 you learned how Punnett squares can be used to predict the probability of genotypes and phenotypes from parent crosses. A **pedigree chart**, also known as a family tree, is a visual representation of the individuals in a family. It can be used to determine the frequency of a trait and the likelihood of inheriting that trait in subsequent generations.

How to draw a pedigree chart

There are some conventions you need to follow when drawing and interpreting pedigree charts. Figure 1.8.1 shows the symbols commonly used.

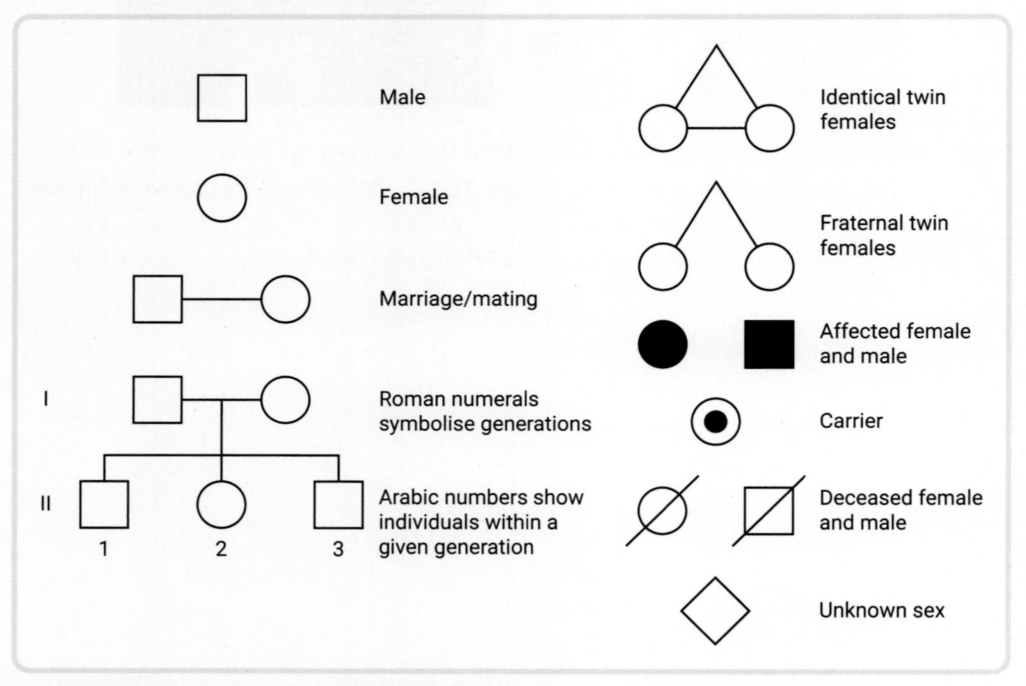

▲ **FIGURE 1.8.1** Common symbols used in drawing and interpreting pedigree charts

How to interpret pedigree charts

Following these conventions allows **geneticists** to interpret pedigrees and predict genotypes and phenotypes of individuals. This information can be particularly important if there is a history of a genetic condition in your family and you are considering having your own children. Analysing your pedigree can help you better understand the chances of passing on a genetic condition.

Alamy Stock Photo/KIKE CALVO

▲ **FIGURE 1.8.2** A family with one child affected with albinism

Albinism is a genetic condition that makes a person unable to produce the protein that makes melanin. Melanin is a pigment that gives skin, eyes and hair their colour. As can be seen in Figure 1.8.2, most people with albinism have pale skin and eye conditions and are sensitive to the Sun.

Consider the pedigree shown in Figure 1.8.3, which demonstrates the inheritance of albinism in a family.

geneticists
scientists who study genetics and inheritance

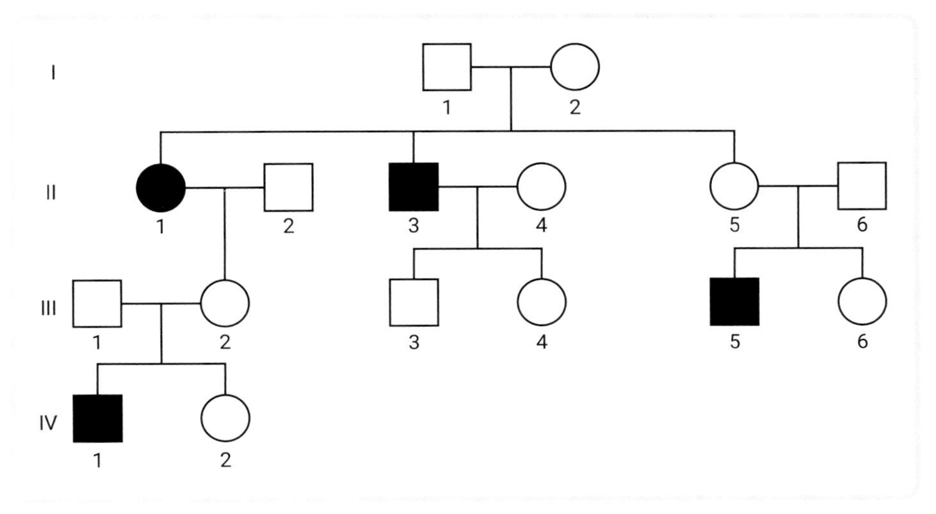

▲ **FIGURE 1.8.3** A pedigree showing the occurrence of albinism in four generations of a family

Here are some facts we can pull from this pedigree by simple observation.

- Generation II individuals 1 and 3, generation III individual 5 and generation IV individual 1 are all affected with albinism.
- Generation I individuals 1 and 2 had three children, two females and a male.
- Generation III individuals 1 and 2 had two children. The first-born son has albinism and the second-born daughter does not.

Determining mode of inheritance

A quick look at the pedigree in Figure 1.8.3 reveals some clues about the **mode of inheritance** of albinism. Albinism is not present in every generation (it is absent in generation I). This is because it skips a generation or, using terms from Mendel's laws of inheritance, the trait is masked in some individuals. This shows us that albinism is a recessive trait.

mode of inheritance
the manner in which a genetic trait or disorder is passed from one generation to the next

Albinism has affected males and females in this pedigree. This indicates that the gene for albinism is carried on the somatic chromosomes, rather than the sex chromosomes. (You can learn more about sex-linked inheritance in Chapter 9 – Biology Extension on Nelson MindTap.) We call this **autosomal** inheritance.

autosomal
the inheritance of genes on the autosome chromosomes

Determining genotype

The next step is to assign genotypes to the individuals of the pedigree.

1 Set a key. For example, A = normal skin pigmentation and a = albinism. The individuals that are shaded on the pedigree have albinism. As albinism is a recessive condition, their genotype must be homozygous recessive, or aa.
2 Work out the parents of each affected individual. As they do not have albinism, they must possess one dominant allele (A). However, because they have a child with albinism both parents must have the recessive allele in their genotype (a). This makes them heterozygotes, or Aa.
3 Finally, there are some individuals in the pedigree chart whose genotype we cannot be sure about; for example, because they have married into the family or have not produced offspring. These individuals are generation II 4, generation III 6 and generation IV 2. They do not have albinism, so must possess at least one dominant allele (A), but there is not enough information to determine if they are heterozygotes (Aa) or homozygote dominant (AA). As you are unsure, you need to annotate these examples with both possibilities, as shown in Figure 1.8.4.

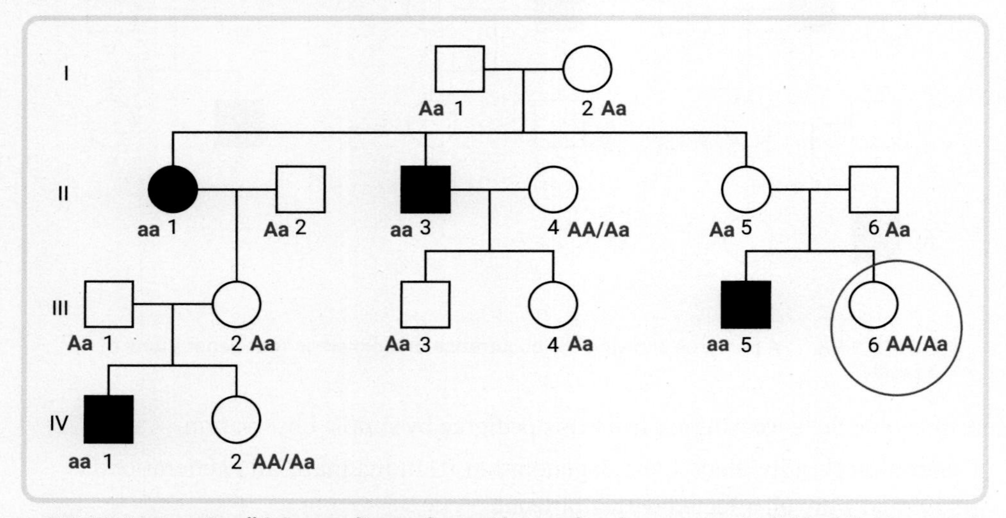

▲ **FIGURE 1.8.4** An albinism pedigree chart with completed genotypes. Individual III 6 is circled, showing her two possible genotypes.

Pedigree charts and autosomal dominant traits

Pedigree charts can also be used to show autosomal dominant traits. Consider the pedigree showing Huntington's disease in Figure 1.8.5. Huntington's disease is a rare **genetic disorder** that results in the progressive breakdown of neurons in the brain, affecting a person's cognitive, social and physical abilities.

genetic disorder
disease symptoms produced when a DNA sequence is different from normal

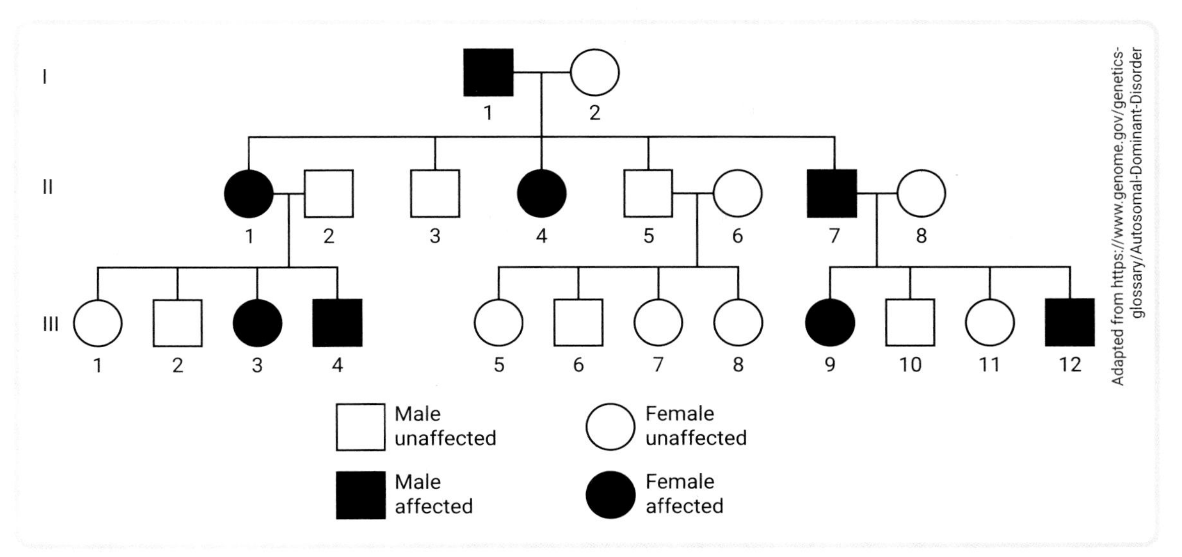

▲ **FIGURE 1.8.5** A pedigree showing the inheritance of Huntington's disease

In this pedigree, you can observe that Huntington's disease appears in every generation, and that every affected offspring has at least one affected parent. This is due to the dominant allele that is expressed in both the homozygous dominant genotype and the heterozygous genotype. Equal numbers of males and females are also affected. The mode of inheritance for Huntington's disease is autosomal dominant.

1.8 LEARNING CHECK

1 **Draw** the correct symbols for the following:
 a male without the trait.
 b female with the trait.
 c a deceased male with the trait.
 d identical twins.

2 Consider the pedigree of a family with near-sightedness shown in Figure 1.8.6. Near-sightedness is a recessive condition. Use the letters N for normal sight and n for near-sightedness to annotate each individual's genotype.
 a How many generations are shown in this pedigree?
 b How many children did individuals 1 and 2 have?
 c How many sets of partners/marriages are shown in this pedigree?
 d What is the sex and genotype of individual 2?
 e What are Jane's possible genotypes?
 f What is the probability that individuals 5 and 6 will have a child that is near-sighted?
 g What is Jane's relationship to individual 1?

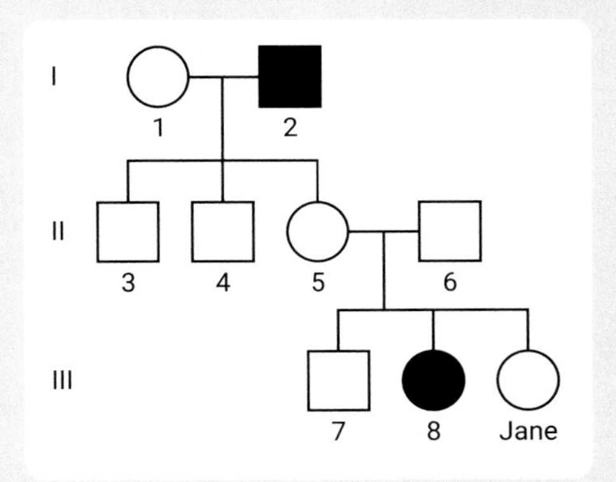

▲ **FIGURE 1.8.6** A pedigree chart showing near-sightedness

Cancer and genetic disorders

BY THE END OF THIS MODULE, YOU WILL BE ABLE TO:

- ✓ define and provide examples of cancer and genetic disorders
- ✓ predict whether a genetic disorder is dominant or recessive based on a pedigree chart
- ✓ investigate the incidences of genetic disorders in families by using pedigree charts
- ✓ explain the role of DNA in cancer and genetic disorders.

Video activity
Cancer and genetics

GET THINKING

Nearly half of all Australian men and women will be diagnosed with cancer by the age of 85. Cancer is a leading cause of death in Australia, with almost 50 000 deaths caused by cancer in 2021. These statistics are taken from the Cancer Council website. What are some positive steps you can take to reduce your risk of cancer?

Cancer is uncontrolled cell division

apoptosis
programmed cell death

cancer
uncontrolled cell division resulting in a growth or tumour

DNA is an incredible molecule! It stores all the information needed to build the proteins in a cell, produces the enzymes needed for normal cellular function, and controls your height and hair colour and even your blood type. But what happens when the code in DNA produces a protein that can be damaging to the health of an individual?

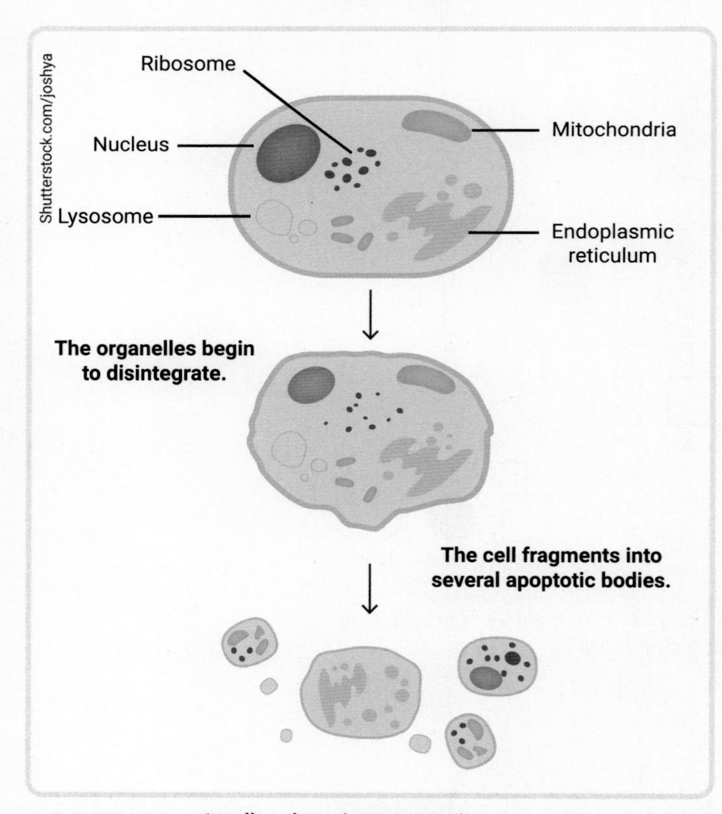

▲ **FIGURE 1.9.1** A cell undergoing apoptosis

In Module 1.3 you learned about mitosis, the cell division that is essential for the growth, replacement and repair of somatic cells. This important process is controlled by genes that help regulate the cell cycle. This ensures that the cell's DNA is copied accurately, that any errors in the DNA are repaired and that each daughter cell receives a full set of chromosomes. The cell cycle has checkpoints during interphase that allow certain genes to check for errors throughout the process. If a cell has a DNA error, it means it can't be repaired. This can lead to programmed cell death, known as **apoptosis** (see Figure 1.9.1). This removes damaged cells from the body.

However, if the body is unable to regulate the cell cycle and the cell division continues unchecked, it can lead to diseases such as **cancer**. Cancer can be defined as uncontrolled cell division. There are many different types of cancer.

9780170463041

Genetic link to cancer

Sometimes faulty genes can increase the risk of getting a specific type of cancer. For example, 5–10 per cent of breast cancers are due to faulty *BRCA1* and *BRCA2* genes. The normal versions of these genes are responsible for repairing cell damage and for keeping breast and ovarian cells growing normally. However, if there is a **mutation** present in these genes, the risk of breast and ovarian cancers increases. These faulty genes can be inherited.

mutation
a spontaneous and permanent change to a DNA sequence

Other cancers that can have a strong family link include melanoma, colon, pancreatic, uterine and prostate cancers. If you have a family history of cancers, your doctor might recommend you undergo genetic testing to see if you have a mutated gene.

Other causes of cancer

Cancers can also be caused by exposure to **carcinogens**. These are cancer-causing agents like UV radiation, chemicals in cigarettes and X-rays. Carcinogens can damage DNA or cause the cell cycle to continue unregulated, leading to cancers. Many studies have shown that exposure to carcinogens can increase the likelihood of developing cancers by up to 70 per cent. Things like smoking, drinking alcohol and leading a sedentary lifestyle can increase the chance of developing cancer, as can be seen in Figure 1.9.2.

carcinogen
an agent that increases the likelihood of developing cancer

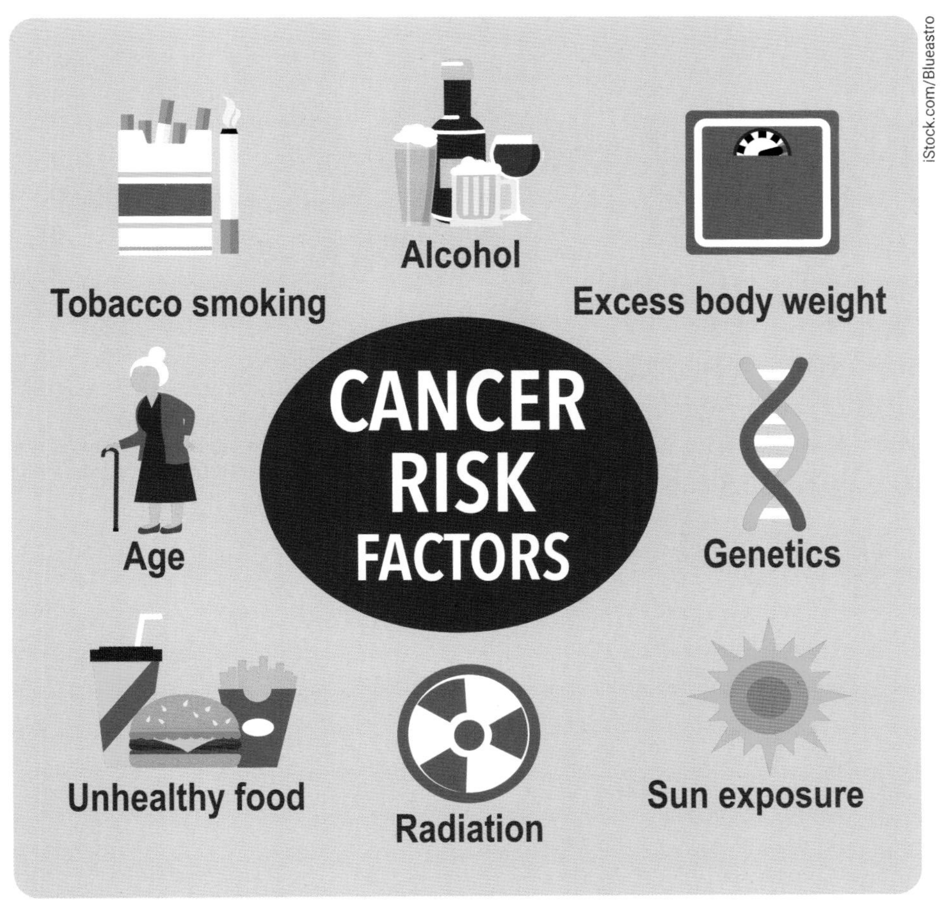

▲ **FIGURE 1.9.2** Factors that can increase the risk of cancer

Mistakes in the duplication and division of chromosomes lead to mutations. A mutation can either be inherited or spontaneously appear in an individual. Mutations can be passed on to new cells produced from the abnormal cell. While some mutations can be beneficial and important in the process of evolution, many mutations result in conditions that are not favourable to the individual or the population. Mutations will be discussed in further detail in Chapter 2.

Genetic disorders

Genetic disorders arise when the DNA sequences are changed, resulting in the sequence producing malfunctioning proteins, or no proteins at all. Genetic disorders can be caused by:

- a mutation in one gene or many genes
- damage to chromosomes
- abnormal chromosome number.

Mutations can also be influenced by environmental factors, including exposure to harmful chemicals or radiation.

Examples of genetic disorders in a single gene include cystic fibrosis, phenylketonuria (PKU), albinism and sickle-cell anaemia. Examples of genetic disorders at the chromosome level include Klinefelter syndrome, Turner syndrome and trisomy 21 (see Figure 1.9.3a and 1.9.3b).

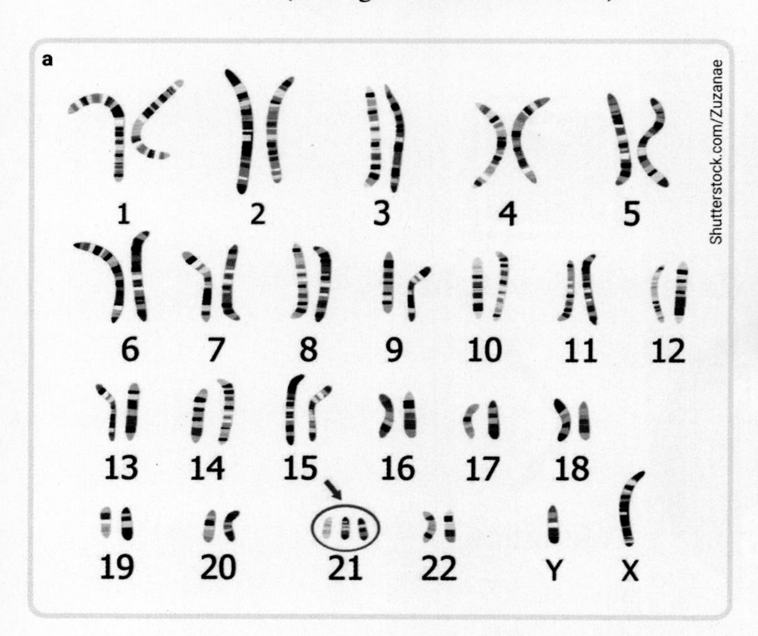

▲ **FIGURE 1.9.3** **(a)** A karyotype of person with Down syndrome, with three copies of chromosome 21 (trisomy 21); **(b)** A boy with Down syndrome and his family.

LEARNING CHECK

1 In your workbook, copy and **match** each term to the correct definition.

Term	Definition
Cancer	An allele that is masked in the heterozygote condition
Mutation	Programmed cell death
Recessive	A spontaneous and permanent change to a DNA sequence
Apoptosis	Uncontrolled cell division

2 Consider the pedigree in Figure 1.9.4, showing the incidence of PKU in a family. PKU is a rare genetic disorder that causes an amino acid called phenylalanine to build up in the body. If left untreated, this can lead to brain damage, intellectual disabilities, behavioural symptoms or seizures.

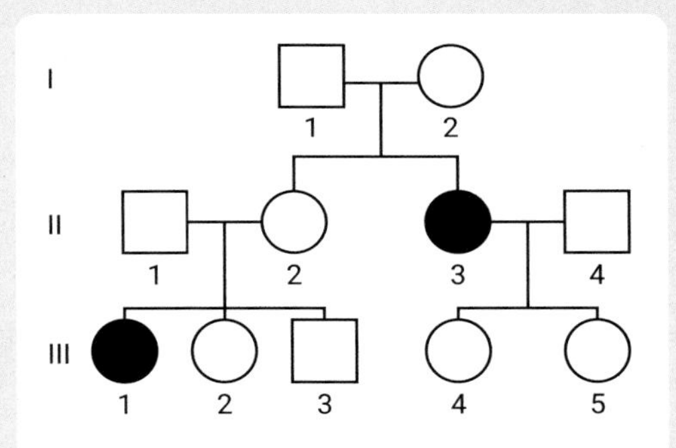

▲ **FIGURE 1.9.4** A pedigree showing the incidence of PKU in a family

- **a** Using evidence from the pedigree, what is the mode of inheritance for PKU?
- **b** **Complete** the genotypes for the following individuals:
  - **i** Gen I1
  - **ii** Gen II3
  - **iii** Gen II2
  - **iv** Gen III5
- **c** **Calculate** the probability of Gen II3 and 4 having a child with PKU.
- **d** What is the probability of Gen III2 being a heterozygote for this condition?

3 Use the Internet to **research** the following cancers. Include the causes (genetic or environmental), the likelihood of developing the cancer, ways to prevent the cancer occurring and any treatments available. Present your findings in a table.

- Breast cancer
- Melanoma
- Prostate cancer
- Cervical cancer
- Bowel cancer

1.10 First Nations Australians' kinship structures

IN THIS MODULE, YOU WILL:

✓ **Explore the significance of First Nations Australians kinship and family structures.**

Kinship is fundamental to First Nations Australians' societal structure. First Nations Australians' cultures have complex societal structures. These determine how people relate to each other socially, ceremonially and spiritually. An important aspect of these structures is they inform who is allowed to marry whom. These structures have existed long before British colonisation. In part, they function to prevent relationships between people who are too closely related. Offspring born to closely related people may inherit detrimental traits. First Nations Australians' kinship systems ensure that such relationships do not occur.

Kinship structures include:

- moiety: two halves of society divided along matrilineal (mother) and patrilineal (father) lines of descent
- totem: the natural resources a person is accountable for, ensuring their protection for passing to the next generation
- skin name: a person's bloodline and connections across generations.

Skin names are the element of kinship systems that indicates a person's bloodline. It also communicates information about how generations are linked and how they should interact. Husbands and wives do not share a skin name, and children do not have the skin name of their parents. The system is sequential in that the name is given based on a person's position in the cycle. Males and females of the same skin name are also distinguished by variations to the skin name.

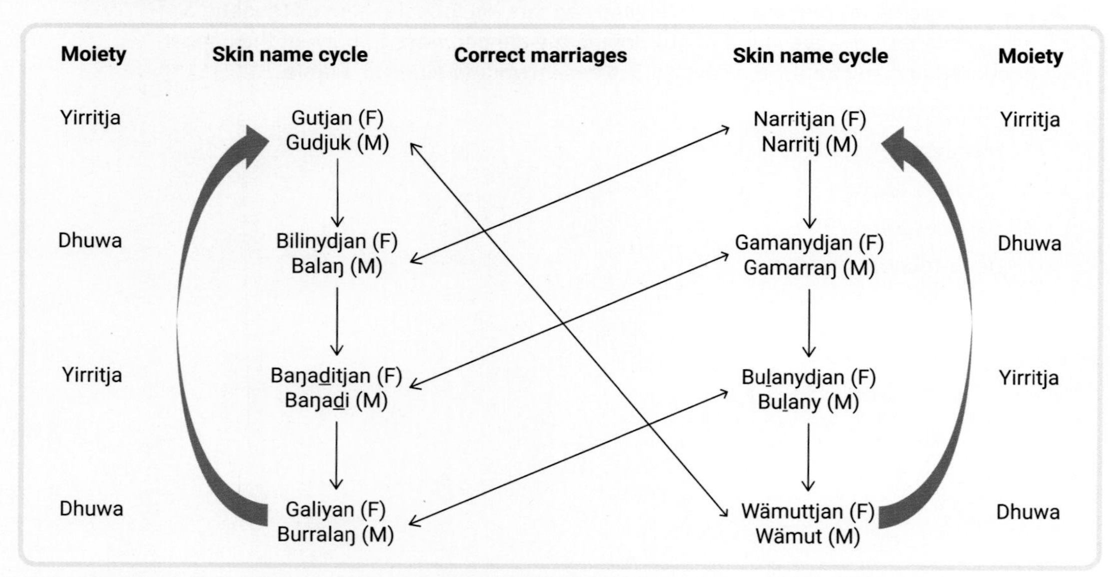

Figure adapted from diagram of Gurrutu skin names cycle, Chris Matthews, 'Indigenous perspectives in maths: Understanding Gurrutu', Teacher Magazine April 27 2020 https://www.teachermagazine.com/au_en/articles/indigenous-perspectives-in-maths-understanding-gurruu

▲ **FIGURE 1.10.1** **The skin name cycles of the Gurrutu system. The skin name cycles are only one component of Gurrutu system.**

The Gurrutu system is a complex societal structure of kinship used by the Yolngu People of north-east Arnhem Land. In the Gurrutu system, the two moieties are *Yirritja* and *Dhuwa*. A person's moiety is always opposite to their mother. Each moiety has four skin groups, and each skin group has a male (M) and female (F) version of the name, as seen in Figure 1.10.1.

The arrows show the sequential cycles of the skin names. For example, children of a person with the skin name *Gamanydjan* would have the skin name *Bulanydjan/Bulany* and their children would take the skin name next in the cycle. The moiety *(Yirritja* [Y]/ *Dhuwa* [D]) changes with each generation.

Correct marriages connect people of opposite moiety. These systems are underpinned by an understanding of the biological principles of heredity and the transmission of heritable characteristics. The diagonal arrows in Figure 1.10.1 show the correct marriages between the two cycles of skin names.

Mapping the Gurrutu system of kinship

☆ ACTIVITY

Method

1 Take 16 icy pole sticks.
   a Write a Gurrutu system skin name on each icy pole stick.
   b Write a Y or a D to indicate the moiety.
   c Colour the icy pole sticks with male skin names one colour and the icy pole sticks with female skin names another colour.
2 Use these icy pole sticks to model the generations of the Gurrutu system.
   a Begin by selecting a skin name from the cycle. Identify the correct marriage of this person.
   b Then identify the skin name of their child. Choose either the male or the female version of the skin name.
3 Continue this process to model the Gurrutu skin name system. Note that in the Gurrutu system, the children take the skin name next in the cycle following matrilineal lines (i.e. the child takes the name next in the cycle from the mother's skin name).

Evaluation

1 What does this show you about the purpose of skin name structure as a societal organisation system?
2 How do such systems prevent the transmission of inheritable harmful traits?
3 How does this system relate to Mendelian genetics?

SCIENCE AS A HUMAN ENDEAVOUR

1.11 The race to discover DNA

BY THE END OF THIS MODULE, YOU WILL BE ABLE TO:

✓ explain how scientific knowledge is validated and subject to peer review

✓ examine the work of Rosalind Franklin and how her publications contributed to the findings of Watson and Crick.

Video activity
Rosalind Franklin

A scientific race

In the 1950s, the challenge to solve the mystery of the structure of DNA became a race, riddled with rivalry and failed partnerships.

- James Watson and Francis Crick were young scientists working at Cambridge University (see Figure 1.11.1 a).
- Rosalind Franklin, Raymond Gosling and Maurice Wilkins were working at King's College, London (see Figures 1.11.1 b and c).

Alamy Stock Photo/Science Photo Library

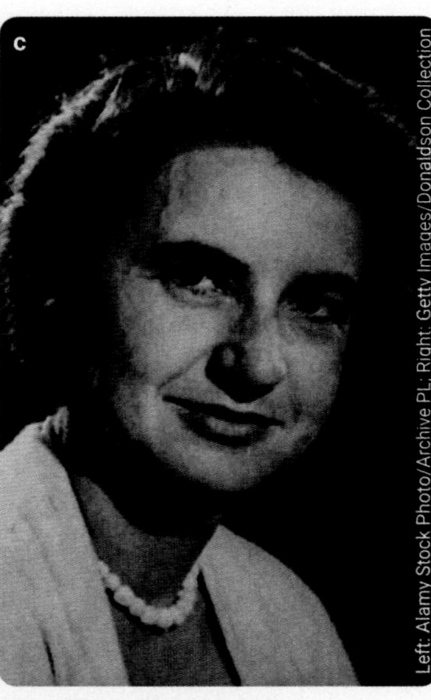

Left: Alamy Stock Photo/Archive PL; Right: Getty Images/Donaldson Collection

▲ **FIGURE 1.11.1** (a) James Watson and Francis Crick with the 3D model of DNA, (b) Maurice Wilkins and (c) Rosalind Franklin

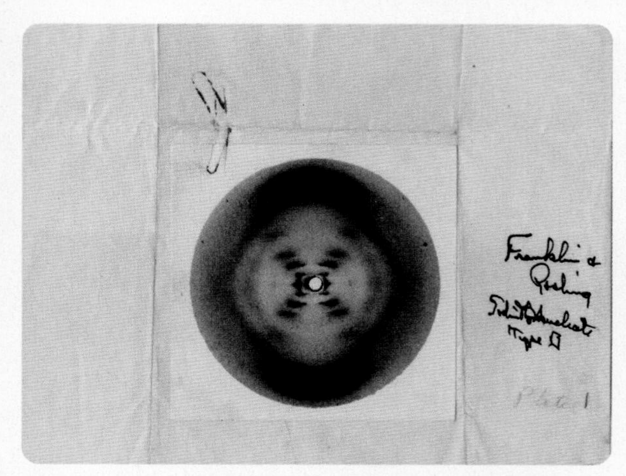

▲ **FIGURE 1.11.2** ***Photo 51*** **showing a clear cross in the middle, indicating a helical structure**

Watson and Crick were passionate about DNA. They were building models to represent how they thought the molecule was structured by trying to follow chemistry rules. However, with insufficient data, their goal was too far out of reach.

Franklin was brought onto the King's College team due to her expertise in using X-ray crystallography to carefully photograph individual strands of DNA. Watson and Crick would later use these photographs to map out the molecule's structure. Gosling, working under the supervision of Franklin, took the famous *Photo 51* (see Figure 1.11.2).

A dissolved partnership

While the partnership between Watson and Crick was successful, Franklin and Wilkins struggled to maintain a cohesive team. Franklin believed the job of determining the structure of DNA was hers alone, and Wilkins had not been consulted about Franklin joining the team of scientists working at King's College. Franklin ultimately chose to leave the team after writing a short paper providing evidence towards the double helix model of DNA.

During a visit to London, Wilkins showed Watson *Photo 51*. It was the final clue they needed to help build the helix model that had so far eluded the Cambridge pair.

The race is over

On 28 February 1953, Watson and Crick unveiled their double helix model. Nine years later, in 1962, Watson, Crick and Wilkins shared the Nobel Prize in Physiology and Medicine. Franklin had died in 1958 and, as the statutes of the Nobel Foundation stipulate that the award may not be given posthumously, they did not formally recognise her for her work.

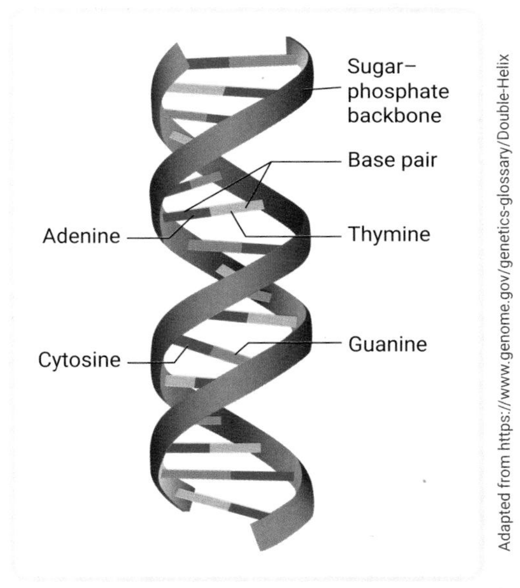

▲ **FIGURE 1.11.3** The double helix model of DNA

1.11 LEARNING CHECK

1. **Name** the team credited with the discovery of the structure of the DNA molecule.
2. **Explain** why scientific knowledge needs to be validated and reviewed by peers.
3. Do you think that the Nobel Prize committee should recognise Rosalind Franklin's work towards the discovery of the structure of DNA?

SCIENCE INVESTIGATIONS

1.12 Modelling DNA

SCIENCE SKILLS IN FOCUS

IN THIS MODULE, YOU WILL FOCUS ON LEARNING AND IMPROVING THESE SKILLS:

- building a model to represent DNA
- extracting DNA from an organism
- using a microscope to examine and draw stages of mitosis and meiosis
- using a model to predict genotypes and phenotypes from parent crosses.

Models are used in science to explain or predict a scientific concept or phenomenon. Models are especially helpful when we want to show or explain something we can't see with the naked eye. For example, we use models to represent very small structures, such as a molecule of DNA.

A good model is clear and interactive and helps communicate ideas and information clearly. Look around your science classroom. There will be models on display that your teacher refers to or passes around for students to use. Can you pick your favourite model? What is it about that model that attracts you? Is it:

- colourful?
- clearly visible in the room?
- simple?
- able to communicate an idea to a range of students with different ages and abilities?

When you make your DNA model, consider the audience and purpose.

- Who are you making your model for?
- What do you want them to understand or learn?

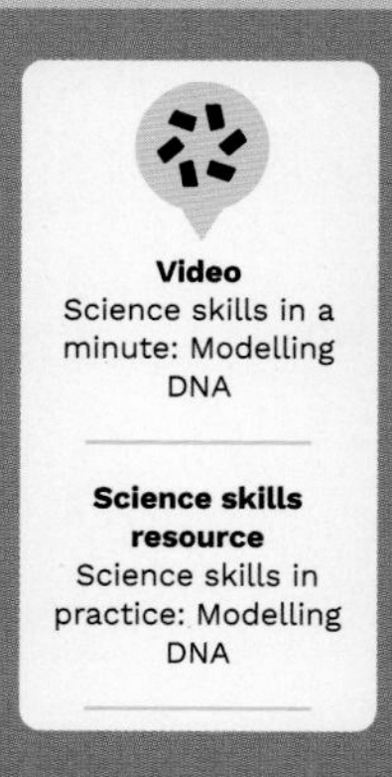

Video
Science skills in a minute: Modelling DNA
Science skills resource
Science skills in practice: Modelling DNA

INVESTIGATION 1: BUILDING A DNA MOLECULE

AIM

To build a model DNA molecule

MATERIALS

- ☑ 4 × 30 cm long pipe cleaners of two different colours
- ☑ 60 assorted beads of six different colours. Plastic beads with holes work best for this activity
- ☑ 30 cm ruler

METHOD

1 Cut one pair of pipe cleaners into strips that are 5 cm in length. You will have 12 short strips in total. You will use 10 of these in the activity (the other two are spares). Leave the other pair of pipe cleaners at their full length (see Figure 1.12.1).

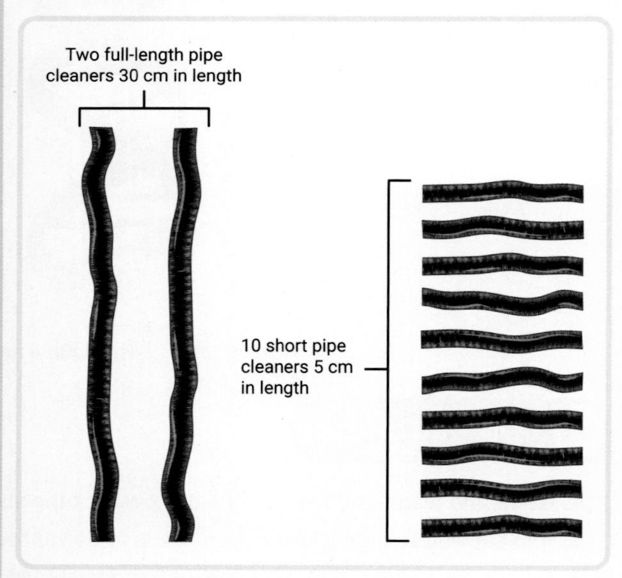

▲ **FIGURE 1.12.1** Step 1

9780170463041

2 Using two different colours of beads, thread the beads in alternating colours down each full-length pipe cleaner. Ensure the two strands match. Leave about 1.5 cm between successive beads (see Figure 1.12.2).

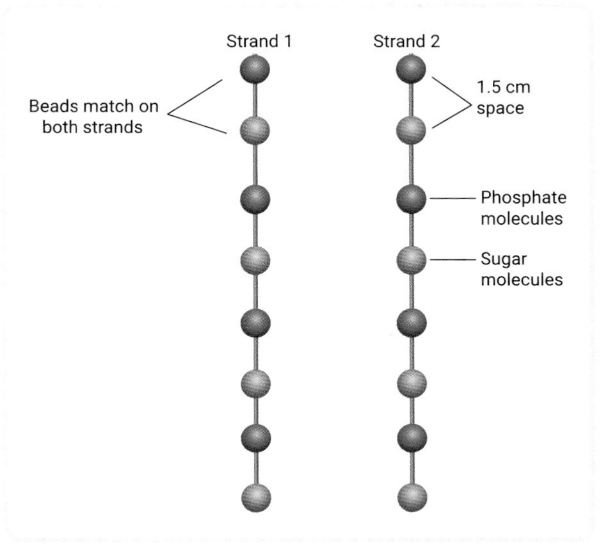

▲ **FIGURE 1.12.2** Step 2

3 Pair up the other four colours of beads so that the same two colours always match. Thread the pairs onto the 5 cm lengths of pipe cleaner (see Figure 1.12.3).

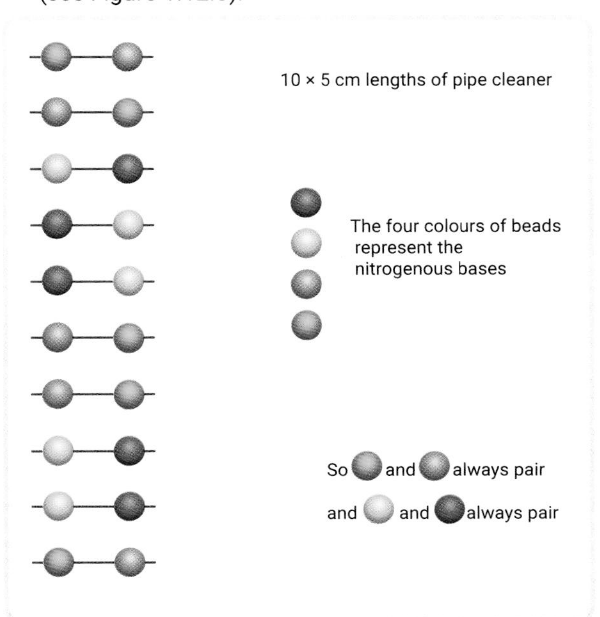

▲ **FIGURE 1.12.3** Step 3

4 Attach the 5 cm lengths to the long pipe cleaner strands. Do this by hooking each end of the 5cm pipe cleaner behind the 'sugar molecule' bead on the long pipe cleaner strand (see Figure 1.12.4). Ensure you are attaching them to the same colour bead on each long strand.

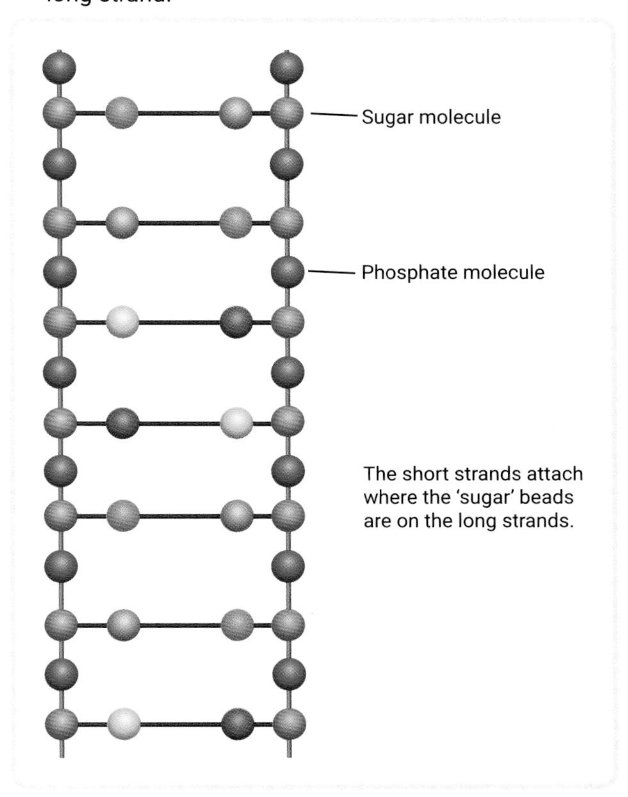

▲ **FIGURE 1.12.4** Step 4

5 Once all the small pieces have been attached, make the long strands into a double helix shape by twisting them in a anticlockwise direction (see Figure 1.12.5).

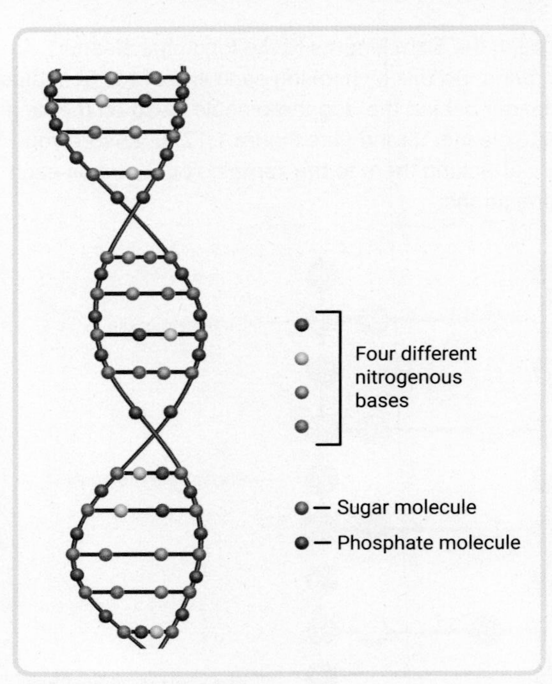

▲ **FIGURE 1.12.5** Step 5

EVALUATION

1 What part of your model represents the sugar and phosphate backbone?

2 How did you represent the nitrogenous base pairs?

3 Which colour bead in your model represents the deoxyribose sugar? Why?

4 **Draw** a labelled diagram of your model in your workbook.

5 **Examine** the other students' models. Are any strands identical? How does this represent DNA in your cells?

INVESTIGATION 2: EXTRACTING DNA

AIM

To extract DNA from an organism

MATERIALS

- ☑ biological tissue containing DNA, such as strawberry, kiwi fruit, banana, wheat germ or onion
- ☑ buffer solution containing water, salt, dishwashing detergent and meat tenderiser
- ☑ zip-lock bag
- ☑ filter funnel with filter paper
- ☑ ice-cold ethanol
- ☑ test tube
- ☑ 250 mL beaker
- ☑ hooked Pasteur pipette or paperclip

METHOD

1 Place the biological tissue into the zip-lock bag. Make a note in your workbook of the size of the biological tissue that you start with. Physically break up the material by massaging it with your hands through the plastic bag for one minute.

2 Place the buffer solution into the zip-lock bag. Continue to massage the bag for another two minutes.

3 Using the filter funnel and filter paper, filter the contents of the zip-lock bag into the beaker.

4 Pour the filtrate into the test tube until it is about a third full.

5 Holding the test tube at an angle, slowly add the ice-cold ethanol to form a layer on top of the filtrate. Add roughly the same volume of ethanol as filtrate, so the test tube is two-thirds full. DNA is insoluble in alcohol, so it will form visible strands where the two liquids meet.

7 Lift the DNA out of the test tube using the hooked Pasteur pipette or paperclip.

EVALUATION

1 What material did you extract the DNA from?

2 What is the purpose of using a buffer solution?

4 Why does the ethanol remain on top of the filtrate in the test tube?

5 **Compare** the volume of the DNA extracted from the original volume of the tissue you started with. **Estimate** the percentage of DNA present in the tissue.

INVESTIGATION 3: MITOSIS AND MEIOSIS UNDER THE MICROSCOPE

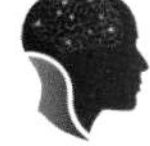

AIM

To identify key stages of mitosis and meiosis in plant cells using a compound microscope

MATERIALS

- ☑ compound microscope
- ☑ microscope slides showing meiosis (e.g. lily anthers)
- ☑ microscope slides showing mitosis (e.g. onion root tip)

Warning

Microscope slides are fragile and can break easily. If they break, please dispose of the glass safely.

METHOD

1 Turn on your microscope and position it so that you can easily look through the lens. Ensure it is on the lowest magnification setting.

2 Carefully place the mitosis slide into position and centre it under the microscopic field. Rotate to the high-power objective and refocus to observe the cells in greater detail.

3 Once you have located the cells, draw a scientific diagram of what you can see in your workbook. Remember to include the magnification.

4 Identify the stage of mitosis each cell is undergoing and label the diagram accordingly.

5 Once your diagram is complete, move your microscope back to the lowest setting and remove the mitosis slide. Replace it with the meiosis slide.

6 Increase the magnification as before and refocus to see the cells in clearer detail.

7 Complete a scientific diagram of what you can see on the slide in your workbook. Identify the stage of meiosis each cell is undergoing and label the diagram accordingly. Remember to include the magnification.

EVALUATION

1 Which phases of mitosis were you able to identify? What are the key processes occurring during each phase?

2 Which phases of meiosis were you able to identify? What are the key processes occurring during each phase?

3 From your observations, how are meiosis and mitosis different?

4 From your observations, how are meiosis and mitosis similar?

5 Why is it important to record the magnification of the microscope?

INVESTIGATION 4: PHENYLTHIOCARBAMIDE ANALYSIS

Phenylthiocarbamide, or PTC, is a natural chemical found in many vegetables, such as kale and broccoli. PTC tastes bitter to some people but is tasteless to others. The ability to taste PTC is due to the presence of a dominant gene. PTC can be purchased from a biological supply company.

AIM

To measure the frequency of the PTC gene in a population

Warning

- **Wear appropriate personal protective equipment.**
- **Follow the laboratory guidelines for the disposal of PTC papers.**
- **Ensure you are aware of any food allergies.**
- **PTC is safe when consumed in small amounts; do not exceed one strip.**
- **Wash your hands thoroughly before and after this activity.**

MATERIALS

☑ drinking water
☑ PTC papers

METHOD

1. Before tasting the strip, drink some plain tap water.
2. Place a strip of PTC paper on your tongue and let it sit for a few minutes.
3. Note any taste or sensation that you feel.
4. On a scale of 1–9, record how bitter the PTC tastes. Use the PTC sensitivity table (see Table 1.12.1) to determine the PTC sensitivity category you belong to.

▼ **TABLE 1.12.1** PTC sensitivity table

Bitterness score	PTC sensitivity	Genotype
Two or less	Low	pp
3 or 4	Mid-level	Pp
5 or more	High	PP

EVALUATION

The PTC gene was discovered in 2003. It contains two common alleles; one is a tasting allele, and one is a non-tasting allele. The ability to taste PTC is a dominant trait.

1. Using the letters P/p, assign genotypes to your group members based on their results in the taste-test.
2. **Collate** the class data of genotypes. Remember to also **identify** the sex of the individual if appropriate.

   An example of a table that could be used to record the frequency is shown in Table 1.12.2.

▼ **TABLE 1.12.2** Example genotype frequency table

	Male	Female	Prefer not to say
PP			
Pp			
pp			

3. Does the class data reflect Mendel's ratios for single gene inheritance? If so, **explain** what this implies about the mode of inheritance.
4. Some foods that naturally contain PTC are broccoli, pepper and cabbage. **Conduct** a further study by asking your classmates if they like or dislike eating those foods. Use the data collected to determine if there are food preferences or avoidances in your class group that may correlate to the presence of the PTC tasting allele.

9780170463041

REMEMBERING

1 **Recall** what DNA stands for.

2 **Explain** why DNA is known as the 'blueprint of life'.

3 **Define**:
   a chromosome.
   b karyotype.
   c autosomal chromosomes.
   d gene.

4 The shape of the DNA molecule is known as a double helix.
   a **Draw** a diagram to represent this shape.
   b **Explain** how *Photo 51* confirmed the shape of DNA.

5 **Name** the four bases found in a DNA molecule and explain how and why they pair up.

UNDERSTANDING

6 **Explain** why it is important that DNA replicates during interphase, before the cell divides.

7 **Examine** Figure 1.13.1, which shows the phases of mitosis.

Then **copy** and **complete** Table 1.13.1, explaining what is occurring during each phase.

▼ **TABLE 1.13.1** Interphase and phases of mitosis

Phase	Description
Interphase	
	The chromosomes form, nuclear membrane breaks down and the centrioles replicate.
Metaphase	
	The chromatids are pulled to opposite poles of the cell.
Telophase	
	The cytoplasm divides and two daughter cells are produced.

8 **Describe** the result of meiosis.

9 **Demonstrate**, using a colour-coded and labelled diagram, how crossing over results in genetically different chromatids.

10 **State** the benefits of First Nations Australians' kinship structures.

11 **Identify** the following statements as true or false. Modify any false statements in your workbook so they read true.
   a Homozygous individuals have two different alleles for one gene.
   b A dominant allele will mask a recessive allele in the heterozygote genotype.
   c The phenotype is the representation of the alleles in an individual.
   d Lowercase letters are used to denote the dominant trait in the genotype.

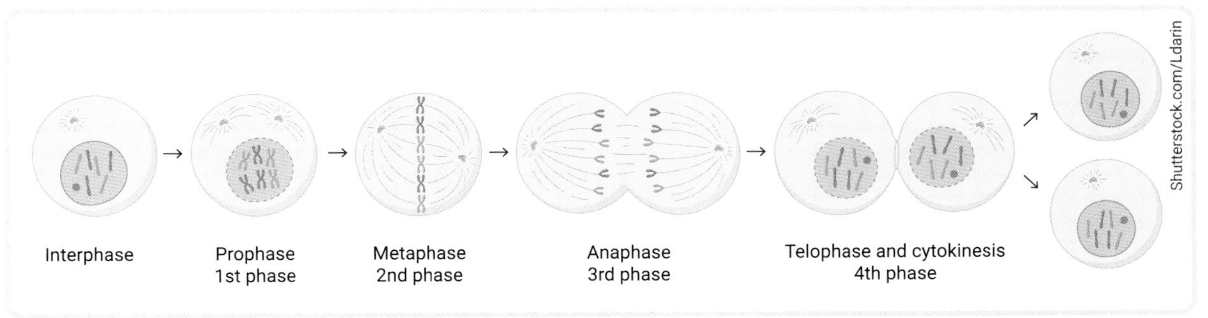

▲ **FIGURE 1.13.1** The process of mitosis consists of four phases and produces two genetically identical daughter cells.

12 **Examine** the karyotype shown in Figure 1.13.2.

a What sex does this karyotype show?

b There is an abnormal number of chromosomes in this karyotype. Identify which chromosome has an abnormal number.

c What is the number of chromosomes found in a normal human body cell?

d What is the number of chromosomes found in a normal human sex cell?

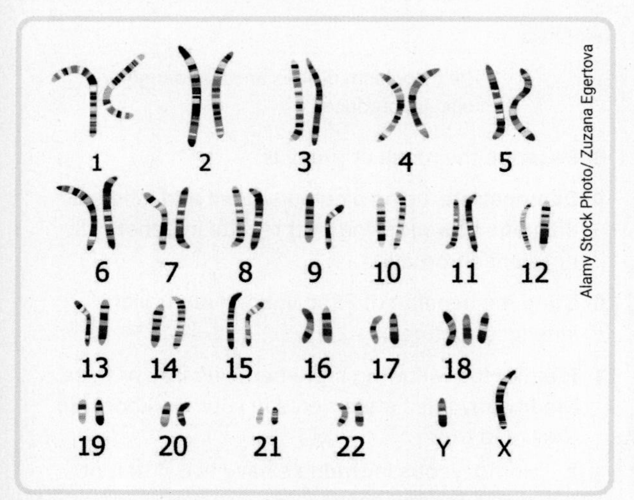

▲ **FIGURE 1.13.2 A karyotype showing an abnormal number of chromosomes**

APPLYING

13 Random assortment is the chance alignment of chromosomes at the metaphase plate during metaphase I. If a cell has a diploid (2*n*) number of 58, **calculate** the total number of combinations of chromosome pairs that can occur.

14 The ability to roll your tongue into a U shape is an autosomal dominant trait. **Justify**, using a Punnett square, how two individuals, homozygous for tongue rolling, are unable to produce offspring who cannot roll their tongue.

15 **Define** 'cancer'.

16 **Give an example** of one type of cancer caused by genetics and one type of cancer caused by environmental factors.

ANALYSING

17 Belle does not have albinism; however, her maternal grandfather did, and her maternal aunt does too. On her father's side, there is no family history of albinism.

a What are the genotypes for Belle's:

i grandfather?

ii mother?

iii father?

b What is the probability of Belle being a heterozygote for albinism? **Justify** your response by showing your working out.

c **Draw** a pedigree of Belle's family, showing the occurrence of albinism. Include all labels and genotypes of every individual mentioned in this question.

d **Classify** the mode of inheritance of albinism.

18 Melanoma is a skin cancer that can develop from exposure to UV rays. **Identify** five steps that you could take to limit your chance of developing skin cancer.

19 During anaphase II, chromosome pair 21 failed to separate into their chromatids in one of the cells.

a **Draw** the resulting daughter cells from this meiotic division.

b **Explain** what will occur if these cells are fertilised with another gamete and name this genetic condition.

EVALUATING

20 **Write** a short paragraph that links the terms nucleotide, nitrogenous base, DNA, genes and chromosomes.

21 **Conduct** a survey with your class to determine the ratio of left or right thumbs on top when fingers are casually interlocked together. Left thumbs on top is dominant to right thumbs on top.

a Does the ratio in your classroom reflect one of Mendel's laws?

b Is there a correlation between each person's dominant writing hand and which thumb naturally rests on top?

BIG SCIENCE CHALLENGE PROJECT

1 Connect what you've learned

In this unit you have learned about DNA, its role in cellular activities, how it is inherited and how the instructions it holds can influence the characteristics of future generations. Using the key terms from each module, create a mind map to show how the information that you have learned is connected.

2 Check your thinking

Explain why recessive genetic conditions such as albinism or cystic fibrosis can be unknowingly passed down from generation to generation.

How do these types of conditions compare to autosomal dominant disorders that may cause premature death, such as Huntington's disease?

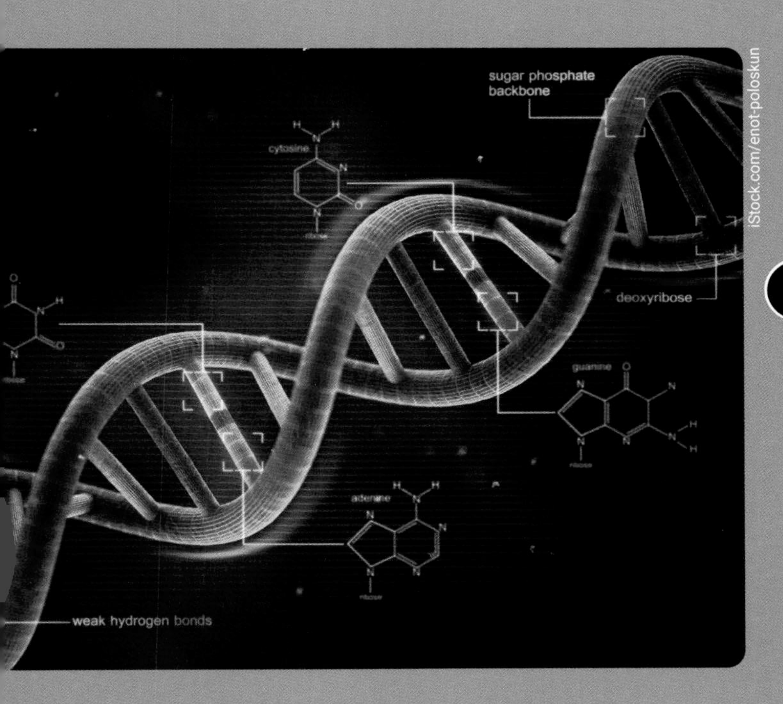

iStock.com/enot-poloskun

3 Make an action plan

Conduct research into a genetic condition of your choice. In your research you will need to determine the:

- mode of inheritance
- frequency of the condition in Australian populations
- signs and symptoms of the disorder
- treatment or prevention of the disorder.

Some suggestions to consider include cystic fibrosis, Huntington's disease, sickle-cell anaemia, melanoma, Duchenne muscular dystrophy, fragile X syndrome, Klinefelter syndrome, neurofibromatosis and von Willebrand disease.

4 Communicate

Present your research about a genetic condition as a flyer that could be used in medical centres to inform patients about the disease. It should be clear and visually communicate your research.

Use your knowledge and understanding to create a pedigree of your own family. Select a physical characteristic, such as earlobe attachment, tongue rolling or blood type, and trace the characteristic through your family as far back as possible. Present your family tree as a poster. Include photographs of your relatives and show the inheritance of your chosen characteristic through the generations.

Evolution

2.1 Variation between individuals (p. 52)
Variation allows for evolution to occur in a species.

2.2 Mutation (p. 54)
Mutations are a major source of variation and can influence evolution.

2.3 History of evolution theory (p. 59)
Many scientists worked on the theory of evolution.

2.4 Natural selection (p. 62)
Natural selection is the process by which organisms evolve and become better suited to their environment.

2.5 Speciation (p. 66)
A new species can form when selection processes act on a population.

2.6 Evolution (p. 70)
Evolution is the gradual change in characteristics in a population over many generations, which leads to different species.

2.7 Evidence for evolution: fossils (p. 72)
Fossils are the physical evidence of previously living organisms and evidence for evolution.

2.8 Evidence for evolution: comparative anatomy (p. 78)
The study of anatomical structures provides evidence for evolution.

2.9 Evidence for evolution: DNA and proteins (p. 82)
The study of DNA and proteins provides biochemical evidence for evolution.

2.10 FIRST NATIONS SCIENCE CONTEXTS: First Nations Australians' physiological responses to the Australian environment (p. 86)
First Nations Australians' long occupation of the Australian continent has resulted in the evolution of physical traits that have been advantageous in particular environments.

2.11 SCIENCE AS A HUMAN ENDEAVOUR: Genetically modified food (p. 89)
Humans have been altering the genetics of organisms for over 30 000 years.

2.12 SCIENCE INVESTIGATIONS: Identifying trends (p. 91)
1. Environmental factors influencing the hatching viability of brine shrimp
2. Observing fossils
3. Bird beaks – an investigation into natural selection
4. Examining homologous structures

BIG SCIENCE CHALLENGE 

Shutterstock.com/TOON WELL

▲ FIGURE 2.0.1 Hominins in order of evolution, ending in *Homo sapiens*

It has taken millions of years for humans to evolve into the species *Homo sapiens*. During that time, our ancestors were influenced by many evolutionary mechanisms that have resulted in us. But are we still capable of evolving? *Homo sapiens'* biological and cultural evolution has resulted in our species not feeling the influence of the environment as strongly as we once did; we can treat diseases that would previously have killed us, and we can even manipulate DNA to aid in our survival.

- **What advances in medical technology have been made to extend the life span of humans?**
- **How do you think these technological advances have affected the evolution of *Homo sapiens*?**
- **Will there be a species after *Homo sapiens*? Or are we the 'ultimate species'?**

 **SCIENCE CHALLENGE ACCEPTED!**

At the end of this chapter, you can complete the 'Big Science Challenge Project #2'. You can use the information you learn in this chapter to complete the project.

Assessments

- Prior knowledge quiz
- Chapter review questions
- End-of-chapter test
- Portfolio assessment task: Science investigation

Videos

- Science skills in a minute: Identifying trends **(2.12)**
- Video activities: Mechanisms of evolution **(2.2)**; Darwin **(2.3)**; Natural selection **(2.4)**; Speciation **(2.5)**; Fossil evidence **(2.7)**; Evolution: the evidence **(2.9)**; GMOs **(2.11)**

Science skills resources

- Science skills in practice: Identifying trends **(2.12)**
- Extra science investigations: Modelling natural selection **(2.4)**; Modelling selection pressures **(2.5)**; Modelling fossilisation processes **(2.7)**

Interactive resources

- Drag and drop: Types of mutation **(2.2)**; Structural evidence of evolution **(2.8)**
- Simulation: Natural selection **(2.4)**
- Crossword: Evolution **(2.6)**

Nelson MindTap

To access these resources and many more, visit: **cengage.com.au/nelsonmindtap**

2.1 Variation between individuals

BY THE END OF THIS MODULE, YOU WILL BE ABLE TO:

- ✓ define 'variation', 'mutation', 'crossing over' and 'independent assortment'
- ✓ explain how mutations and meiosis contribute to variation.

Quiz
Why is there evolution?

GET THINKING

Imagine walking through a rainforest or snorkelling above a coral reef. How many different organisms can you see? If you take a closer look and only focus on one type of plant or one type of fish, do you notice small differences between them? Perhaps not every leaf is the same shape, or the fish have slightly different markings. These differences are known as variation. Without variation, the process of natural selection cannot occur, and there is a greater risk of extinction. In your workbook, write down your explanation of why variation is important. After reading this module, come back and see if you need to adjust your understanding.

Variation between individuals

Have you ever noticed that even within a group of people of similar age or ethnicity, no one looks the same? Even within a family, siblings with the same parents can look similar or quite different. The reason behind this is **variation**. Variation can be described as the differences seen in the phenotypes that are the expression of genotypes of individuals of the same **species**.

variation
a difference in characteristics due to different genes

species
a group of organisms capable of reproducing under natural conditions to produce fertile offspring

evolution
the gradual change in characteristics of a species over many generations resulting in a new and different species

population
a group of individuals of the same species living in the same place at the same time

Why is variation so important? The key here lies in understanding the processes of change that result in **evolution**. Imagine if every individual within a species was genetically identical. What would happen to the species if the environment was no longer favourable to individuals with this specific genetic make-up? The species would become extinct.

An example is the small **population** of cheetahs in Africa. The cheetahs show characteristics of inbreeding and are highly vulnerable to infectious diseases carried by domestic cats. Cheetahs struggle to cope with these pressures partly because they lack the genetic variation to respond to them effectively.

Variation in a species has several sources. Meiosis is a primary source whereby the process of cell division of germline cells produces daughter cells that are genetically different from the parent cell, as shown in Figure 2.1.1.

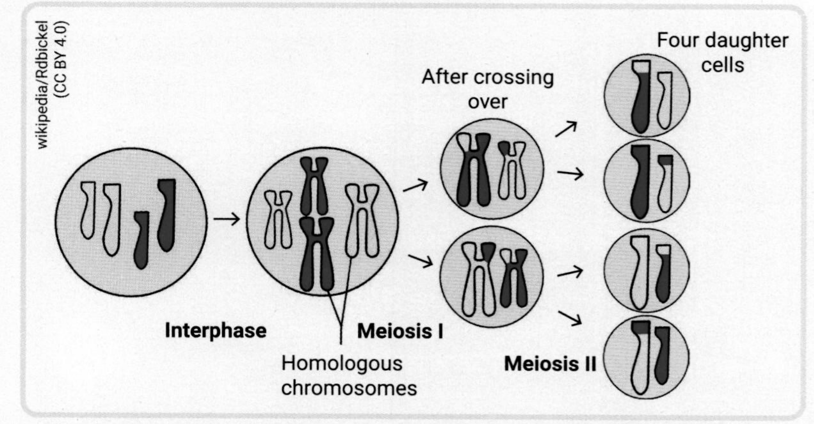

▲ **FIGURE 2.1.1** The result of meiosis is four genetically different haploid daughter cells.

Variation in meiosis

Crossing over during prophase I results in the recombination of alleles between non-sister chromatids of homologous pairs of chromosomes. The non-sister chromatid arms entangle and, at the chiasma, can detach and re-attach, exchanging genetic material. The amount of genetic material exchanged during crossing over varies, so the resulting recombination of alleles can be small or extensive.

9780170463041

Independent assortment occurs during metaphase I. This involves the random lining up of the chromosome pairs at the cell's equator (Figure 2.1.2). The paternal or maternal chromosomes line up randomly on either side of the equator. So, when the chromosomes split during anaphase I, there is a random chance of paternal and maternal chromosomes occurring in the resulting daughter cells.

independent assortment the random lining up of maternal and paternal chromosomes during metaphase I

The number of combinations in which chromosomes can line up at the equator is 2^n, where n is the haploid number of the organism. For example, humans have a haploid number (n) of 23, making the total number of possible combinations of paternal/maternal chromosomes $2^{23} = 8\,388\,608$.

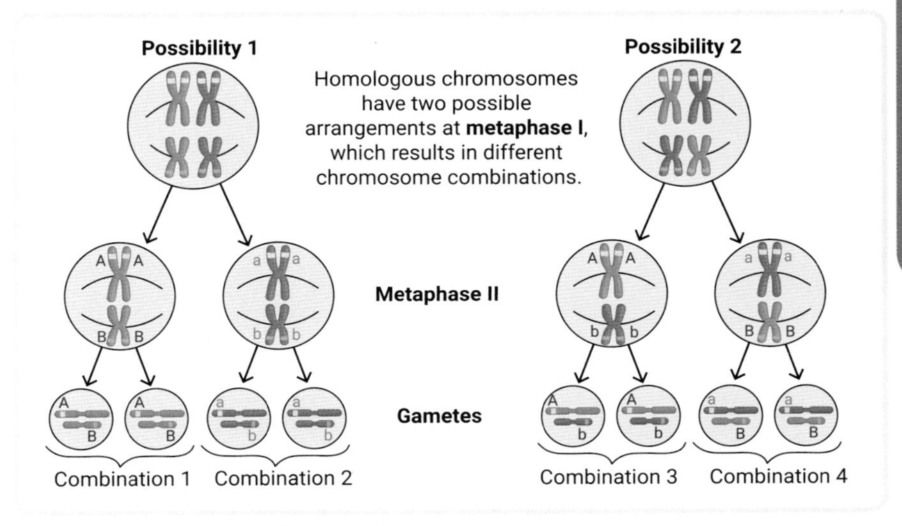

▲ **FIGURE 2.1.2** The possible combinations of chromosomes lining up during metaphase I

Other sources of genetic variation

Random fertilisation and random mating are other sources of variation in species that reproduce sexually. The different types of gametes produced through meiosis and the fact that most animal species mate with multiple individuals result in a wide variety of allele combinations in the offspring.

Mutations are spontaneous and random changes in the DNA sequence. If a mutation is inherited, it can change how frequently an allele appears in a population, resulting in many individuals with affected phenotypes. We will look at mutations in more detail in Module 2.2.

2.1 LEARNING CHECK

1. **List** four sources of variation in a species.
2. **Calculate** the number of combinations in which the chromosomes can line up during metaphase I if the diploid number of an organism is 14.
3. **Examine** the paired homologous chromosomes in Figure 2.1.3. Write out all the possible genotypes found in the gametes if crossing over occurred at the marked chiasma.

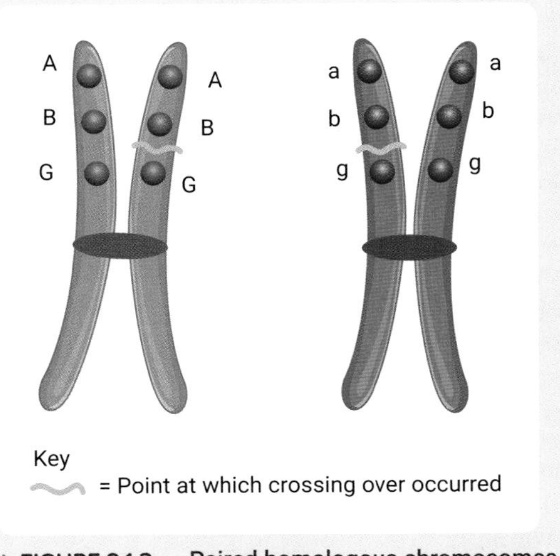

▲ **FIGURE 2.1.3** Paired homologous chromosomes

2.2 Mutation

BY THE END OF THIS MODULE, YOU WILL BE ABLE TO:

✓ define and provide examples of mutation, mutagens, genes and chromosomal mutations
✓ describe the differences between gene and chromosomal mutations
✓ recognise the changes to the DNA sequence resulting from the types of mutations within these categories.

Video activity
Mechanisms of evolution

Interactive resource
Drag and drop: Types of mutation

GET THINKING

What comes to mind when you think of the word 'mutant'? Perhaps the X-Men are the basis of your understanding of a mutant. In biology, a mutant is an organism that is physically different because of a change in its genes. Using this definition, write down some examples of mutants.

Mutations are a significant contributor to variation in a population and are important mechanisms of change in evolution.

Mutations

genome
an organism's full set of DNA

mutagen
an agent that increases the likelihood of mutation

Even though mutations are spontaneous and random, exposure to certain chemicals or agents can increase the likelihood of a mutation in the **genome.** These agents can be physical or environmental and are known as **mutagens**. Mutagens can be further classified by their origin and effects on the DNA.

Physical, chemical and biological mutagens

Physical mutagens include ultraviolet and gamma radiation. These create free radicals that cause unusual bonds between bases in a DNA molecule. This prevents specific proteins from being produced. Figure 2.2.1 demonstrates how UV radiation can cause a mismatch in a sequence of DNA.

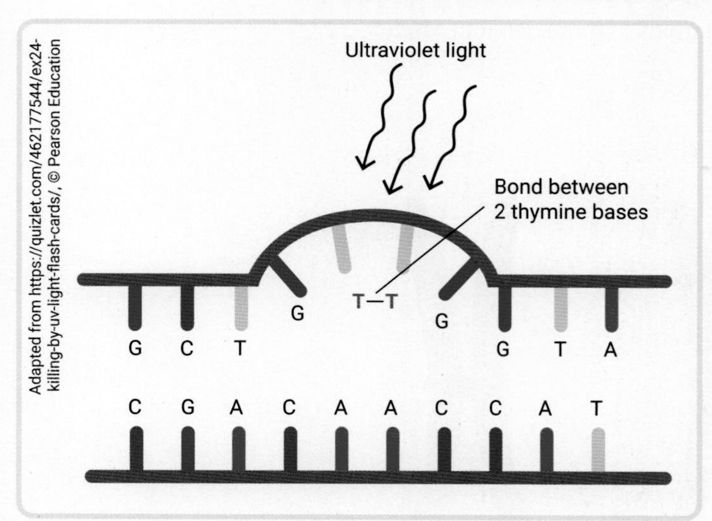

▲ **FIGURE 2.2.1** UV radiation causes a mismatch in a sequence of DNA, with two thymine bases (Ts) bonding with each other instead of with their complementary base, adenine (A).

Chemical mutagens are found in toxic compounds; for example, mustard gas. Exposure to these chemicals cause the DNA double strands to break, affecting the cells' ability to divide and multiply. Chemical mutagens are commonly used in chemotherapy to prevent cancerous cells from reproducing.

Biological mutagens are bacteria or viruses that can cause changes in the DNA sequence. An example is the human papillomavirus, which can result in cancerous cells in infected tissues, commonly the cervix or throat.

Gene mutations

Mutations can be categorised as gene or chromosomal.

Gene mutations occur in a single gene because of changes to the nucleotide sequence. Occasionally they occur during DNA replication. The **insertion**, **deletion** or **substitution** of a single nucleotide results in a point mutation and can result in changes to the protein.

In Chapter 1, we looked at how genes consist of triplets of nitrogenous bases (A, T, G and C) that code for a specific protein. Keeping this in mind, consider the sequence of words: THE CAT ATE THE RAT. Think of each word as representing a triplet of nucleotides in DNA and the sentence representing a sequence of DNA. Changing just one letter, such as the fifth letter, A, to an O, makes the sequence THE COT ATE THE RAT. You can still read the new sentence, but it no longer makes sense. This is what happens in a substitution mutation. Although the gene can still produce a protein, it may not be the protein required for normal cell function.

An insertion occurs when an additional nucleotide is inserted into a gene sequence. For example, adding a second C in the fifth position in the original sentence gives us THE CCA TAT ETH ERA T.

A deletion occurs when a nucleotide is removed from a gene sequence. For example, removing the sixth letter in the sentence gives us the sequence of THE CAA TET HER AT.

Both an insertion and a deletion result in a frameshift mutation. This is when the sequence of the nucleotides is read differently, which interferes with protein production.

Figure 2.2.2 offers a visual depiction of the three different types of gene mutations.

gene mutation a change in the DNA sequence in one or more genes

insertion the addition of a nucleotide into a DNA sequence

deletion the removal of a nucleotide from a DNA sequence

substitution the swapping of a nucleotide within a DNA sequence

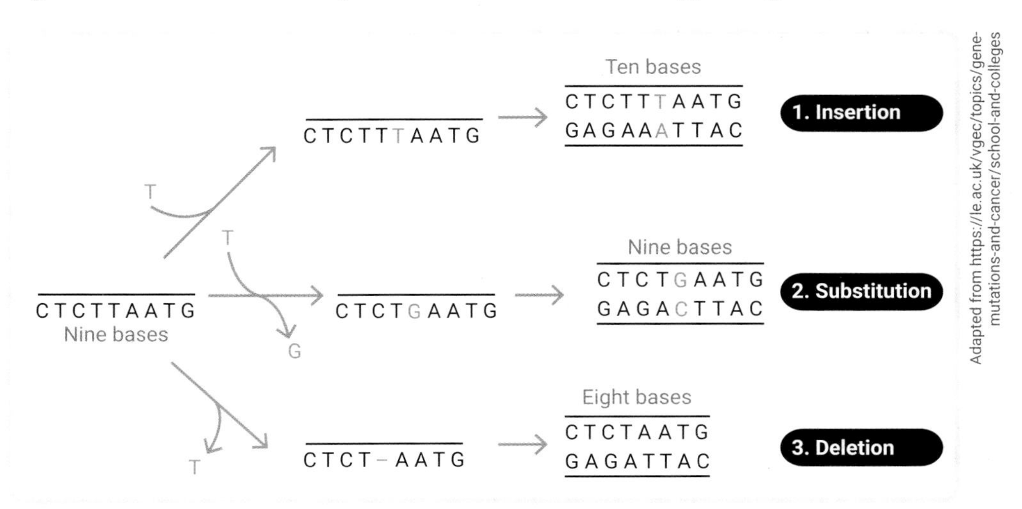

▲ **FIGURE 2.2.2** The three different types of gene mutations

Gene mutations may result in a dysfunctional protein, which can lead to a genetic condition or a cancerous growth.

Chromosomal mutations

chromosomal mutation
a change in the number of chromosomes or an arm of a chromosome

Chromosomal mutations occur at the chromosome level and involve more than one gene. They occur because of an error in the cell division during mitosis or meiosis.

Deletion mutations occur when part of a chromosome is removed. As a result, the genes that part of the chromosome should carry cannot be read, and no proteins are produced. An example in humans is cri-du-chat syndrome, where a portion of chromosome 5 is deleted (see Figure 2.2.3).

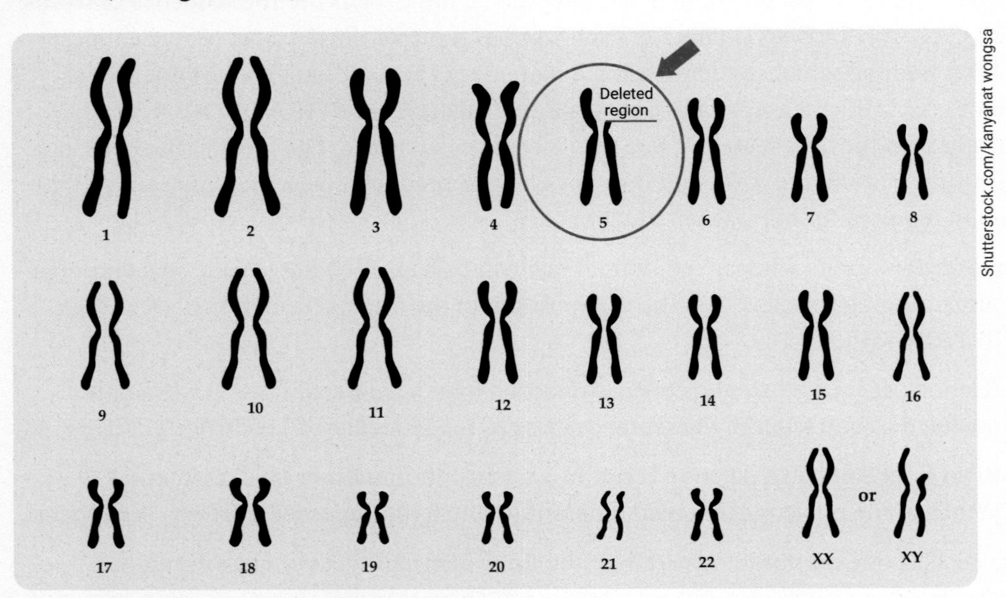

▲ **FIGURE 2.2.3** Cri-du-chat syndrome (or cat-cry syndrome) occurs when part of chromosome 5 is deleted.

inversion mutation
a chromosomal mutation in which part of a chromosome is reversed end-to-end

translocation
the result when part of a chromosome detaches and reattaches to a different chromosome

Inversion mutations occur when the arm of a chromosome breaks off and reattaches upside down. While no genes are lost, the sequence of the nucleotides is different, and functional proteins may not be produced. For example, haemophilia A results when a portion of the blood coagulation factor VIII is dysfunctional, and blood cannot clot.

Translocation results when part of a chromosome detaches and reattaches to a different chromosome pair. This results in part of a chromosome going missing or an extra part of a chromosome in the nucleus. In males, the translocation of the *SRY* gene from a Y chromosome onto an X chromosome can occur when sperm is produced during meiosis. If sperm with this X chromosome fertilises an egg, the individual will develop male characteristics, despite not having a Y chromosome.

Figure 2.2.4 offers a visual depiction of the types of chromosomal mutations.

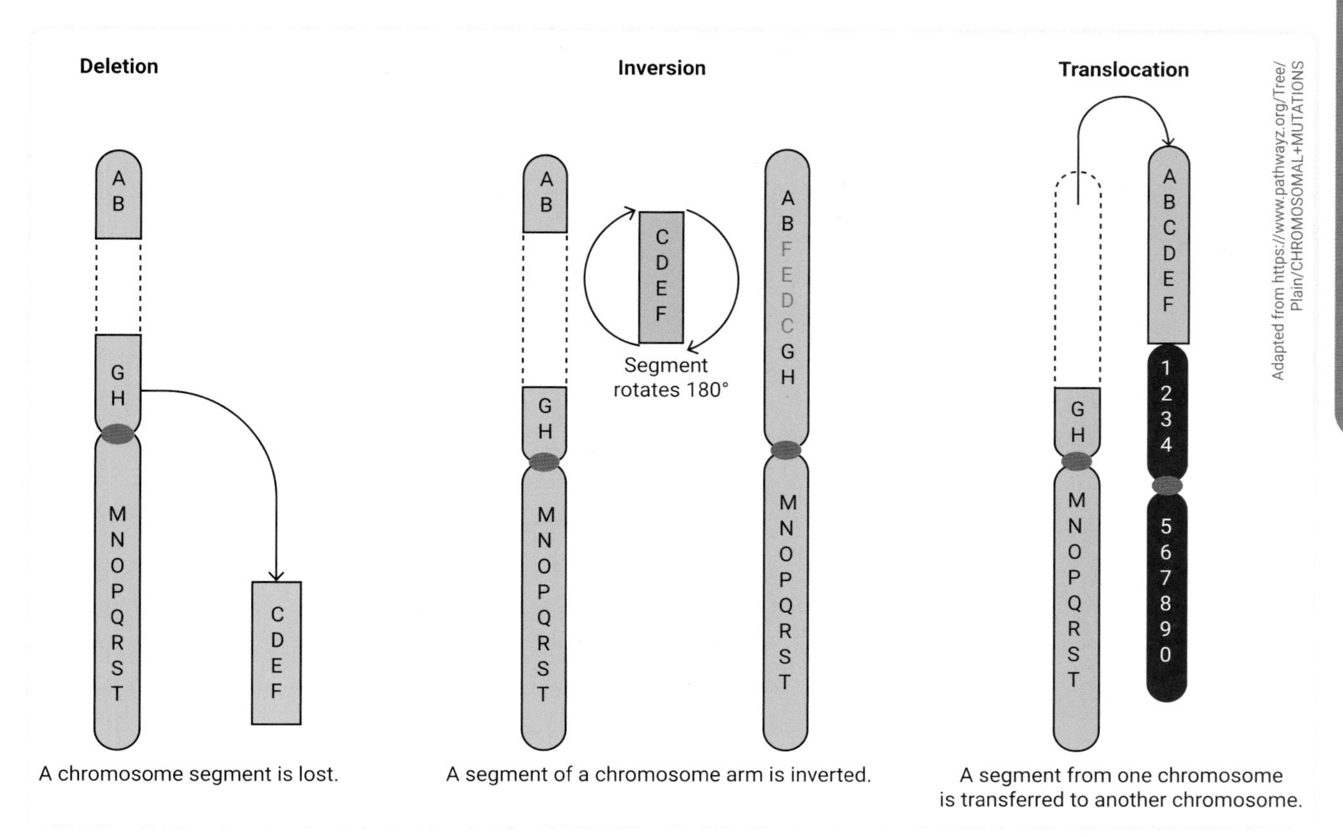

▲ **FIGURE 2.2.4** The types of chromosomal mutations

aneuploidy
having additional or missing chromosomes

Another type of chromosomal mutation is non-disjunction. This results from the incorrect separation of the chromosome or chromatid pairs during anaphase I or anaphase II in meiosis. The resulting daughter cells (gametes) display **aneuploidy**, an unusual number of chromosomes. An example of aneuploidy is trisomy 21, or Down syndrome, a genetic disorder that results when the 21st chromosome pair do not separate as normal, and the resulting zygote has three number 21 chromosomes (see Figure 2.2.5).

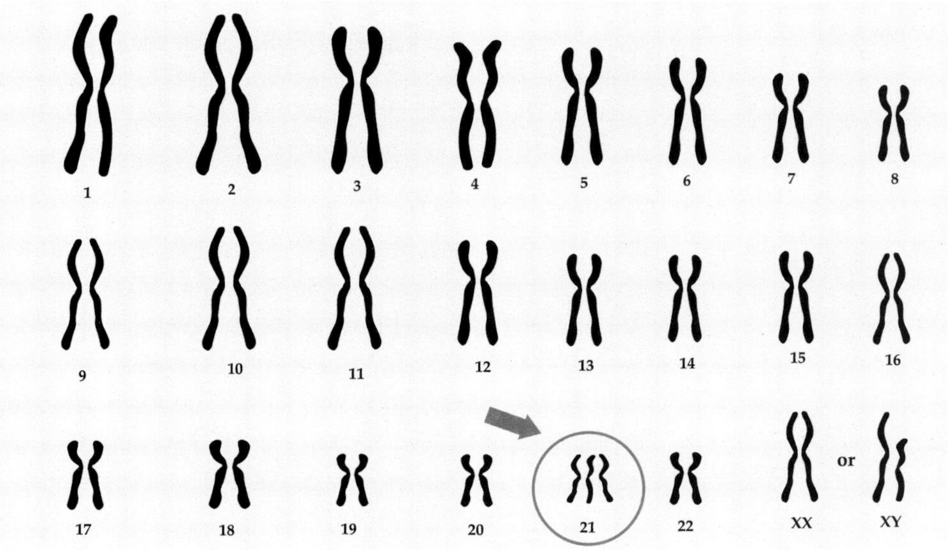

Shutterstock.com/kanyanat wongsa

▲ **FIGURE 2.2.5** Down syndrome occurs when there are three number 21 chromosomes (trisomy 21).

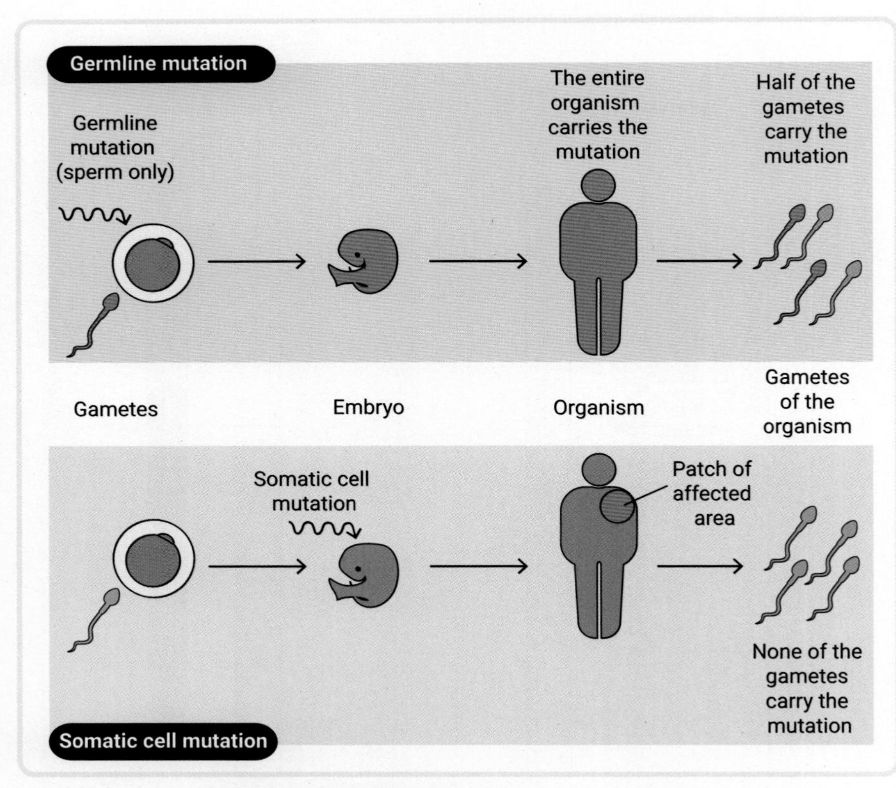

▲ **FIGURE 2.2.6** Somatic and germline mutations

Somatic and germline mutations

Many mutations occur in somatic cells rather than genes or chromosomes and so are not passed on to offspring. These are called somatic mutations. However, if the mutation occurs in the gametes (sex cells), these can be passed on and are deemed **germline mutations** (see Figure 2.2.6).

From an evolutionary perspective, somatic mutations do not impact a species' ability to evolve because the mutation ends when the individual dies. However, germline mutations can affect the offspring and subsequent generations. If the mutation is beneficial to the survival of the individual, it can be maintained in the population. If it is not beneficial, it may reduce in frequency or be removed over time.

germline mutations mutations that occur in sperm or ova

2.2 LEARNING CHECK

1 **Explain** the difference between mutation and mutagen.
2 A mutation resulted in a person being unable to make a protein to assist in blood clotting. The condition this person has is called haemophilia.
   a **Classify** haemophilia as a gene mutation or a chromosomal mutation and **justify** your choice.
   b The offspring of this parent also had blood that did not clot properly. Using this information, **classify** haemophilia as either a somatic or a germline mutation.
3 Consider the DNA sequence TAC–GCA–AAA–CGA–GTC–ATT.
   Rewrite the DNA sequence after the following mutations occurred.
   a Deletion of the 5th nucleotide in the sequence
   b Insertion of adenine (A) after the 7th nucleotide in the sequence
   c Substitution of every thymine (T) with a guanine (G)
4 Sometimes, a nonsense mutation results from the deletion or insertion of a nucleotide. Use the Internet to research what 'nonsense mutation' means and how it might affect the protein produced by that gene.

2.3 History of evolution theory

BY THE END OF THIS MODULE, YOU WILL BE ABLE TO:

✓ describe Darwin's observations and inferences on natural selection.

GET THINKING

Look at Figure 2.3.1.

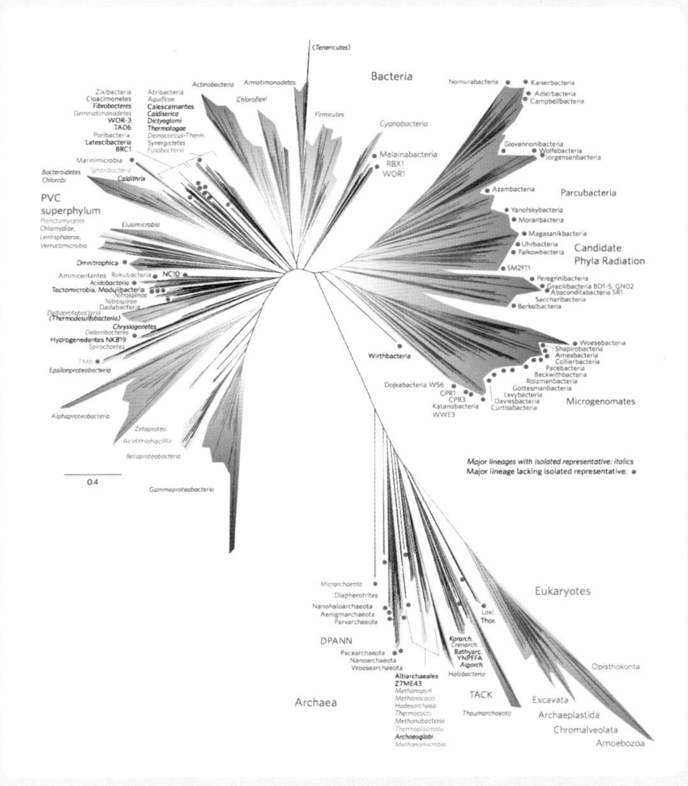

▲ **FIGURE 2.3.1** The universal tree of life

Upon first glance, it resembles a bird in flight. However, on closer inspection, it is a detailed diagram representing the hypothesis that all life descended from the last universal common ancestor. It is known as the 'tree of life' and is used both as a metaphor and as a research tool to help understand the evolution of life and the relationships between organisms, living and extinct. In your workbook, write two or three sentences in your own words to describe how the tree of life could represent evolution.

Video activity
Darwin

You might have heard the expression '**survival of the fittest**' used to explain evolution, but what does it actually mean?

survival of the fittest
the idea that individuals with the best suited characteristics will survive, reproduce and pass their traits on to the next generation

Charles Darwin—the father of evolution

In 1831, a young naturalist by the name of Charles Darwin left England for South America on the HMS *Beagle* as part of a survey of the South American coastline (see Figure 2.3.2).

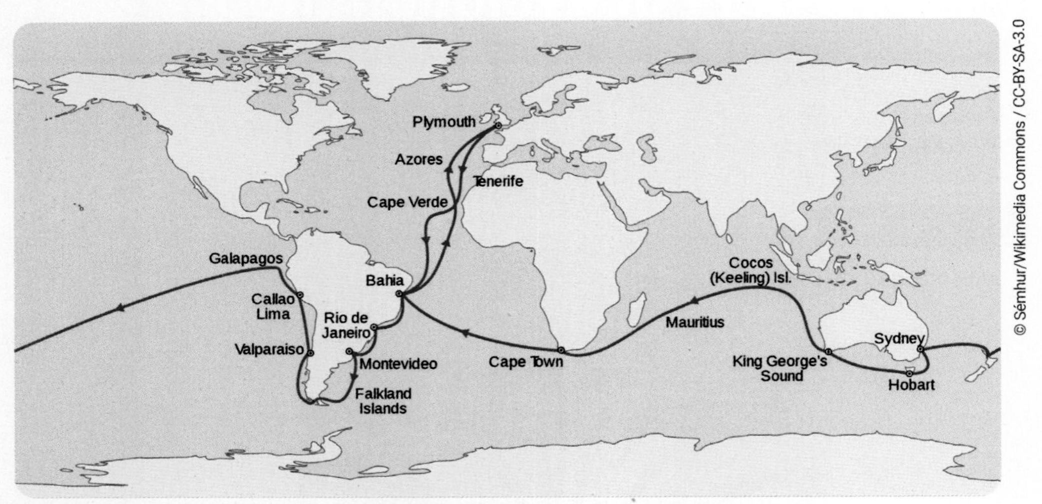

▲ **FIGURE 2.3.2** The route of the HMS *Beagle* voyage from 1831 to 1836

natural selection
the process in which an environmental factor acts on a population, resulting in some individuals being more likely to survive and reproduce

During the nearly 5-year voyage, Darwin studied many countries' geology and natural history and gathered an immense private collection of specimens. He took careful notes in his journals and sent many specimens to Cambridge University for further study. On his return to England in 1836, Darwin worked as a self-funded 'gentleman scientist', free to explore his own collections and publish his research. Darwin's notes on the tortoises and the finches of the Galápagos Islands (see Figure 2.3.3), coupled with the information given to him by a renowned ornithologist, allowed Darwin to speculate on the possibility that 'one species does change into another'.

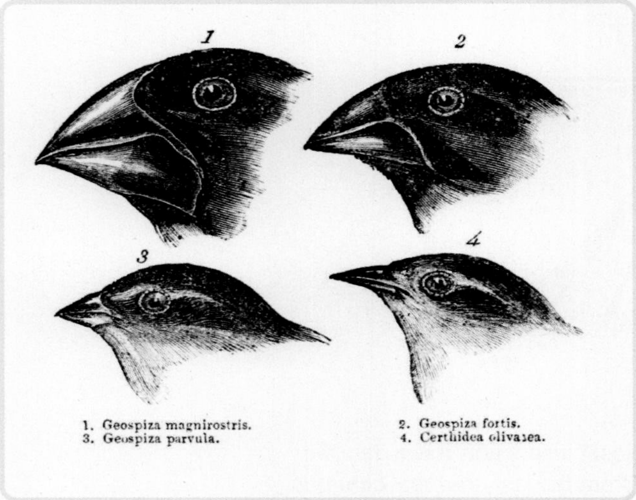

1. Geospiza magnirostris. 2. Geospiza fortis.
3. Geospiza parvula. 4. Certhidea olivacea.

▲ **FIGURE 2.3.3** Darwin's finches and the variety of beak shapes observed in the Galápagos Islands

Darwin felt no urgency to publish his ideas until another British naturalist and explorer, Alfred Russel Wallace (see Figure 2.3.4), wrote about his observations on species distribution in the Malay Archipelago. Incredibly, Wallace independently conceived a very similar theory of evolution through **natural selection**. Wallace wrote a paper, 'On the Law which has Regulated the Introduction of New Species', which was read by Darwin's colleague. This led to Wallace's paper being jointly published with some of Darwin's writings in 1858. The positive reception from the scientific community encouraged Darwin to publish *On the Origin of Species by Means of Natural Selection or the Preservation of Favoured Races in the Struggle for Life* in 1859.

▲ **FIGURE 2.3.4** (a) Alfred Russel Wallace and (b) Charles Darwin

Darwin's key observations

On the Origin of Species is considered the foundation of evolutionary biology. The book presents a body of evidence that explains the diversity of life from a **common ancestor**. Darwin explains this descent with modification through a branching pattern of evolution, as seen in Figure 2.3.5.

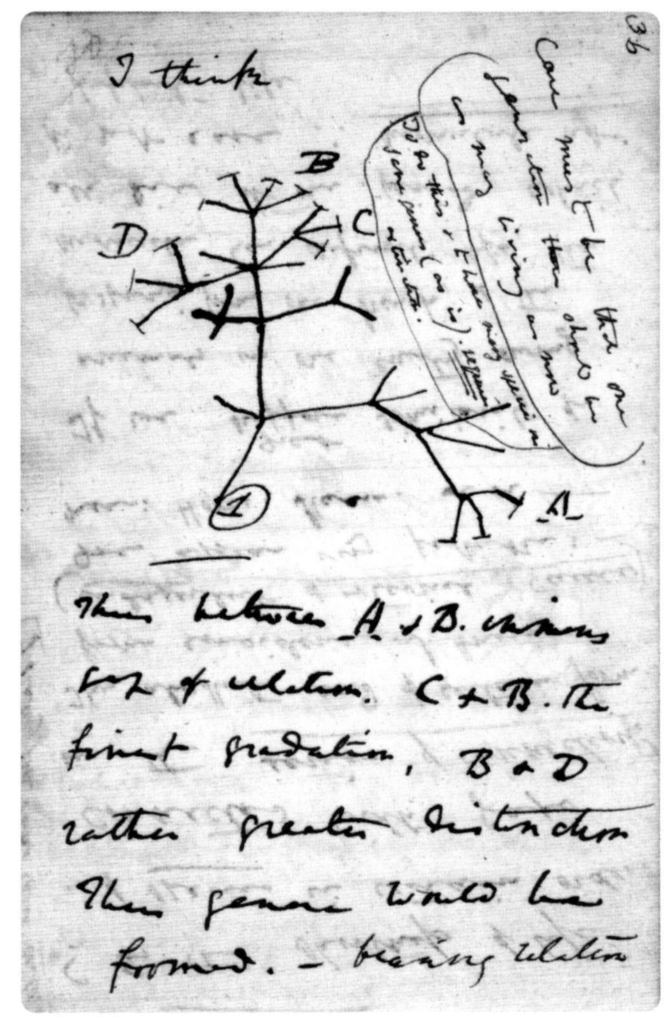

▲ **FIGURE 2.3.5** Darwin's first evolutionary tree sketch, complete with 'I think'

In his book, Darwin identified his key observations:

- traits are inherited
- all species produce more offspring than will survive to reproduce
- all members of a species show variation.

These observations gave rise to Darwin's two main inferences.

- There is a **struggle for existence** due to high reproduction rates and limited resources in the environment.
- The organisms with the traits to help them survive best will produce more offspring, passing on the genes that helped them survive. Over generations, the population will have more individuals with the genes best suited to their environment. This is coined survival of the fittest.

This is the mechanism of evolution known as natural selection.

2.3 LEARNING CHECK

1 **Define** 'natural selection'.

2 a Who was Charles Darwin?

  b What were his key observations made on the HMS *Beagle* voyage?

  c **List** the two inferences he made from these observations.

3 Use the Internet to find information about Charles Darwin's scientific findings and **create** a timeline from the beginning of his journey on the HMS *Beagle* to the publication of *On the Origin of Species*.

common ancestor
the ancestor that two or more descendants have in common

struggle for existence
the competition between individuals for required resources such as food, water or space

2.4 Natural selection

BY THE END OF THIS MODULE, YOU WILL BE ABLE TO:

- ✓ define and describe the processes of natural selection and artificial selection
- ✓ define the term 'selective agent' and describe the effect of a selective agent on a population
- ✓ compare and contrast natural selection and artificial selection.

selective agent
the environmental factor acting on the population

gene pool
the total amount of genetic material available in a population

GET THINKING

Many scientists accept the theory of evolution because of the law of natural selection. In science, we often describe theories and laws to explain observations about our natural world. Using the Internet, write definitions of 'theory' and 'law'. Using these definitions, write a short statement to explain the concept of evolution to a person who does not yet know about or understand the concept.

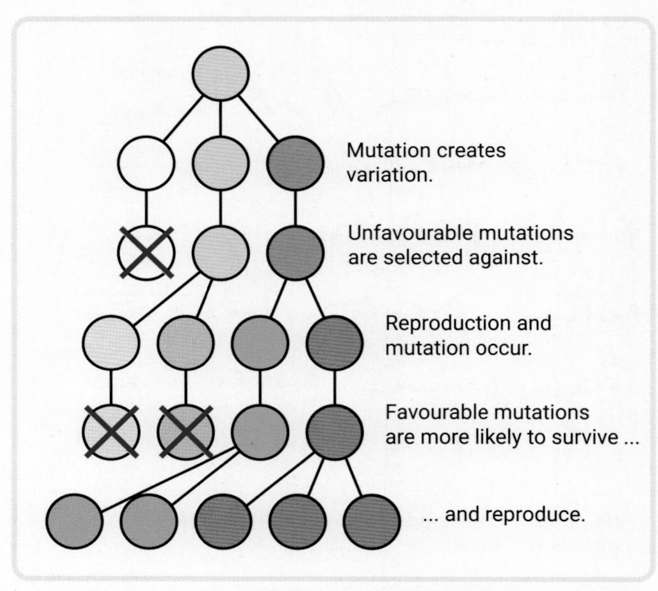

▲ **FIGURE 2.4.1** A representation of natural selection

Natural selection as a mechanism of evolution is now widely regarded as the clearest explanation for the diversity of organisms on Earth. A visual representation of natural selection is provided in Figure 2.4.1.

Peppered moths: a natural selection case study

Natural selection can be seen in peppered moths. Peppered moths exhibit variation in the colour of their body and wings, ranging from a light-coloured body with black speckles on the wings to a variant form in which both the body and wings are entirely black, as seen in Figure 2.4.2. Both these populations lived in rural England and in the cities prior to the Industrial Revolution.

During the Industrial Revolution, the cities became much dirtier due to the pollution produced by engines. The walls of buildings became covered in dark soot. The light-body variant of the peppered moth was unable to camouflage against the darkened walls, and the **selective agent**, predatory birds, could easily find and eat the light-coloured moths, removing them and their alleles from the **gene pool**. The black moths were hidden against the black walls, allowing them to survive, reproduce and pass the black allele on to the subsequent offspring. Over time, the population of peppered moths in industrial areas became, predominantly, the black form variant.

▲ **FIGURE 2.4.2** The two variants of the peppered moth on a tree. The light-body variant is well camouflaged on this pollutant-free tree.

Natural selection in humans

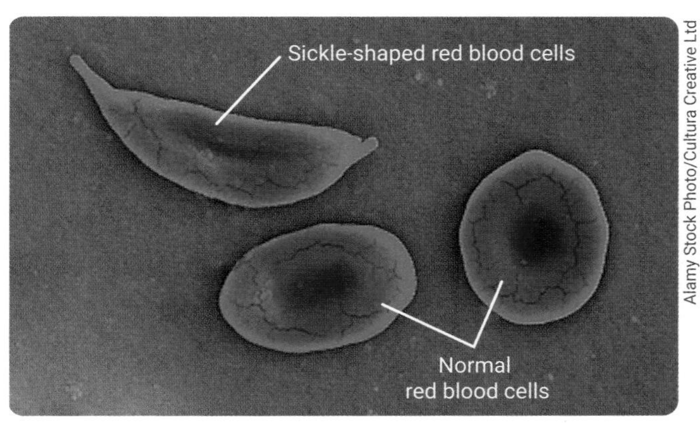

▲ **FIGURE 2.4.3** Normal red blood cells and sickle-shaped red blood cells

Natural selection can also be seen in human populations. Sickle-cell anaemia is a genetic disorder caused by a mutation in the gene coding for the beta-haemoglobin chain. Individuals with two recessive alleles have sickle-cell anaemia, a condition that results in red blood cells having a sickle shape and being unable to carry sufficient oxygen.

Individuals with one allele for the disease have the sickle-cell trait but suffer no major ill effects, provided they live at sea level. Living above sea level, where there is less oxygen in the atmosphere, causes some of the red blood cells to become sickled, which can become stuck in the blood vessels, creating a clot. The difference between normal red blood cells and sickle-shaped red blood cells is shown in Figure 2.4.3. The advantage to humans is seen in malaria zones, where the heterozygotes have a resistance to malaria, survive and reproduce, passing the sickle cell allele on to the next generation.

Artificial selection

Darwin also observed farmers selectively breeding stock to produce desired characteristics in their offspring. He saw this as an analogy for natural selection, whereby humans were the selective agent choosing the best traits to be passed on to subsequent generations. This is called **artificial selection**.

artificial selection
the process whereby humans breed organisms for desired traits

Artificial selection is a faster mechanism of evolution than natural selection, which takes many generations. Artificial selection is driven by human choice, while natural selection is environmentally driven.

Humans have selected desired characteristics in plants and animals since the Neolithic age, 10 000 years ago. For example, humans have selected animals such as sheep, goats and cows for breeding based on their particular traits, such as good milk production, thicker wool or more meat, so these desirable traits were passed on to the next generations. Belgian Blue cattle, for instance, are bred to produce twice the amount of meat (see Figure 2.4.4).

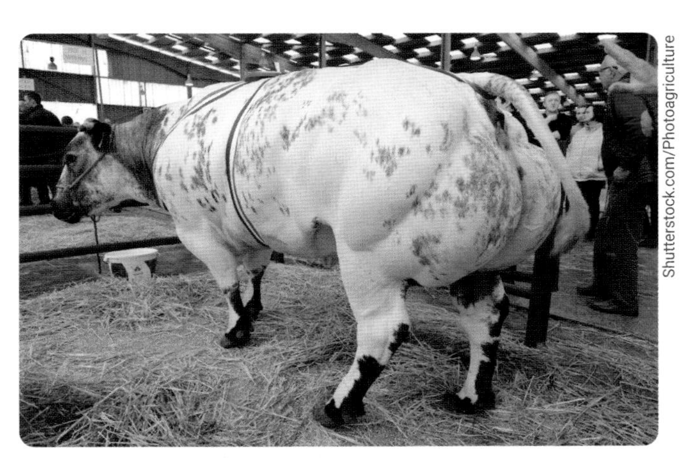

▲ **FIGURE 2.4.4** Belgian Blue cattle are bred to produce twice the amount of meat.

Video activity
Natural selection

Interactive resource
Simulation: Natural selection

Extra science investigation
Modelling natural selection

Today, the results of artificial selection are clearly seen in domestic cats and dogs, with the careful selection and breeding of traits in individuals used to produce a wide variety of breeds. For example, cattle dogs are selected for their hard work, loyalty and innate ability to herd and Siamese cats are selected for their light grey fur and blue eyes (see Figure 2.4.5).

Shutterstock.com/Sari Oneal

▲ **FIGURE 2.4.5** Over thousands of years, humans have selected traits in domesticated animals to suit their needs.

Humans also use artificial selection in plants. Many of the fruits and vegetables you purchase from the supermarket look nothing like their ancestral wild forms. *Brassica oleracea* (see Figure 2.4.6) is the wild form of cabbage and has been carefully cultivated over many years to produce vegetables such as cabbage, cauliflower, broccoli, kale, kohlrabi and brussels sprouts (see Figure 2.4.7).

wikimedia/MPF

▲ **FIGURE 2.4.6** *Brassica oleracea*, commonly known as wild cabbage, grows in many parts of Europe.

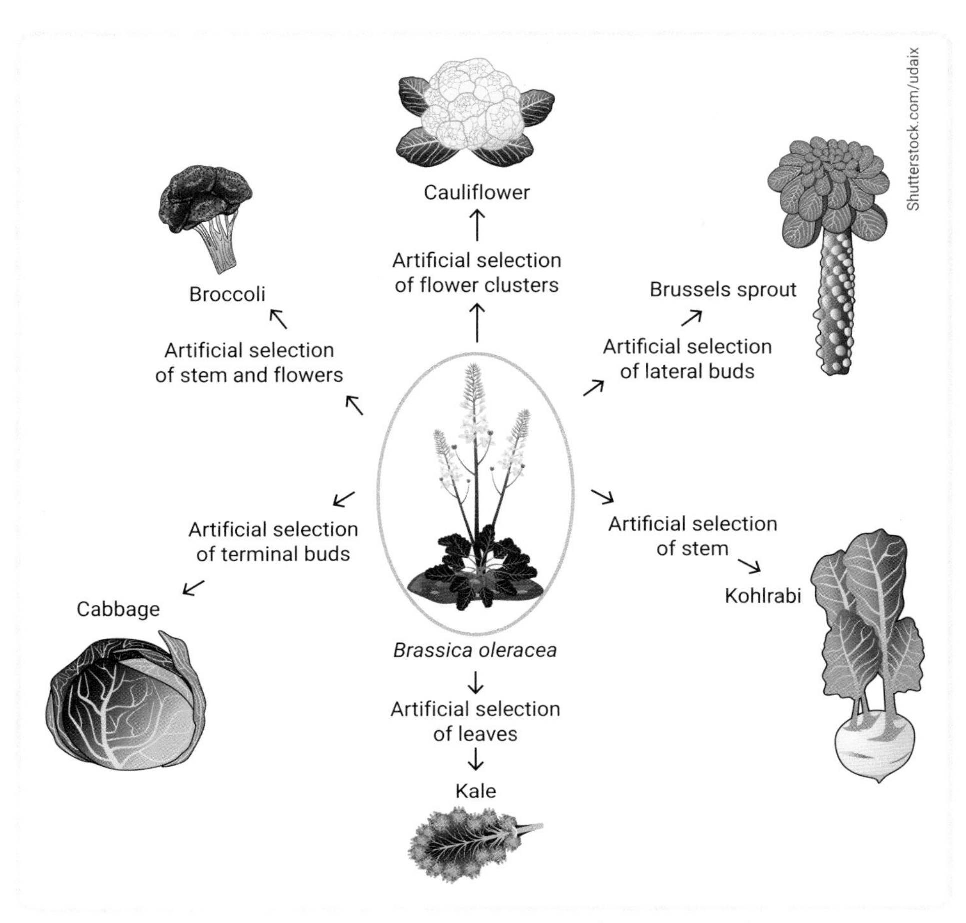

▲ **FIGURE 2.4.7** Many vegetables we eat today originate from *Brassica oleracea*.

2.4 LEARNING CHECK

1 **Define**:
   a natural selection.
   b artificial selection.
2 Considering the case study of the peppered moth:
   a **identify** the two variants.
   b **name** the selective agent acting on the moths.
   c **describe** how the different variants of moth in the city regions were affected by the selective agents and explain what impact this had on their gene pool.
3 Antibiotic-resistant bacteria are an increasingly common problem. Many bacterial infections (e.g. 'golden staph') no longer respond to antibiotics. Using your understanding of natural selection, **explain** how antibiotic-resistant bacteria have formed.
4 **Draw** a Venn diagram to compare and contrast the processes of natural and artificial selection.

2.5 Speciation

BY THE END OF THIS MODULE, YOU WILL BE ABLE TO:

✓ describe the process of speciation to form a new species.

Video activity
Speciation

Extra science investigation
Modelling selection pressures

GET THINKING

In 2021, 23 species of animals and plants were declared extinct worldwide. Largely, these extinctions resulted from human impacts on habitat, including climate change. However, Professor Andrew Pask, a researcher from the University of Melbourne, is working towards making the Tasmanian tiger de-extinct. The technology he will be using is at the heart of the project to bring the woolly mammoth back from extinction by 2027.

What animals or plants would you bring back from extinction if you could? What do you think will need to change in our global environment to support the de-extinction of these animals and plants?

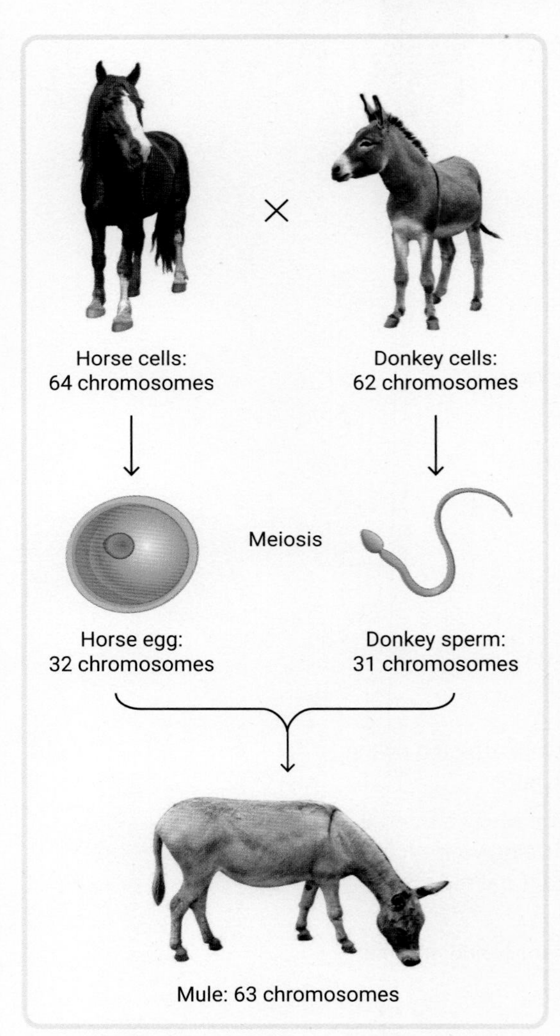

▲ **FIGURE 2.5.1** A cross between a horse and a donkey results in an infertile hybrid mule.

What is a species?

As humans, we have the biological name *Homo sapiens*. This means we belong to the genus of *Homo* and are part of the species *sapiens*. There are many other species in the *Homo* genus that existed before us and that are now extinct. What determines a species, and why is it important in the process of evolution?

Biologically, a species is defined as a group of individuals that breed under natural conditions to produce fertile offspring. For example, the domestic cat belongs to the species *Felis catus*. While there are many different breeds of cats, they can all produce fertile offspring if they mate. If members of different species mate, any successful offspring will likely be infertile.

Consider the hybrid mule, as shown in Figure 2.5.1. A mule is produced when a horse (*Equus caballus*) reproduces with a donkey (*Equus asinus*). Even though the horse and donkey are from the same genus, their different species names indicate that they would not be able to produce fertile offspring if they mate. This is because each species has a different and incompatible number of chromosomes, resulting in offspring with an odd number of chromosomes.

Speciation

The process of **speciation** is when two populations become so genetically different that they can no longer breed with each other to produce fertile offspring. Speciation relies on variation and **selection pressures** on a population. For speciation to occur, variation must exist within a population. That's because different phenotypes allow for selective agents to act on the population, providing an advantage to the individuals that are the 'most fit' for survival in that specific environment.

For speciation to occur, there must also be a degree of **isolation** between the original population and the break-away group(s). Isolation prevents the movement of alleles between population groups, often described as 'gene flow'. Isolation can be classified as geographical or sociocultural in human populations (see Table 2.5.1). When isolation occurs over an extended period, it can lead to the groups becoming reproductively isolated, meaning the different populations cannot reproduce with each other.

speciation
a process in which two groups become so genetically different they can no longer breed with each other under natural conditions to produce fertile offspring

selection pressure
the effect the selective agent has on the population

isolation
a mechanism or barrier to separate breeding populations

▼ **TABLE 2.5.1** Examples of the different ways a population can become isolated

Geographical isolation	Sociocultural isolation in human populations
Large bodies of water, such as the ocean, lakes or rivers	Language
Mountain ridges	Religion
Canyons	Economic status
Deserts	Education
Icebergs	Sexual selection

As groups become increasingly isolated due to limited gene flow, different selective agents begin to act on the separated populations. These selective agents provide an advantage to the individuals with the phenotype best suited to a specific environment. Over many generations, the separated gene pools will become significantly different from each other. Once individuals in the different gene pools can no longer produce fertile offspring, they are considered different species. This process is demonstrated in Figure 2.5.2.

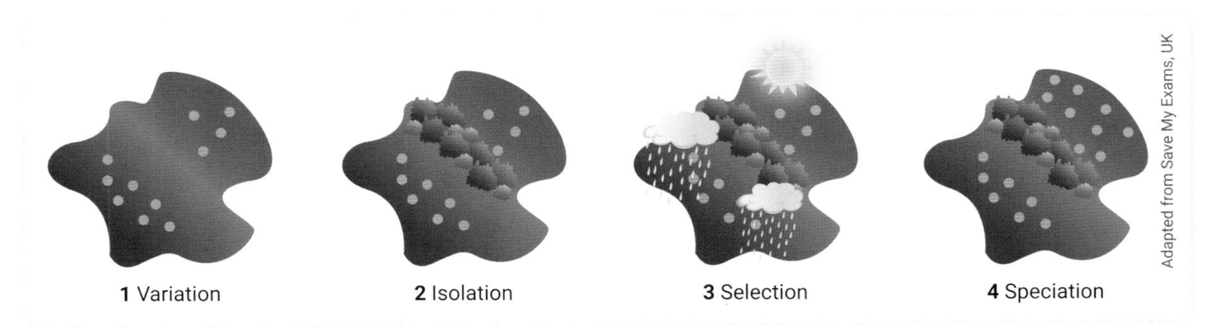

▲ **FIGURE 2.5.2** The process of speciation

Types of speciation

allopatric speciation speciation due to a barrier

Allopatric speciation is defined as speciation from a geographical barrier, resulting in reproductive isolation. When the barrier is removed, the two groups can no longer breed because they are now genetically distinct species. Charles Darwin saw this on the Galápagos Islands while studying the islands' finches, as shown in Figure 2.5.3.

sympatric speciation speciation due to reproductive isolation

Sympatric speciation is defined as speciation occurring without a geographical barrier. It is more commonly seen in plants as they self-fertilise more easily. An example is the cultivated wheat plant species of emmer, durum and common wheat.

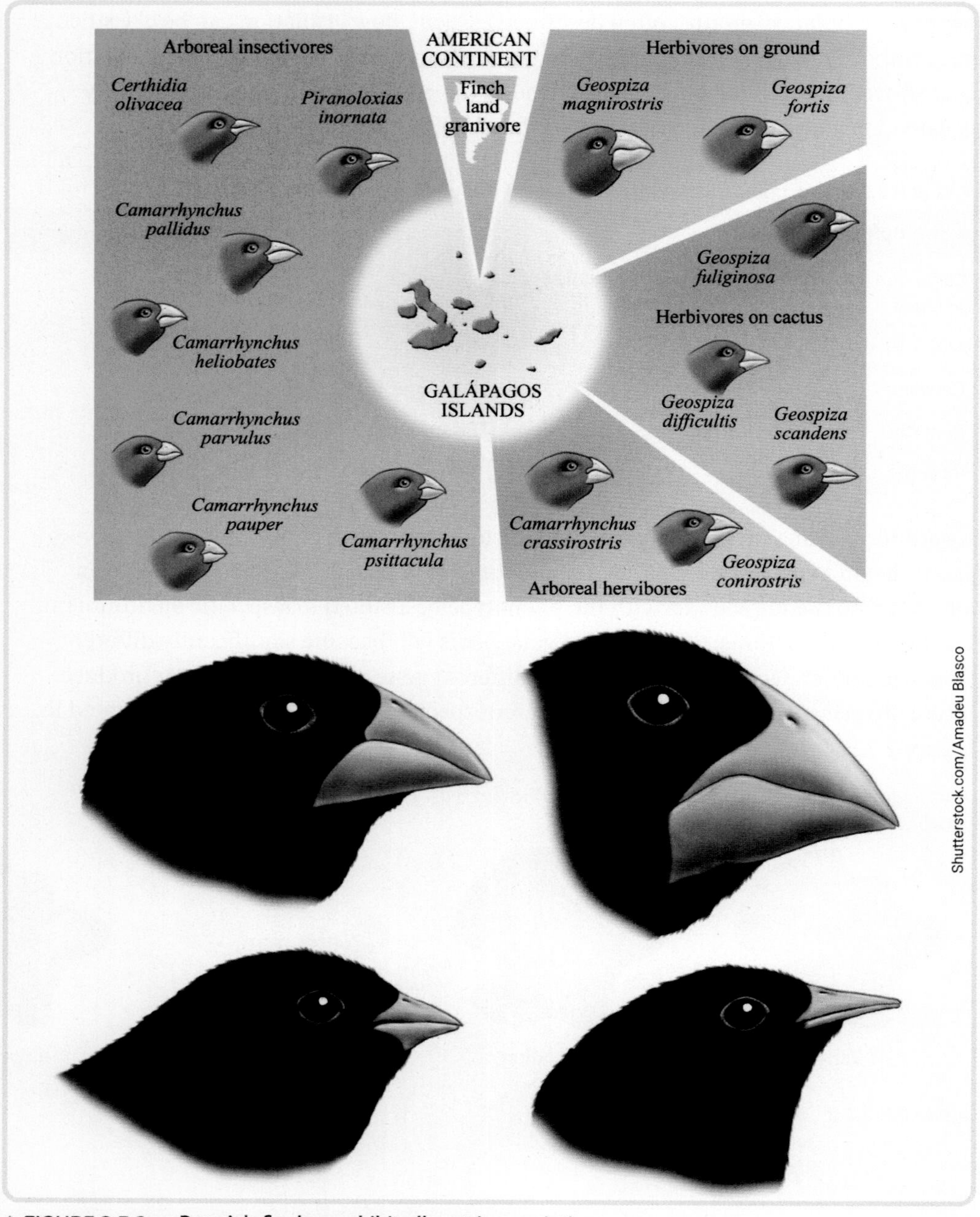

▲ **FIGURE 2.5.3** Darwin's finches exhibit allopatric speciation

2.5 LEARNING CHECK

1 **Explain** why a mule is not considered a species.
2 **Outline** the process of speciation.
3 **Compare** allopatric speciation with sympatric speciation.
4 Tigons and ligers are hybrid great cats (Figures 2.5.4 and 2.5.5). They are the offspring from breeding tigers and lions in captivity. Despite being hybrids, there have been successful offspring when tigons or ligers are bred.
   a Tigons and ligers have a chromosome number of 38. **Explain** why these hybrids are not sterile, unlike the mule.
   b Despite tigons and ligers being capable of breeding, it is a discouraged practice. Why do you think conservationists do not advocate for this breeding?

Shutterstock.com/videohouse

▲ **FIGURE 2.5.4** A liger

Alamy Stock Photo/ONE WORLD IMAGES

▲ **FIGURE 2.5.5** A tigon

5 **Create** a table to summarise the main speciation processes for Darwin's finches.

2.6 Evolution

BY THE END OF THIS MODULE, YOU WILL BE ABLE TO:

- ✓ define 'evolution' and describe the mechanisms that result in evolution
- ✓ define 'gene drift' and 'allele frequency'
- ✓ justify the theory of evolution using specific examples and evidence.

Interactive resource
Crossword: Evolution

GET THINKING

Pingelap is a small island in the South Pacific, sometimes described as the 'colour blindness' island. Between four and 10 per cent of its inhabitants carry the gene for total colour blindness. In comparison, the incidence of total colour blindness is 0.003 per cent globally. Why do you think there is such a high frequency of the colour blindness gene in this population?

Shutterstock.com/FOTOGRIN

▲ **FIGURE 2.6.1** A population of flamingos

Mechanisms of evolution

Evolution can be defined as the gradual change in the characteristics of an organism from earlier forms over many generations. However, an individual does not evolve. The population evolves.

A population is a group of individuals capable of breeding who are living in the same place at the same time, such as the group of flamingos shown in Figure 2.6.1. The genetic information the population possesses can be collectively called the gene pool.

Evolution mechanisms such as natural selection and mutation affect individual organisms in a gene pool. However, it is the collective change in **allele frequency** that results in the population evolving.

allele frequency
the measure of how common an allele is in a population

Allele frequency

Allele frequency is a measure of how common an allele is in a population. It can be calculated by determining how many times the allele appears in the population and dividing by the total number of copies of the gene.

$$\text{Frequency of allele A} = \frac{\text{number of copies of allele A in population}}{\text{total number of copies of gene in population}}$$

If a gene pool is small, the frequency of some alleles might be disproportionality high when compared with a larger population. Smaller gene pools tend to undergo greater change when the mechanisms of evolution are applied to them.

Genetic drift

When small populations of a species are isolated, the small number of individuals with rare genes may fail to transmit them, leading to the disappearance of the gene and the rise of a new species. This effect is known as **genetic drift**, or the Sewall Wright effect after the US geneticist who proposed the concept.

genetic drift
the change in allele frequency seen in small populations due to chance events from one generation to the next

Genetic drift can be modelled using coloured balls (see Figure 2.6.2). Consider a population of 100 individuals represented by 50 red and 50 black balls. The gene is represented by the colour, and the alleles are red and black. If you were to place the 100 balls in a bag and randomly select 10 individuals to form the gene pool, the selection of red to black is unlikely to represent the 50:50 ratio seen in the original group. If those individuals are then allowed to breed to bring the new population back up to 100, the resulting allele frequency after one generation will not match the frequency that existed in the original population.

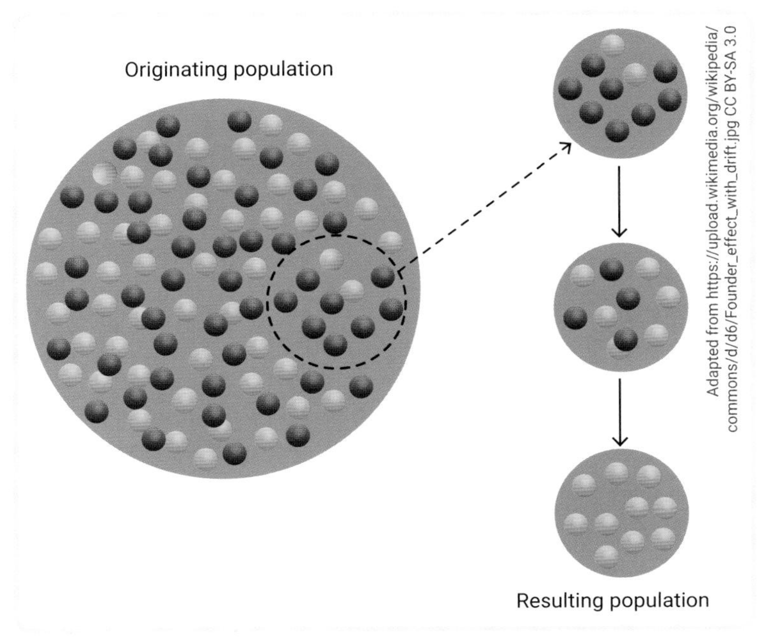

▲ **FIGURE 2.6.2** The Sewell Wright effect of genetic drift due to allele frequency changes in small populations

Genetic drift is seen in populations that are populating a new area or repopulating after an event that resulted in the death of many individuals. The bushfires on Kangaroo Island in 2020 reduced the koala population by up to 90 per cent. Twelve koala orphans were moved to a wildlife park in the Adelaide Hills, where they will form a new population, free of diseases including chlamydia and koala retrovirus. This new population will be used to help build koala populations in the rest of Australia. However, the Kangaroo Island sub-sample was bred from another small sample in the 1920s. Due to genetic drift, these koalas have testicular aplasia, meaning they have only one testicle.

2.6 LEARNING CHECK

1. Read the following statements and **determine** if they are true or false. Rewrite any false statements so that they read correctly.
   a. Evolution is a quick change in characteristics resulting in a new species.
   b. A gene pool is all available genes in a population at any one time.
   c. Allele frequency can change randomly in large populations.
2. A scientist called Lamarck incorrectly hypothesised that giraffes developed long necks by stretching towards branches on trees. By continually doing this, they stretched their necks and their offspring were born with long necks as well. Using your understanding of natural selection, speciation and evolution, **explain** why this hypothesis is incorrect.

2.7 Evidence for evolution: fossils

BY THE END OF THIS MODULE, YOU WILL BE ABLE TO:

- ✓ define and provide examples of fossils, absolute dating and relative dating
- ✓ describe fossil formation
- ✓ explain how fossils can be used as evidence for evolution.

Video activity
Fossil evidence

Extra science investigation
Modelling fossilisation processes

GET THINKING

In 1974, American palaeoanthropologist Donald Johanson discovered hundreds of small bone fragments belonging to an individual female hominin of the species *Australopithecus afarensis*. The fossil was dated to 3.2 million years and nicknamed 'Lucy'. She was 40 per cent complete and an important fossil discovery. Examine Figure 2.7.1. In your workbook, identify the characteristics of the Lucy skeleton that led scientists to believe she could stand upright and walk bipedally (on two feet).

▲ **FIGURE 2.7.1** The Lucy specimen

What are fossils?

fossil
any preserved trace of a once-living organism

Fossils are a lot more than just dinosaur bones! Fossils can be defined as any preserved trace of an organism. This definition includes bones, footprints, burrows, casts, moulds and even faeces. For example, the footprints of sauropod shown in Figure 2.7.2 are a type of fossil. Fossils provide important evidence of evolution.

▲ **FIGURE 2.7.2** Sauropod footprints dated to 130 million years ago in West Kimberley, Western Australia

Fossils are generally hard to make, find and preserve. Most of what was living has decayed, has been destroyed or is inaccessible. As such, there are large gaps in the **fossil record**, which can only be filled using assumptions and inferences based on the discoveries on either side of the gap.

fossil record
the list of fossil finds, their classification and their age

Fossil formation

The conditions required to make a fossil of an organism include:

- the organism must have been buried rapidly under sediment
- the organism must have had hard parts, such as bones, teeth or a shell
- the sediment within which the organism is buried must be alkaline (pH of more than 7) to preserve the hard parts
- the organism must have been left undisturbed for an extended period.

When an organism dies, sediment from the river or at the bottom of lakes or the ocean will quickly cover the organism's body. This prevents scavenging and decay from destroying the organism. The rapid burial also limits oxygen, further slowing decomposition. The process of fossilisation is shown in Figure 2.7.3.

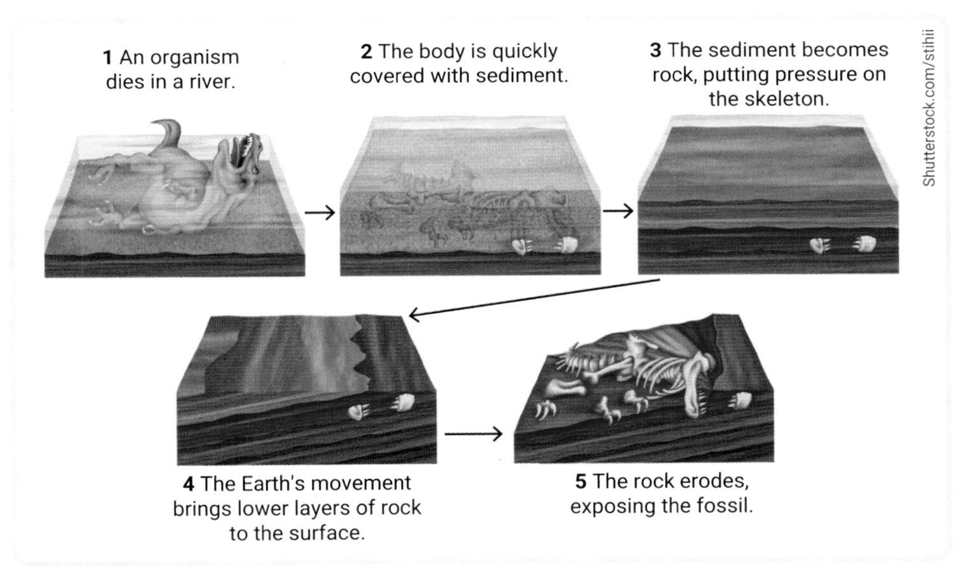

▲ FIGURE 2.7.3 The process of fossilisation

▼ TABLE 2.7.1 The different types of fossils

Type of fossil	Description	Example	Image of fossil
Original fossils	Formed when the chemical composition remains, similar to when the organism was alive. The fossils are recognisable.	Small insects trapped in amber or whole-body fossils	
Carbon trace fossils	Formed when a single layer of leaves or seeds is pressed between layers of sediment. The fine details of the fossil are preserved as a black carbon image.	Commonly plant fossils	
Replacement fossils	Formed when the organic material is replaced with silica, petrifying the fossil specimen.	Petrified wood	
Trace fossils	Formed from the trace of an animal rather than the animal itself. These fossils provide information about the organism's environment, diet and life.	Burrows, tracks or faeces (also known as coprolites)	The Laetoli footprints in Olduvai Gorge, Tanzania

Absolute dating methods

Once a fossil has been discovered, it needs to be classified and dated to place it on the fossil record. Dating fossils can be done in two ways: **absolute dating** or **relative dating**.

Absolute dating determines the actual age of the fossil in years. This is achieved by using **radiometric dating**, where age can be calculated based on the decay of radioactive isotopes found in the fossil. Carbon-14 dating and potassium–argon dating are examples of radiometric dating.

Carbon-14 dating is useful for determining the actual age of an organic fossil because all living things contain carbon-14. This type of dating is used when the age of the fossil is less than 70 000 years.

Potassium–argon dating is an example of absolute dating suitable for older fossils at least 200 000 years old.

absolute dating
determining the age of a fossil in years

relative dating
determining if a fossil or rock is older or younger than another

radiometric dating
a dating method that measures the decay of radioactive isotopes to determine the age of fossils

Relative dating

Relative dating is used to organise fossils on a scale to determine if a fossil is older or younger than another fossil, without determining an exact age. Examples of relative dating are **stratigraphy** and the use of **index fossils**.

Stratigraphy is the study of strata, or rock layers (see Figure 2.7.4). It uses the superposition principle, which states that layers that are lower in strata are older than layers above (see Figure 2.7.5). Fossils that are found in different layers can then be relatively dated as older or younger than each other.

stratigraphy
comparing strata or layers of rock to determine the relative age of fossils

index fossil
a fossil that can be used to compare the relative age of rock strata from different locations

iStock.com/salajean

▲ FIGURE 2.7.4 Rock layers

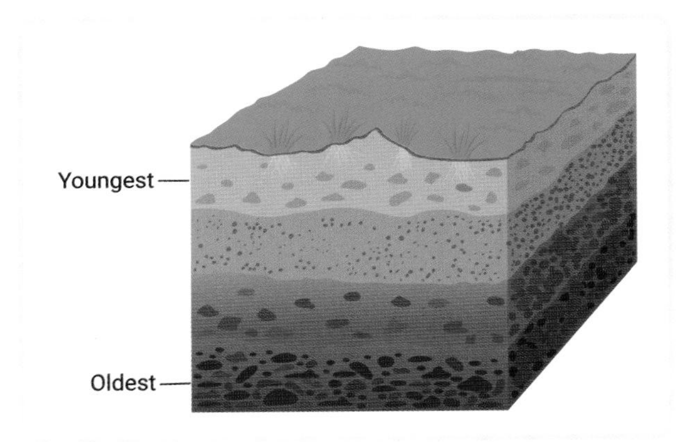

▲ FIGURE 2.7.5 The principle of superposition in rock strata

Index fossils are fossils that can be classified to a species level and are geographically widespread over a short time frame. Identical fossils were formed at the same time at the same location, so finding them widespread across Earth helps reinforce the movement of Earth's crust at various times in geological history. Fossils found alongside index fossils can then be relatively dated or correlated to strata from different areas of the world. Index fossils are frequently marine organisms, such as trilobites and ammonites (see Figure 2.7.6).

▲ **FIGURE 2.7.6** Common index fossils: (a) trilobite and (b) ammonite

Transitional fossils are important to provide further evidence of evolution. The *Archaeopteryx* was a 'bird-like' dinosaur that bridges the gap between the non-avian dinosaurs and birds, as it had both feathers and scales. It was first discovered in 1861, and there have been a further 11 body fossil specimens discovered (see Figure 2.7.7).

▲ **FIGURE 2.7.7** The *Archaeopteryx*: a transitional fossil between dinosaurs and birds

2.7 LEARNING CHECK

1 **Define:**
   a fossil.
   b absolute dating.
   c relative dating.
   d index fossils.
2 Consider the following organisms. Determine which one(s) would most likely form a fossil. **Explain** your choices.
   - Earthworm
   - Clam
   - Bird
   - Jellyfish
   - Sea urchin
   - Leech
3 In your workbook, copy and **match** the type of fossil to its description.

Type	Description
Trace	• Minerals replace the organic material in the hard parts of an organism
Carbon film	• Unchanged parts of an animal or plant
Original	• A thin black deposit of carbon that shows the fine details of fish scales or plant leaves
Replacement	• The footprint, trail, burrow or faeces of an organism

4 The fossil record is incomplete. **Describe** three reasons why there are gaps in the fossil record.
5 **Explain** why a fossil dated to 100 000 years cannot have been dated using carbon-14.
6 **Compare and contrast** absolute and relative dating methods.

2.8 Evidence for evolution: comparative anatomy

BY THE END OF THIS MODULE, YOU WILL BE ABLE TO:

- ✓ describe how comparative anatomy can be used as evidence for evolution
- ✓ define and provide examples of homologous structures and vestigial organs
- ✓ explain how embryology is used to provide evidence for evolution
- ✓ explain how comparative anatomy indicates common ancestry.

Interactive resource
Drag and drop: Structural evidence of evolution

GET THINKING

The appendix in humans has long been thought of as a remnant organ that does not serve a purpose. New research shows that the role of the appendix is to store beneficial bacteria for good gut health. During early development, the appendix plays a role in maturing white blood cells and producing antibodies for the immune system. Humans have a collection of these remnant organs; for example, the muscles that control the outer ear (pinna). Conduct some research and write a short statement about why these muscles may become 'useful' again in time.

Fossils provide great evidence for evolution as they show the gradual change in characteristics of organisms. As you learned in Module 2.7, fossils can be hard to make, find and classify, so looking for other structural pieces of evidence that do not require fossilisation to support evolution is useful.

Structural evidence for evolution

comparative anatomy the study of the body structures of organisms to show the adaptive changes made from a common ancestor

homologous structures body parts that can be found in a range of organisms that have similar structures but different functions

vestigial organs organs that are retained in a species despite no longer being functional as they were in the ancestral species

embryology the study of the early stages of development

Comparative anatomy is the study of body structures of organisms to show the adaptive changes that occur when species share a common ancestor. Examples of comparative anatomy include:

- **homologous structures**
- **vestigial organs**
- **embryology.**

Homologous structures are body parts that can be found in a range of organisms but have different functions to suit the organism's way of life. An example is the vertebrate forelimb. Look at Figure 2.8.1.

In the vertebrate animals shown, the forelimbs show similar structure, called the pentadactyl limb ('penta' means five and 'dactyl' means fingers). However, each limb is adapted for different purposes. For example, a bat's pentadactyl limb is modified for flying, and a human limb is modified for grasping and using tools. The similar structure indicates a common ancestor between all vertebrates. All physical structures are encoded by DNA, which is inherited across generations. The instructions to build the pentadactyl limbs had to exist in a common ancestor, and, over time, adaptations were collected, so each organism is best suited to its environment.

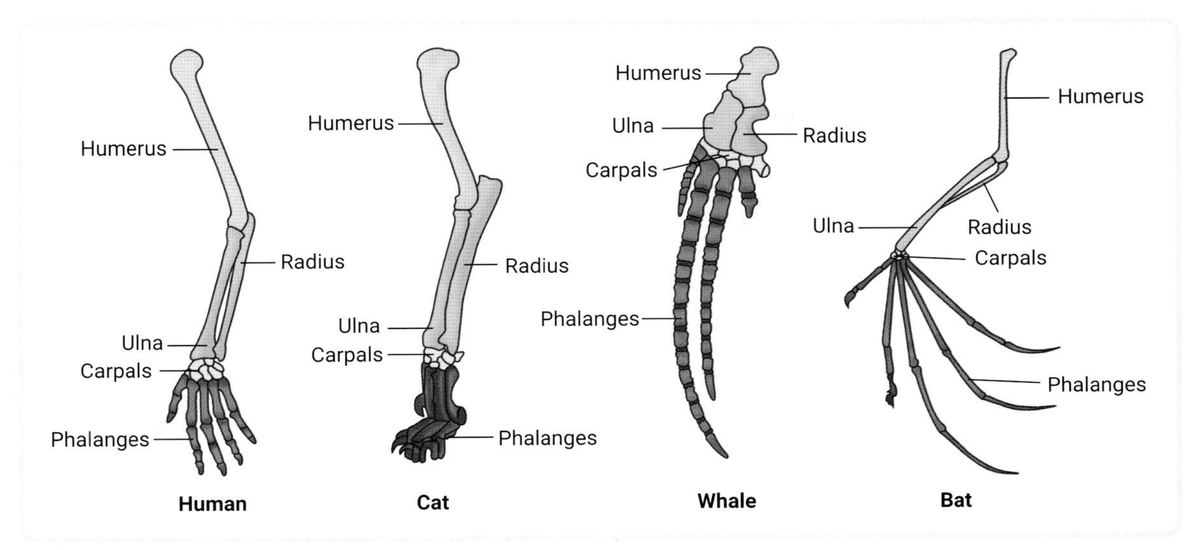

▲ **FIGURE 2.8.1** The vertebrate forelimb structure in four different vertebrates

Analogous structures are features of species that are similar in function but not in structure. An example is the wings of an insect and a bird (see Figure 2.8.2). Both animals require wings to survive in their environment, but the underlying structure of the wing is very different, and there is not a recent common ancestor between insects and birds.

analogous structures
structures in different species that are anatomically different but have the same or similar function

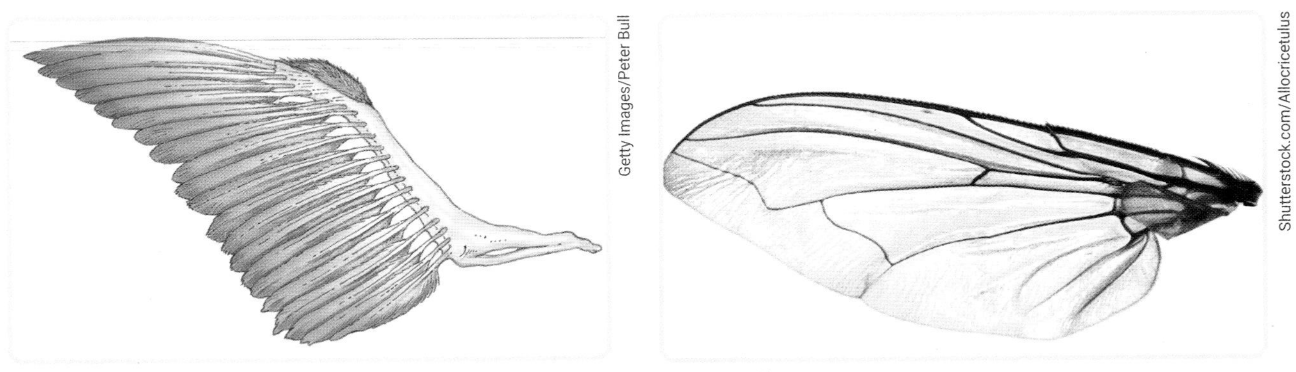

▲ **FIGURE 2.8.2** Bird wings and insect wings have similar functions but different structures.

Vestigial organs

Vestigial organs are the organs in living organisms that no longer serve a purpose. However, these organs were once useful in common ancestors. The DNA that codes for the structures is still present and is passed on to future generations, indicating common ancestry. The structures tend to be of reduced size, so they do not require as much energy in the living organism. Over time, some organisms are born without these structures.

Humans have a few examples of vestigial organs, which are an evolutionary leftover from our primate ancestors (see Figure 2.8.3).

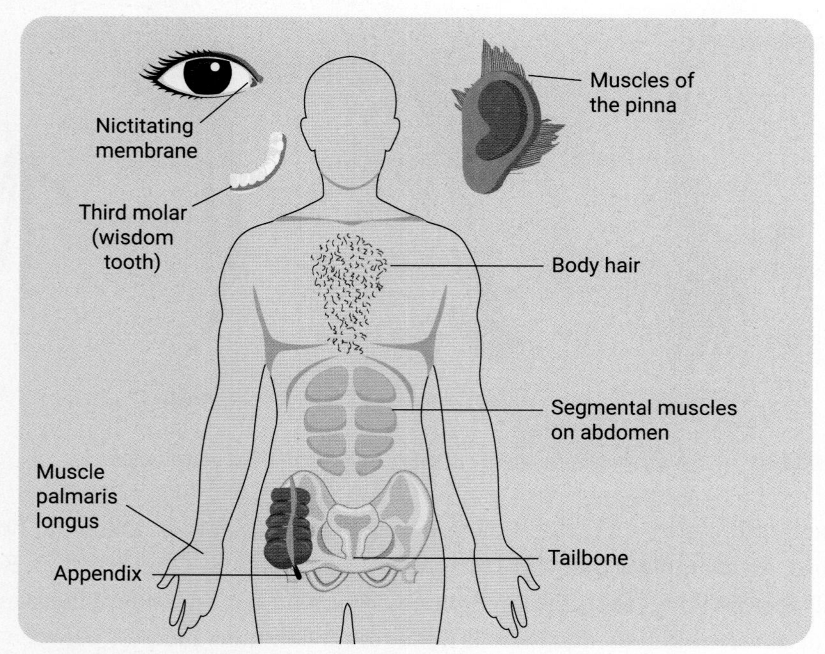

▲ **FIGURE 2.8.3** The vestigial organs present in humans

An easy one to check is the presence of the muscle palmaris longus (see Figure 2.8.4). This muscle in the wrist and hand was used to help primates move around the trees. Place your hand flat on a surface and touch your pinkie finger to your thumb. If you see a raised band in your wrist, that is the vestigial muscle. It may not be present in both wrists, and about 14 per cent of humans do not have the muscle anymore.

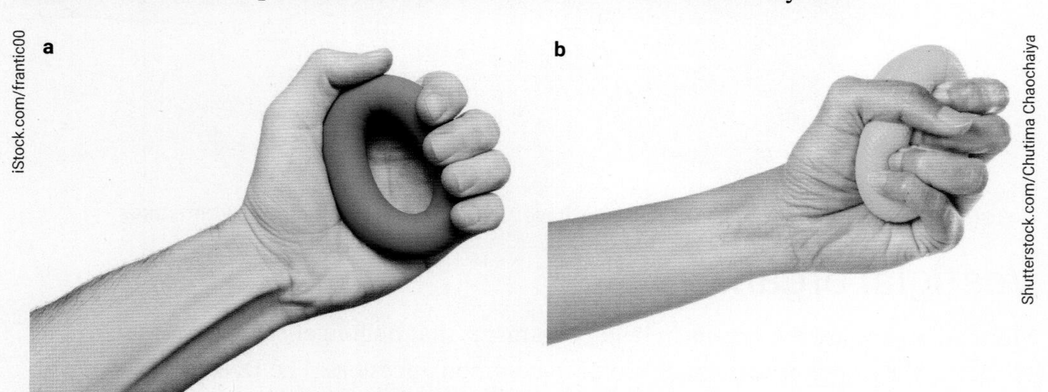

iStock.com/frantic00

Shutterstock.com/Chutima Chaochaiya

▲ **FIGURE 2.8.4** The (a) presence and (b) absence of the palmaris longus muscle

Embryology

Embryology is the study of embryos and can be used to show structural evidence for evolution. An embryo is the name given to an organism early in its development. During this time, the body plan is determined based on the instructions coded in DNA. Many vertebrate embryos show a similar developmental pattern and can be indistinguishable from each other in the early stages of development.

All vertebrate embryos have gill slits, although this feature is lost or changed in the adult form (see Figure 2.8.5). All human embryos have a tail that becomes the tail bone during later development. The similarity in embryological development between different species implies common ancestry.

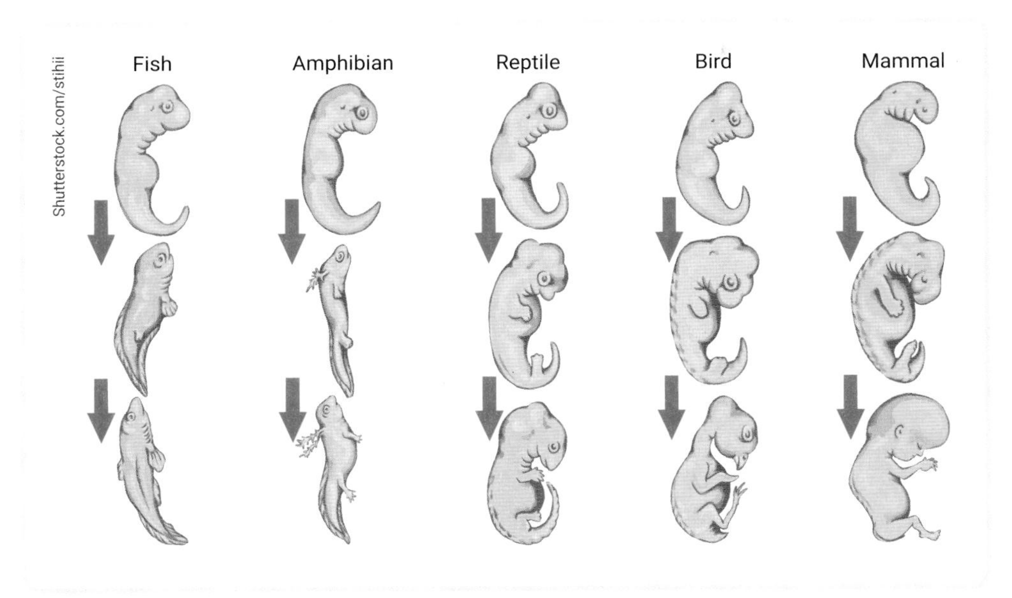

▲ **FIGURE 2.8.5** The development of vertebrate embryos

Comparative anatomy provides further evidence for evolution that can be seen in living organisms, not just fossils.

2.8 LEARNING CHECK

1. **Define** 'comparative anatomy' and provide three examples.
2. **Define** 'vestigial organs'.
3. Use the Internet to identify the vestigial organ in whales. **Explain** why whales have this vestigial organ.
4. In South-East Asia, there are flying lemurs: primates with a flap of skin between their front and back paws that allows the animals to glide from tree to tree. In Australia, the sugar glider displays a similar webbing to help it move between trees (see Figure 2.8.6). **Identify** if this is an example of homologous or analogous structures and explain your choice.

▲ **FIGURE 2.8.6** (a) A flying lemur from Madagascar and (b) an Australian sugar glider

2.9 Evidence for evolution: DNA and proteins

BY THE END OF THIS MODULE, YOU WILL BE ABLE TO:

- ✓ define and give examples of ubiquitous proteins and bioinformatics
- ✓ predict how recent the common ancestor is based on the similarity of the DNA or proteins
- ✓ explain how phylogenetic trees model evolution.

Video activity
Evolution: The evidence

GET THINKING

Humans and chimpanzees, including the bonobos, share 98.8 per cent of the same DNA. Yet we look very different, have different behaviours and live in very different environments. The 1.2 per cent difference equates to about 35 million differences in the nucleotides, and we understand that the genes in chimpanzees operate at different levels compared to the same genes in humans. Using the Internet, research how similar humans are to rats, pigs and bananas. How does comparing our genetic code help scientists understand evolution? Write a sentence to explain the importance of DNA to evolution.

So far, you have learned about structural evidence that supports the theory of evolution. Homologous structures, vestigial organs and embryology imply that these structures are similar because of shared DNA from a common ancestor.

biochemical evidence evidence of evolution based on the fact the same enzymes are found in the cells of most organisms

Biochemical evidence for evolution

Biochemical evidence of evolution uses the premise that proteins, particularly enzymes, are found in the cells of nearly all life on Earth. To understand biochemical evidence for evolution, let us first revisit what DNA is and how it controls the characteristics of an organism.

DNA holds the sequence of nucleotides (or base pairs) that code for amino acids, the building blocks of proteins (see Figure 2.9.1). Proteins are essential for cell processes and characteristics and, ultimately, the traits of the whole organism. DNA is inherited equally from two parents, and offspring will express the traits of the parents in their phenotype.

Natural selection, as a mechanism for evolution, states that organisms with the traits best suited for their environment will survive and reproduce, passing on the DNA for advantageous or neutral changes to the next generation, resulting in the gradual change of the species.

This means that the DNA and proteins can also indicate relationships between species and be used to show common ancestry.

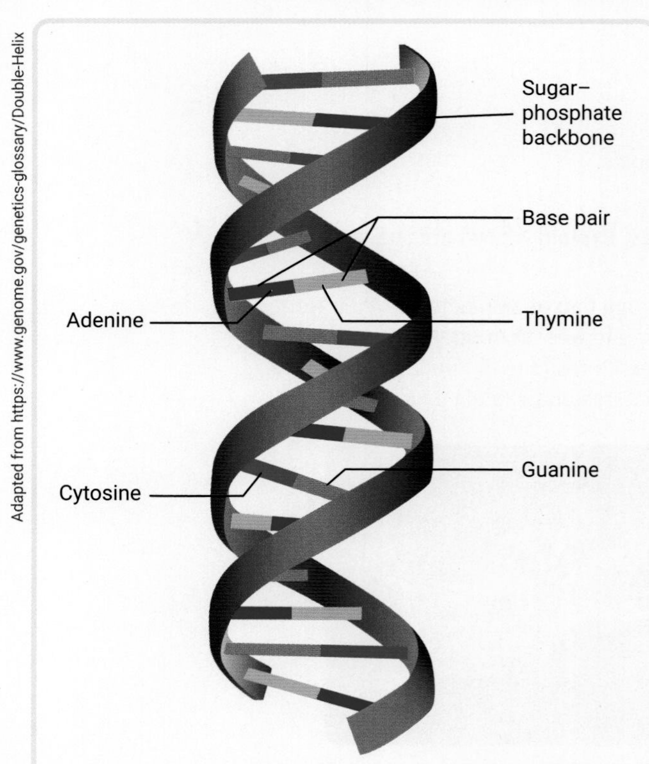

▲ **FIGURE 2.9.1** A DNA molecule showing base pairs: guanine and cytosine, and adenine and thymine.

Bioinformatics

Bioinformatics is the use of technologies to collect and analyse biological data, including genetic code. It is necessary because the genome of an organism can be billions of base pairs in length.

Certain proteins are used in bioinformatics because they can be found in the cells of nearly all living things and perform the same function. These are known as **ubiquitous proteins**. An example is cytochrome C, which is a protein necessary in the mitochondria to synthesise adenosine triphosphate during cellular metabolism. It is made up of 100–104 amino acids and is largely unchanged between species (see Table 2.9.1).

bioinformatics
the use of technology to collect and analyse biological data, such as the sequence of amino acids or proteins

ubiquitous proteins
proteins that are found in nearly all organisms and carry out the same function

▼ **TABLE 2.9.1** Comparing the amino acid sequence in cytochrome C in humans to other organisms

Organism	Difference in amino acid sequence
Chimpanzee	0
Rhesus monkey	1
Rabbit	9
Pig	10
Dog	10
Penguin	11
Horse	12
Moth	24
Yeast	38

Note: Data taken from Margoliash, E. and Finch, W.M. 1967. Construction of phylogenetic trees. Science 155: 279–284.

Comparing the sequence of amino acids in the protein of two organisms indicates the closeness of the relationship between the organisms. The more differences seen in the sequence, the more time has passed since divergence from the common ancestor. Conversely, the more similar the sequence of amino acids, the more recent the common ancestor.

The alpha and beta chains in haemoglobin are another example of ubiquitous proteins. Haemoglobin is found in red blood cells. It binds with oxygen to enable red blood cells to transport oxygen to the body's cells. When comparing the amino acid sequence of both the alpha and beta chains, we can observe there are no differences in the sequence between chimpanzees and humans, and only one difference in each chain comparing humans to gorillas.

Shutterstock.com/GUDKOV ANDREY

Shutterstock.com/Claire E Carter

▲ **FIGURE 2.9.2** (a) Chimpanzees and humans have identical amino acid sequences in alpha and beta chains in haemoglobin. (b) By comparison, humans and gorillas have only one difference in each chain.

phylogenetic tree
a diagram representing lines of evolutionary descent from a common ancestor

Data on the relationship between organisms can be represented pictorially as a **phylogenetic tree**. Phylogenetic trees are diagrams that show the evolutionary descent of different species from a common ancestor. In the tree, two species are drawn closer together with a more recent common ancestor if they are more related. Less related species are drawn further apart with a more distant common ancestor.

Phylogenetic trees are used to show the best hypothesis about how a set of species evolved from a common ancestor. They can be drawn in various ways but follow a basic pattern.

1. The species of interest are found at the tips of the branches.
2. Each node, or branching point, represents a point of divergence from the common ancestor.
3. Each branch represents the series of ancestors leading up to the species at the end.
4. The trunk of the tree is the common ancestor between all species.

An example of a phylogenetic tree based on the data from Table 2.9.1 can be seen in Figure 2.9.2.

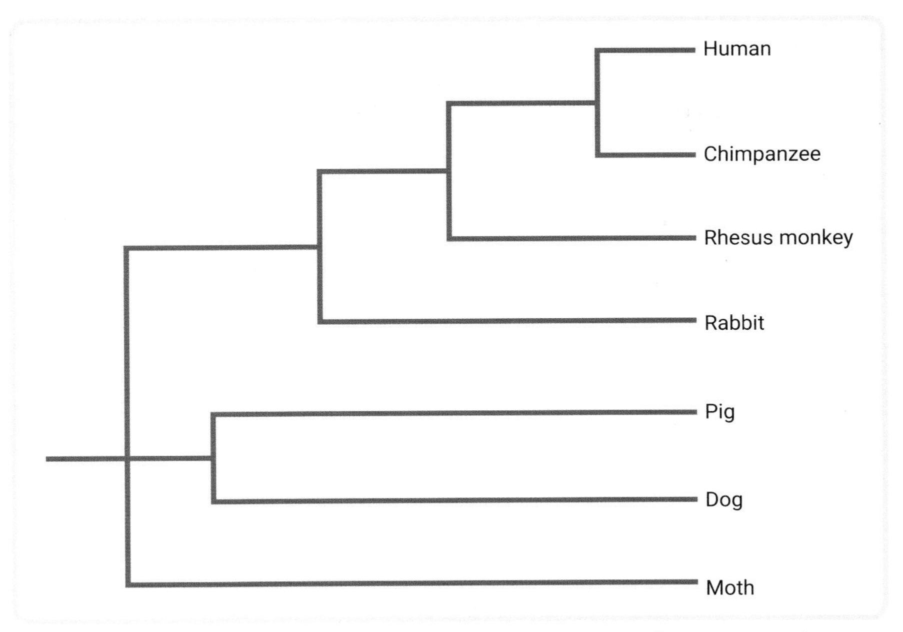

▲ **FIGURE 2.9.2** A phylogenetic tree showing the relatedness of organisms using amino acid sequencing from cytochrome C

DNA sequences can also be used to show relatedness. Three nucleotides code for one amino acid, and there are 64 different combinations to code for the 20 amino acids produced in an organism. More than one sequence of nucleotides may code for the same amino acid. An example is the amino acid valine, which can be coded for in the DNA as CAA, CAG, CAT or CAC.

As such, there could be a difference in the DNA sequence that is not represented by a change in the amino acid sequence. Comparative genomics uses bioinformatics to compare sequences of nucleotides to determine relatedness between species and can result in a more accurate determination of relatedness.

2.9 LEARNING CHECK

1 **Explain** how DNA provides evidence for evolution.
2 **Define** 'ubiquitous proteins' and provide three examples.
3 Use the data in Table 2.9.1 to **explain** how closely related humans are to chimpanzees compared to a penguin and a moth.
4 DNA analysis compares the nucleotide sequence, and protein analysis compares the amino acid sequence of organisms. Which is more accurate? **Justify** your choice.

FIRST NATIONS SCIENCE CONTEXTS

2.10 First Nations Australians' physiological responses to the Australian environment

IN THIS MODULE, YOU WILL:

✓ explore the cultural, environmental and genetic influences on the structural and physiological adaptations of First Nations Australians

First Nations Australians' adaptations

Humans have anatomical and physiological variability that enable us to adapt and thrive in the environments we inhabit. The theory of evolution by natural selection explains how advantageous traits can become more dominant in a population. Our environments, histories and cultures influence the appearance of such genetic characteristics.

First Nations Australians' long occupation of the Australian continent has resulted in the evolution of physical traits that confer advantages to living in certain diverse environments. Early European colonists recorded observations of First Nations Australians, noting their remarkable physiological and anatomical capabilities such as throwing accuracy, visual acuity (the ability to distinguish shapes and details of objects at a distance), running speed and endurance. First Nations Australians who were observed to have such advantageous capabilities were also often sought by European colonisers for particular roles. For example, some First Nations Australians were considered excellent whalers due to their extraordinary eyesight. One record notes that a renowned Aboriginal whaler of the early 1800s could see a whale with the naked eye that was invisible to others even when they used a telescope.

Joe Sambono

▲ **FIGURE 2.10.1** Many genetic and environmental factors affect people's athletic abilities, such as stride distance.

Factors affecting physical traits

People have always been interested in observing and measuring the extreme limits of the human body. We celebrate incredible athletic achievements in events such as the Olympics. Some physical traits that are beneficial for particular skills are controlled by both genetic and environmental factors. Height is an example of such a trait. Height is strongly influenced by genetic factors, but environmental factors also play critical roles. Cultural factors play a role, such as food traditions that demonstrate a complex understanding of the nutritional requirements for a complete diet to be obtained for an individual to attain maximal height. Other cultural influences include the societal structures that consider roles and responsibilities of individuals, patterns of movement and daily or seasonal routines. Historical factors have also affected the attainment of physical traits, such as the impact of diseases, population loss, access to medicines and the immigration of other groups of people.

When used in reference to a person's height, the term 'genetic potential' describes the maximum height an adult could reach if other factors did not contribute. A sophisticated understanding of the interrelated factors that influence genetic potential allowed the remarkable genotypes of some groups of First Nations Australians to be fully attained.

A person's height phenotype (the observable characteristic) is a result of all factors, genetic, environmental, cultural and historical, combined. Anthropologists calculate the height of individuals who lived in the past by measuring bone length, particularly the femur, or thighbone. This bone generally accounts for a quarter of an individual's height and is the longest bone in the human body. These measurements provide information about an individual's physical height trait in consideration of the cultural, historical and environmental influences of the time.

First Nations elite athletes

In Australia, First Nations Australians are well represented in elite sports. The scientific and sporting communities acknowledge that genetic factors can contribute to athletic performance. However, like a person's height phenotype, aspects of athletic performance can also be influenced by other factors, including:

- diet and nutrition
- training
- socioeconomic factors that affect access and proximity to specialised resources and facilities, such as the need to relocate from family or community to access to sports facilities and coaches
- individual ambition and determination.

While First Nations Australians comprise about 3.5 per cent of the total Australian population, they are often represented in higher proportions in elite sports. Research is investigating whether the athletic abilities of some First Nations Australians are connected to some of the incredible physiological and anatomical attributes that First Nations Australians have accumulated over millennia. These attributes have been shaped by cultural factors, including a sophisticated understanding of the nutritional requirements for an individual's optimal diet and nutrition and tens of thousands of years of living on the Australian continent. These factors may today contribute, with other influences such as training, dedication and hard work, to the success of many First Nations Australians in professional sports.

Scientists are now identifying genes that may influence physiological traits. A variation in the *ACTN3* gene that results in a deficiency of the protein alpha-actinin-3 was found to occur less often in sprint athletes, despite being found in about 20 per cent of the general population. Research suggests that the alpha-actinin-3 protein is critical for the function of fast-twitch muscle fibres that produce bursts of strength or speed. In contrast, the genetic variant that results in alpha-actinin-3 protein deficiency was higher in endurance athletes than in the general population. Scientists are investigating whether the loss of this protein benefits endurance athletes.

☆ ACTIVITY

Calculating your stride distance

In this activity, you will determine your stride length and speed over 100 m.

Materials

- timer
- trundle wheel

Method

1 Measure 100 m using a trundle wheel.
2 Time how long it takes you to walk the 100 m distance. Count the number of steps you took. Record your data in a table.
3 Repeat the measurements at a running pace. Count the number of steps you took.
4 Calculate your stride length by dividing the distance by the number of steps you took. How much did your stride length change from walking to running?
5 Calculate your speed by dividing the distance by the time it took for you to walk and run.

Photo Shelter/Michael Amendolia

▲ FIGURE 2.10.2 A fossilised human footprint from the Willandra Lakes region

Evaluation

Fossil footprint impressions found in the Willandra Lakes region in western New South Wales provide an opportunity to explore historical human anatomy and physiology. The tracks of human footprints, dated at about 20 000 years old, were studied to gain an insight into humans living in this arid region of Australia (see Figure 2.10.2).

Some of the data recorded in relation to this fossil is given in Table 2.10.1.

▼ TABLE 2.10.1 Measurements relating to the fossilised footprints discovered in Willandra Lakes

Track number	Demographics	Foot length (cm)	Stride length (m)	Speed (km/h)
T8	Adult, male	29.5	3.73	37.3
T15	Adult, female	24.0	2.61	26.1
T16	Adult, male	24.4	2.90	32.6
T6	Juvenile, 9 years	20.5	1.17	10.5

1 How do your results compare with the data from the fossilised footprints?
2 What activities might have required running speeds of 37.3 km/h?
3 What does this tell you about human life in the Willandra Lakes region 20 000 years ago?
4 Look up the Olympic 100 m records online. How do the fossil results compare with elite sprinters today?

2.11 Genetically modified food

BY THE END OF THIS MODULE, YOU WILL BE ABLE TO:

- ✓ investigate key factors that contribute to scientific knowledge and practices being adopted more broadly by society
- ✓ understand why agricultural practices have changed to include the widespread use of genetically engineered crops.

Humans have been altering the genetics of organisms for over 30 000 years. Although our ancestors had no concept of genetics, they were able to influence the DNA of organisms through artificial selection. You learned about this in Module 2.4. While artificial selection is not what we typically consider **genetic engineering** today, it is the precursor to modern processes.

Genetically modified organisms, or GMOs, refer to a modern process of altering the genetics of organisms.

genetic engineering
the deliberate modification of an organism's DNA

genetically modified organism
an organism whose genes have been altered in the laboratory to produce a desired trait

The history of GMOs

In 1973, scientists Herbert Boyer and Stanley Cohen together engineered the first successful genetically engineered organism (GE organism). They developed a method to cut out a gene for antibiotic resistance from one bacterium and paste it into another, making the GE organism resistant to antibiotics. One year later, another pair of scientists, Rudolf Jaenisch and Beatrice Mintz, followed a similar procedure and introduced foreign DNA into mouse embryos.

Video activity
GMOs

Despite the incredible possibilities of these new techniques, there was immediate concern about the possible consequences on human health and Earth's ecosystems. By the middle of 1974, a universal prohibition of GE projects was observed, allowing experts time to meet and consider the next steps for the safety of GE experiments. The meeting, known as the Asilomar Conference of 1975, resulted in a set of guidelines with defined safety and containment regulations.

The cooperation seen at the Asilomar Conference gave government bodies around the world confidence to support GE research and launched a new era of modern genetic modification.

GMOs have the green light

In 1980, the US Supreme Court allowed ownership rights over GMOs, giving large companies the incentive to rapidly develop GMO tools that were both useful and profitable. In 1982, the US Food and Drug Administration approved the first human medication produced by a GMO. Bacteria had been genetically engineered to produce human insulin, allowing the hormone to be purified, packaged and prescribed to diabetes patients as the drug Humulin.

The first experiments with food crops that had been genetically modified began in 1987. After five years of extensive health and environmental testing, the Flavr Savr tomato was the first food crop to be approved for commercial production. These tomatoes were modified to include a gene that inhibits a natural tomato protein, thereby increasing the firmness and shelf-life of the Flavr Savr variety (see Figure 2.11.1).

In 1995, the first pesticide-producing crop was approved, and in rapid succession, herbicide-resistant crops were also engineered. This made it easier for farmers to control unwanted plants in their fields.

Scientists have also genetically engineered crops to increase their nutritional value. In 2000, Golden Rice was developed to combat vitamin A deficiency, which is estimated to kill over 500 000 people each year (see Figure 2.11.2).

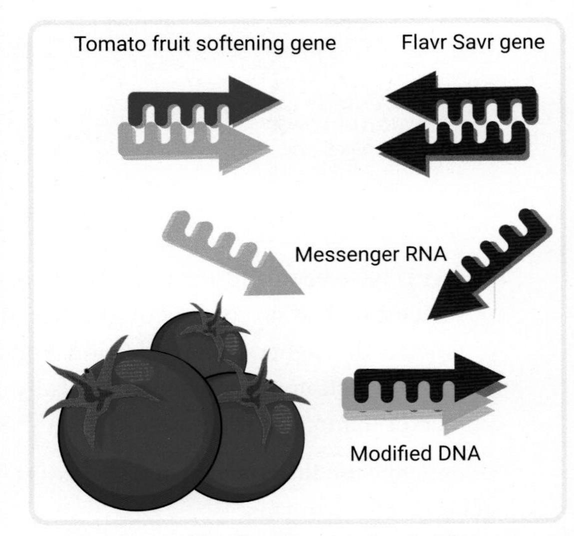

▲ **FIGURE 2.11.1** The Flavr Savr tomato has been genetically engineered to delay rot.

The future of GMOs

A recent review of over 150 studies concluded that genetically modified technology has increased crop yields and farmers' profits over the past 20 years. Today GMOs of soybeans, maize and cotton have been associated with a 22 per cent increase in yield, a 37 per cent decrease in pesticide use and a 68 per cent increase in farmer profits, despite the increase in the cost of GM seeds.

The United Nations has predicted that by 2050 humans will need to produce 60 per cent more food than we currently do to meet the needs of the global population. Innovative approaches, including GMOs, will be required to solve this problem.

▲ **FIGURE 2.11.2** A comparison of normal rice and Golden Rice

2.11 LEARNING CHECK

1. **Define** 'genetic engineering'.
2. **Develop** a timeline to show the development of genetically modified organisms.
3. Using the Internet, **research** GMOs that have been developed in the areas of:
   a. drought-resistant plants.
   b. enhanced growth in animals.
   c. disease-resistant crops such as the Rainbow papaya.
4. What do you think the future of GMO technology is going to be? Write down some of your ideas and share your thoughts with your classmates.

SCIENCE INVESTIGATIONS

2.12 Identifying trends

SCIENCE SKILLS IN FOCUS

IN THIS MODULE, YOU WILL FOCUS ON LEARNING AND IMPROVING THESE SKILLS:

- observing and recognising different types of fossils
- investigating environmental factors influencing species' success
- identifying trends and patterns seen in homologous structures.

Conducting practical investigations allows you to collect, organise and process data. This, in turn, enables you to interpret the data to form conclusions. Hands-on practicals provide an opportunity to see what cannot be observed in nature or what may take many generations to occur. When analysing data, scientists are usually looking to identify three main things:

- patterns
- trends
- relationships.

Patterns

Scientists look at data to see if something repeats over time. Patterns in scientific data give us information about the past and help us make predictions. For example, when we look at fossils in strata, we can see patterns around the world in the distribution of index fossil fossils.

Trends

A trend is when data shows movement or change in a particular direction, such as a population of koalas decreasing over time. Usually, scientists need a great deal of data to be confident that it is showing a trend. An example of a trend in evolution is the slow increase in the size of the brain cavity in the skulls of our early human ancestors over many millions of years.

Relationships

A relationship is a trend that has a mathematical correlation between two or more variables. For example, the more it rains, the higher the water level in a rain gauge. Another example could be a reduction in the consumption of junk food resulting in an improvement in cardiovascular health.

INVESTIGATION 1: ENVIRONMENTAL FACTORS INFLUENCING THE HATCHING VIABILITY OF BRINE SHRIMP

AIM

To understand how environmental factors can affect the hatching viability of brine shrimp and consider how this relates to the theory of natural selection

BACKGROUND

Brine shrimp (*Artemia salina*) is a species of small crustaceans found in saline environments, specifically salt lakes, worldwide. Brine shrimp do not inhabit oceans due to the high presence of predators. Brine shrimp can avoid predators by living in very high saline environments that other aquatic life cannot. This makes them an excellent model for the study of natural selection and adaptations.

Salinity levels within the environment greatly affect the population growth of brine shrimp. In this investigation, you explore how different saline level environments affect hatching viability. To do this, you will attempt to hatch cysts in four different salt concentrations and measure the hatching viability by counting how many nauplii (brine shrimp larva) emerge from the cysts.

Warning

- Wear appropriate personal protective equipment.
- Know and follow all regulatory guidelines for the disposal of laboratory wastes.
- Wash hands thoroughly before and after working with any organic materials.
- To dispose of brine shrimp, pour them down the sink. Alternatively, you may like to keep them for further observation.

Video
Science skills in a minute: Identifying trends

Science skills resource
Science skills in practice: Identifying trends

MATERIALS

- ☑ brine shrimp eggs (cysts)
- ☑ 4 Petri dishes
- ☑ 3 saltwater solutions (0.5%, 1.0% and 2.0%)
- ☑ fine brush
- ☑ 4 microscope slides
- ☑ double-sided tape
- ☑ magnifying glass
- ☑ permanent marker
- ☑ graduated cylinder
- ☑ distilled water

METHOD

Preparing the cysts for hatching (day 1)

1. Using a permanent marker, label four Petri dishes: 0%, 0.5%, 1.0%, 2.0%.
2. Form your hypothesis.
3. Using the graduated cylinder, measure 30 mL of each saline solution and pour it into the appropriately labelled Petri dish.
4. Collect four microscope slides. Measure and cut four 1.5 cm strips of double-sided tape and gently adhere one of them to each of the microscope slides.
5. Lightly touch the fine brush to the side of the dish containing the brine shrimp eggs. Collect 20–30 eggs on the brush. Do not collect too many eggs because you will be required to count them.
6. To adhere the eggs to the double-sided tape, lightly press the brush onto the tape on the first microscope slide. Repeat this step for the remaining three microscope slides.
7. Using a magnifying glass, count the number of eggs on the first slide. Copy Tables 2.12.1 and 2.12.2 into your workbook and record this information in Table 2.12.1.
8. Once the eggs have been counted, place this slide into the 0% salt solution Petri dish, ensuring that you place the slide with the tape side facing up.
9. Count the eggs on each slide and place them in the respective salt solutions. Record the egg count information in the corresponding row in Table 2.12.1.
10. Place the Petri dishes under a light bank for 24 hours at room temperature.

Data collection (day 2 and day 3)

1. After 24 hours, examine the contents of each Petri dish with the magnifying glass. You should see that some brine shrimp have hatched and are swimming in the salt solution. Record the number of eggs, the number of dead or partially hatched eggs and the number of swimming brine shrimp in Table 2.12.1.
2. After 48 hours, examine the contents of the Petri dishes again and record your observations in Table 2.12.1. Calculate the hatching viability of each dish at 48 hours by dividing the number of shrimp swimming by the initial number of eggs in the Petri dish. Round up your calculations to the nearest hundredth and add this information to the class results table (Table 2.12.2).
3. Draw a line graph that shows the sample means from the class results.

SOUTHERN Biological

RESULTS

▼ TABLE 2.12.1 The hatching viability of brine shrimp in varying levels of salinity

% NaCl	0 hours	24 hours			48 hours			
	Eggs	Eggs	Dead or partially hatched	Swimming	Eggs	Dead or partially hatched	Swimming	Hatching viability percentage
0%								
0.5%								
1.5%								
2%								

▼ TABLE 2.12.2 The hatching viability of brine shrimp in varying levels of salinity; results of different student groups (add rows as required)

Class group	Hatching viability at salinity:			
	0%	0.5%	1.5%	2%
1				
2				
3				

EVALUATION

1 Which Petri dish had the highest hatching viability? Which had the lowest? Suggest possible reasons for these results.

2 Based on your data and the class data, is there enough evidence to conclude that environments of different salinities affect the hatching viability of brine shrimp?

3 What is the selective agent in this study?

4 Imagine a hypothetical scenario wherein the salinity of the water lived in by wild brine shrimp dropped to 0.5% and remained that way for a decade. If we repeated this experiment in a decade, do you think the hatching viability percentages would change? Explain your answer.

5 What other conditions may affect the hatching viability of brine shrimp? Design an experiment to investigate another environmental factor that may affect hatching viability.

INVESTIGATION 2: OBSERVING FOSSILS

AIM

To observe different types of fossils

MATERIALS

- ☑ a range of fossils consisting of original, carbon film, replacement, casts, moulds and amber-preserved fossils
- ☑ hand lens
- ☑ access to the Internet

METHOD

1. Examine a fossil. Describe its appearance and suggest what kind of organism it was and what it may have looked like when it was alive.
2. Identify the way the fossil was formed.
3. Using the Internet, research the name of the organism and its age (how long ago it was formed).
4. Repeat steps 1–3 for each fossil.

RESULTS

Record your observations in a table. Remember to give your table a title.

EVALUATION

1. After determining the age of each fossil, create a timeline and place your fossils along it.
   - **a** Are there any gaps in your fossil record? Suggest reasons why this might have occurred.
   - **b** Research the geological timescales and overlay these eras and epochs on your timeline.
   - **c** Research when *Australopithecus afarensis* first appeared in the fossil record and add them to your timeline.
2. Explain how fossils are used as evidence for evolution.

INVESTIGATION 3: BIRD BEAKS – AN INVESTIGATION INTO NATURAL SELECTION

AIM

To investigate how environmental factors can influence the shape of bird beaks over time

MATERIALS

- ☑ 6 varieties of confectionary
- ☑ tools to represent bird beaks: spoon, 2 chopsticks, toothpick, icy pole stick
- ☑ takeaway food container
- ☑ stopwatch
- ☑ sheet of A3 paper

Warning

This activity uses edible confectionary. If that is not an option, replace the confectionary with coins, marbles, toothpicks, small pieces of gravel, pieces of string or table tennis balls.

METHOD

1. Predict which of the tools listed in the materials section (representing a bird's beak) would be best for picking up each type of confectionary.
2. Place the confectionary into the takeaway food container. In 30 seconds, use one tool to collect as much food as possible. Copy Table 2.12.3 into your workbook and record the number of pieces you were able to pick up.
3. Replace the confectionary in the takeaway food container.

4 Repeat Step 2 with each tool and record the totals in Table 2.12.3.

5 Apply a selective agent to your food collection; for example, one of your foods is now toxic, and if you touch it with your 'beak', that beak can no longer collect food. Repeat steps 2–4 and record your totals.

6 Apply a second selective agent to your food collection; for example, another of your foods is now toxic.

7 Extension: using only the tools that can still collect food, cover your left eye and try to collect as much food as possible in 30 seconds.

RESULTS

▼ **TABLE 2.12.3** 'Bird beak' results

Tool	Trial 1	First selective agent applied	Second selective agent applied	Closed left eye
Spoon				
Chopsticks				
Toothpick				
Icy pole stick				

EVALUATION

1 Define 'natural selection'.

2 Compare columns 2 and 3 in your table.
   a Which tool(s) were still able to collect food?
   b What impact did the selective agent have on collecting food?

3 Compare columns 3 and 4 in your table.
   a Which tool(s) were still able to collect food?
   b What impact did the second selective agent have on collecting food?

4 Compare the results in column 5 to the results in columns 2–4.
   a What affect did closing your left eye have on the effectiveness of the tools?
   b Which of the three selective agents had the greatest impact on collecting food? Use the results to explain which type of beak would be the most likely to be passed on to the next generations.

5 Each tool represents a different bird beak. Using the Internet, identify which tool best matches a bird's beak and determine what that bird's diet is like.

INVESTIGATION 4: EXAMINING HOMOLOGOUS STRUCTURES

AIM

To compare different vertebrate skeletons

MATERIALS

- ☑ skeletons of a human, cat, bird, frog, bony fish and lizard
- ☑ if skeletons are not available, X-ray images from the Internet

METHOD

1. Copy Table 2.12.4 into your workbook.
2. Carefully observe each skeleton and complete Table 2.12.4 by recording which animals are most alike and which animals are least alike for the named structures.

RESULTS

▼ **TABLE 2.12.4** A comparison of homologous structures

Structure	Most alike	Least alike
Skull		
Vertebral column		
Ribs		
Front limbs		
Rear limbs		
Pelvic girdle		

EVALUATION

1. Which two animals are the most alike, and which two are the least alike?
2. One of these animals does not live on land. Describe how its skeleton is different from those of the land-dwelling organisms.
3. Most of these animals have similar skeletons, with many of the same bones in the same places. Name this piece of evidence of evolution and describe how this observation came to be.
4. Build a phylogenetic tree based on the similarities recorded in Table 2.12.4. On it, show when the features of scales, feathers, fur/hair and leathery eggs would have appeared.

9780170463041

2 REVIEW

REMEMBERING

1 What is the difference between a gene mutation and a chromosomal mutation?

2 **Identify** an example of a physical mutagen, a chemical mutagen and a biological mutagen.

3 **Draw** a representation of a:

- **a** deletion mutation.
- **b** translocation mutation of chromosome 5.
- **c** substitution mutation.

4 **Define** 'fossil'. Name and provide an example of the four different types of fossils.

5 **Define** 'stratigraphy' and the principle of superposition.

UNDERSTANDING

6 **Explain** why there are gaps in the fossil record.

7 **Describe** the process of fossil formation in sedimentary rock.

8 On his trip to the Galápagos Islands, Charles Darwin observed the giant tortoises and how their shell shape differed between different islands. On one island, the vegetation was lush, green and well watered. The tortoises here had rounded shells. On another island, the available water was less, and the vegetation grew higher on bushes rather than grasses. These tortoises had a flat shell with a peak at the front of the shell. **Explain** these observations of tortoise shells using Darwin's two inferences.

9 **Describe** how different environmental factors can work together with genetic factors to culminate in higher rates of representation by minority groups in some sports.

10 **Examine** the three strata shown in Figure 2.13.1.

- **a** Determine the youngest layer and outcrop.
- **b** Determine the oldest layer and outcrop.
- **c** Explain what conditions must be achieved for a fossil to be classified as an index fossil.
- **d** Propose how index fossils would be useful in dating strata from different areas in the world.

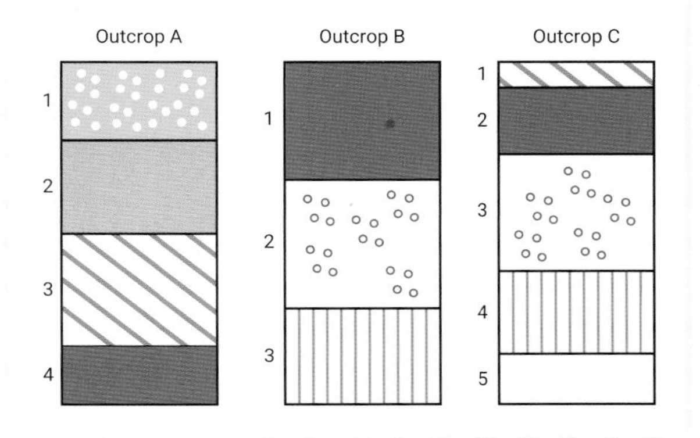

▲ **FIGURE 2.13.1** Strata from three different outcrops

APPLYING

11 **Recall** the calculation to determine the combinations that can be formed when chromosomes line up in metaphase I.

- **a** Using this calculation, determine the combination of chromosomes that can be formed during meiosis in a human.
- **b** Some organisms, such as aphids, reproduce through asexual reproduction, resulting in genetically identical offspring. Propose what would happen to a population of aphids if a new selective agent was introduced into the population.

12 Sharks and dolphins both live in the ocean and require fins or flippers to help them swim.
Research whether shark fins and dolphin flippers are homologous or analogous structures and justify your reasoning.

13 Consider the following information on the domesticated dog, *Canis familiaris*.

- The ancestor of the domesticated dog is an ancient extinct wolf.
- Domestication of the dog began 15 000 years ago, before the development of agriculture.
- All current dog breeds are of the same species.
- Pedigree dogs, such as German shepherds and French bulldogs, are expensive and have many genetic conditions such as hip dysplasia.

- **a** **Identify** what traits the wolves may have displayed for humans to want to domesticate them.

b **Explain** what the humans would have done for the wolf to become tame.

c **Define** 'species'.

d What do you think the term 'pedigree' means?

e **Explain** how artificial selection has increased the frequency of genetic diseases in pedigree dogs.

EVALUATING

14 The amino acid sequence of cytochrome C of five different species of vertebrates was analysed. Table 2.13.1 shows the number of differences in the sequence between each pair of species.

a Using the data in Table 2.13.1, **create** a phylogenetic tree to reflect the evolutionary relationships of the organisms.

b **Identify** the organism that is the least related to the others.

c In your phylogenetic tree, indicate when feathers and fur may have appeared in the ancestral forms.

▼ **TABLE 2.13.1** The number of differences in the cytochrome C sequence of five different species

	Wild horse	Black mamba snake	Red jungle fowl	Emperor penguin	African wild donkey
Wild horse	0	21	11	13	1
Black mamba snake		0	18	17	20
Red jungle fowl			0	3	10
Emperor penguin				0	12
African wild donkey					0

15 Antibiotics are used to treat bacterial infections and have been prescribed to patients since 1910. However, many strains of bacteria no longer die upon exposure to antibiotics. They have become resistant or 'superbugs'.

a Using the principle of natural selection, **explain** how these antibiotic-resistant bacteria came to be.

b **Propose** a change to the use of antibiotics to reduce the occurrence of antibiotic-resistant bacteria.

c Can you think of a way to use DNA technology to develop a new antibiotic?

16 *Archaeopteryx* is a transitional fossil, showing the characteristics of a reptile and a bird. **Explain** why transitional fossils occur and their importance in providing evidence for evolution.

17 **Compare and contrast** artificial selection and natural selection.

CREATING

18 **Draw** a concept map of all the glossary terms in this chapter. Make as many links as possible between them.

BIG SCIENCE CHALLENGE PROJECT #2

1 Connect what you've learned

Consider this statement: 99.9% of everything that has ever lived on Earth has died. Apply your understanding of natural selection and evolution to that statement.

2 Check your thinking

Use the Internet to research the following predecessors to modern humans. For each extinct hominin, identify the temporal distribution, their geographical distribution, their cranial capacity and a summary of any cultural advances scientists have evidence of. Display your findings in a table.

- *Australopithecus afarensis*
- *Paranthropus robustus*
- *Homo habilis*
- *Homo erectus*
- *Homo neanderthalensis*

3 Make an action plan

Imagine you are an anthropologist working on a small island in the Malay Archipelago, and you discover some fossilised bones that look like *Homo habilis*. On closer examination, you can see that they are much younger than the temporal distribution of *Homo habilis* – they are much closer in age to *Homo sapiens*. You also know the accepted geographical distribution of *Homo habilis* has never extended out of Africa.

What dating technique would you have used to determine the age of the fossils? What other evidence (structural and biochemical) would you need to collect to determine if this is a new species or a transitional species?

4 Communicate

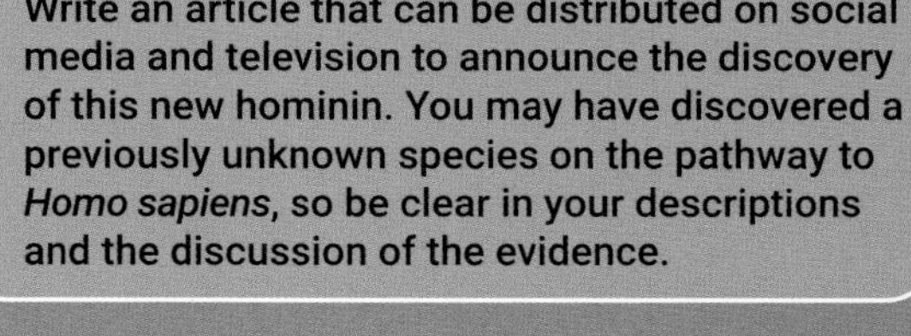

Write an article that can be distributed on social media and television to announce the discovery of this new hominin. You may have discovered a previously unknown species on the pathway to *Homo sapiens*, so be clear in your descriptions and the discussion of the evidence.

Shutterstock.com/TOON WELL

The structure and properties of chemicals

BIG SCIENCE CHALLENGE

Shutterstock.com/ChameleonsEye

▲ FIGURE 3.0.1 (a) A bridge made of steel, (b) a plastic shopping bag, and (c) a copper electrical wire with plastic coating

Have you ever wondered why objects are made from certain materials? Why are bridges made from steel and shopping bags made from plastic? Why is electrical wiring made from copper with plastic around the outside?

The properties of a material are directly related to its structure. When making new products, engineers and designers carefully consider the chemical structure of materials and the properties they have.

- **What properties are needed for the bridge, shopping bag and electrical wiring?**
- **What do you think causes materials to have particular properties?**
- **Can you identify something on your desk that you use because it has particular properties?**
- **What properties does this object have that make it useful?**

#3 **SCIENCE CHALLENGE ACCEPTED!**

At the end of this chapter, you can complete the Big Science Challenge Project #3. You can apply the knowledge and skills you learn in this chapter to complete the project.

Assessments
- Prior knowledge quiz
- Chapter review questions
- End-of-chapter test
- Portfolio assessment task: Science investigation

Videos
- Science skills in a minute: Justifying conclusions **(3.9)**
- Video activities: Atom structure: electron shells **(3.2)**; Introduction to the periodic table **(3.3)**; Introduction to chemical bonding **(3.4)**; Mendeleev and the periodic table **(3.8)**

Science skills resources
- Science skills in practice: Justifying conclusions **(3.9)**
- Extra science investigation: Comparing metal properties **(3.6)**

Interactive resources
- Label: Electrons and shells **(3.2)**; Predicting ions **(3.5)**
- Drag and drop: Types of bonding **(3.4)**; Metal properties **(3.6)**
- Simulation: Build an atom **(3.1)**

To access these resources and many more, visit: **cengage.com.au/nelsonmindtap**

3.1 Review of atomic structure

BY THE END OF THIS MODULE, YOU WILL BE ABLE TO:

✓ describe the subatomic particles in atoms
✓ describe the structure of an atom in terms of its subatomic particles.

Video activity
What is an atom?

Interactive resource
Simulation: Build an atom

GET THINKING

In chemistry, numbers and chemical symbols are often used to describe information about chemicals and atoms. Find examples of numbers and symbols being used in this module and suggest what the numbers might mean.

Subatomic particles

You may recall from Year 9 that **atoms** are the smallest particles of matter that form a chemical **element**. Atoms have a central **nucleus** that contains two types of **subatomic particles: protons** and **neutrons**. Protons are positively charged, and neutrons have no charge. Negatively charged **electrons** are found in **electron shells** around the nucleus.

Protons and neutrons are about the same size. Electrons are much smaller, approximately $\frac{1}{1840}$ the size of a proton or neutron (see Table 3.1.1).

atom
a particle of matter made up of protons, neutrons and electrons

element
matter consisting of atoms with the same number of protons in their nucleus

nucleus
the structure in the centre of an atom containing protons and neutrons

subatomic particle
a particle found inside an atom, such as a proton, a neutron or an electron

proton
a particle in the nucleus of an atom with a positive charge

neutron
an uncharged particle found in the nucleus of an atom

electron
a negatively charged particle in an atom; it moves in space around the nucleus

electron shell
a level around a nucleus containing electrons of the same energy

isotopes
atoms of an element with the same number of protons but a different number of neutrons

▼ **TABLE 3.1.1** Subatomic particles

Subatomic particle	Symbol	Location in the atom	Charge	Relative mass
Proton	p	Nucleus	+1	1
Neutron	n	Nucleus	0	1
Electron	e	In shells around the nucleus	–1	$\frac{1}{1840}$

Atoms, elements and isotopes

There are over 100 different elements. These include many substances you should be familiar with, such as hydrogen, magnesium, copper and oxygen. An element contains atoms with the same number of protons. Different elements have atoms with different numbers of protons. For example, all carbon atoms have six protons, and all oxygen atoms have eight protons.

An element has atoms with the same number of protons, but it can have forms with different numbers of neutrons. These are called **isotopes**.

In Figure 3.1.1 you can see three types of carbon atoms that we refer to as carbon isotopes. Each carbon isotope has six protons; however, the number of neutrons in the nucleus differs.

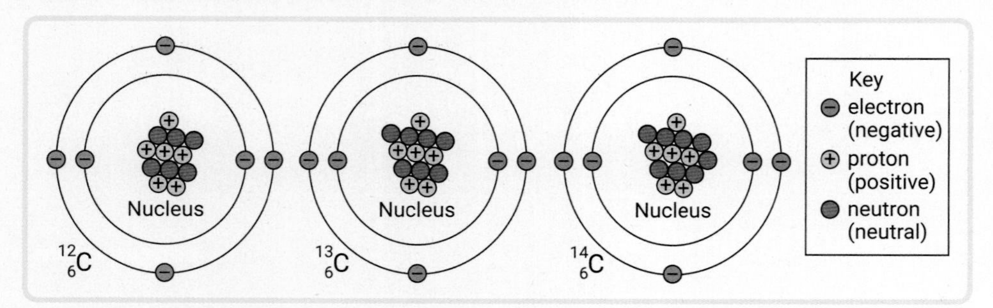

▶ **FIGURE 3.1.1** The structure of carbon atoms

9780170463041

Atomic number and mass number

The number of protons in an element is represented by its **atomic number (Z)**. From Table 3.1.2, you can see that carbon has an atomic number of six. As the number of protons is constant in an element, the atomic number for that element is always the same. Every element has a unique atomic number.

atomic number (Z) the number of protons in a nucleus; the same for every atom of the same element

▼ **TABLE 3.1.2** Isotopes of carbon

Isotope	Number of protons	Number of neutrons	Total number of particles in the nucleus
Carbon-12	6	6	12
Carbon-13	6	7	13
Carbon-14	6	8	14

The atomic number also represents the number of electrons in an atom. All atoms are electrically neutral. They have no overall charge. The number of positive protons is balanced by the number of negative electrons. This means the number of protons and electrons in an atom is equal.

mass number (A) the total number of protons and neutrons in the nucleus of an atom

The **mass number (A)** is the total number of protons and neutrons in the nucleus. The mass number is used with the element name to describe the isotopes. For example, a carbon atom with six protons and six neutrons has a mass number of 12. This isotope is carbon-12.

The atomic number and mass number can be placed next to the element symbol.

Mass number ——— 12
Atomic number ——— 6

$^{12}_{6}C$

▲ **FIGURE 3.1.2** Notation showing the mass number and atomic number for carbon-12

☆ **ACTIVITY**

Modelling subatomic particles in isotopes

Use three different colours of plasticine to create small spheres. Each colour will represent a different type of subatomic particle. Lithium is an element with the atomic number 3. Make models of lithium-6 and lithium-7 that show all subatomic particles. Put the electrons in a circle around the nucleus for now. You will learn more about the placement of electrons in the next module.

a Create a table showing the number of each subatomic particle in each isotope.

b How are the two isotopes similar ? How are they different?

3.1 LEARNING CHECK

1 **Identify** the number of protons, neutrons and electrons for the following.

a $^{23}_{11}Na$ **b** $^{39}_{19}K$

2 Subatomic particles have similarities and differences. This could be about their size, charge or location.

a **Identify** a similarity between any two of the subatomic particles.

b **Identify** a difference between any two of the subatomic particles.

3.2 Bohr's model of the atom

BY THE END OF THIS MODULE, YOU WILL BE ABLE TO:

✓ explain the Bohr model and its limitations
✓ write electron configurations for the first 20 elements.

Video activity
Atom structure: electron shells

Interactive resource
Label: Electron and shells

GET THINKING

This module describes the Bohr model of the atom. We use models to represent complex scientific ideas or objects we cannot observe directly. What scientific models do you know from your studies so far? How can models be helpful to your learning?

The development of the Bohr model

spectroscope
an instrument used to split light into its component colours

emission spectrum
the pattern of lines formed from the movement of electrons between energy shells

Niels Bohr (1885–1962) was a Danish physicist who proposed one of the atomic models you learned about in Year 9. Bohr studied light that was emitted from elements when they were heated. When this light passes through a **spectroscope**, the light is separated into an **emission spectrum** (plural: spectra). A spectroscope works like a prism that breaks white light into a rainbow of colours.

The emission spectrum of hydrogen is shown in Figure 3.2.1. All samples of hydrogen that are heated emit light that gives the same emission spectrum. The same four coloured lines (violet, blue, blue-green and red) always appear in the same pattern.

Alamy Stock Photo/ Phil Degginger

▲ **FIGURE 3.2.1** The emission spectrum of hydrogen

Bohr examined spectra from hydrogen atoms and used what he observed to propose his model of the atom. His model was an attempt to explain how electrons were arranged around the nucleus.

The Bohr model

The Bohr model, shown in Figure 3.2.2, is based on a series of assumptions made by Bohr.

- Electrons have energy and exist in a circular orbit around the nucleus in an electron shell.
- All electrons in the same electron shell have the same amount of energy.
- Electrons in shells that are further away from the nucleus have more energy.

excited
the state of an atom or electron when it absorbs energy

Bohr proposed that electrons in atoms would be arranged in electron shells and would stay in those shells unless the atom was heated and became **excited**. If the atom were heated, the electrons would absorb this energy and move to a higher energy level further away from the nucleus. This electron would now be unstable.

9780170463041

Unstable electrons would release energy by returning to their original electron shell. The energy would be released as light. The amount of energy released would be related to how many electron shells the electron moved up and down (see Figure 3.2.3).

Bohr's model was able to explain why the emission spectrum pattern of hydrogen was always the same. Bohr performed calculations to predict features of the emitted light. His calculations were based on his assumptions about the electron shells and the movement of electrons. The experimental evidence supported his proposed model and, thus, it was accepted by the scientific community.

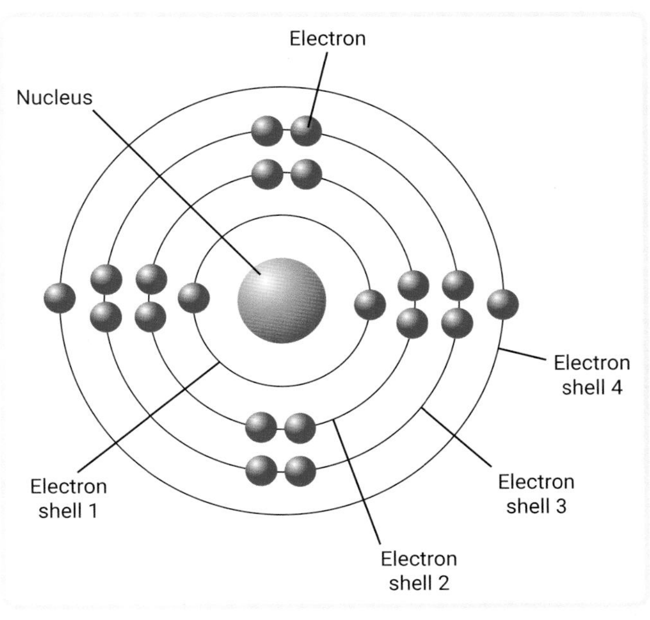

▲ **FIGURE 3.2.2** The Bohr model of the calcium atom

Limitations of the Bohr model

While the Bohr model was accepted as the most accurate model at the time, it still was not perfect. The model had limitations that led to further development of the atomic model. It could not explain:

- all the lines in the emission spectra of elements other than hydrogen
- why some lines were more intense than others (in hydrogen, the purple line is very faint, the red line is very bright)
- other features of emission spectra that had been observed by scientists.

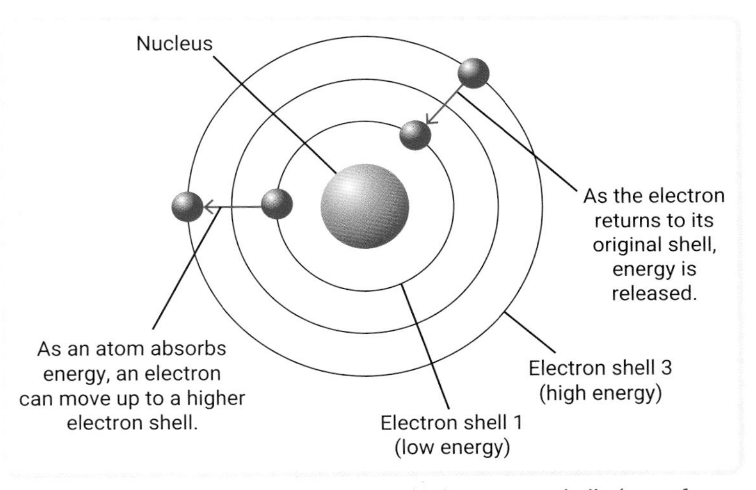

▲ **FIGURE 3.2.3** Electrons can move to higher energy shells (away from the nucleus) when an atom absorbs energy. Electrons release energy when they move down electron shells (towards the nucleus).

Electron configuration

The arrangement of electrons in **energy shells** is called the **electron configuration**. The electron configuration of any atom can be determined by knowing the set of rules that apply.

The first thing to know is how many electrons can fit into each shell. You can find the maximum number of electrons that can fit into a shell by applying the formula $2n^2$, where n represents the number of the electron shell (see Table 3.2.1). We will use this and other rules to predict the electron configuration for the first 20 elements in the **periodic table**.

energy shells
regions of space around a nucleus that contain electrons of the same energy

electron configuration
the arrangement of electrons in electron shells in an atom

periodic table
a method of arranging elements by increasing atomic number

▼ **TABLE 3.2.1** The maximum number of electrons in each electron shell

Shell number (n)	Maximum number of electrons
1	$2 \times 1^2 = 2$
2	$2 \times 2^2 = 8$
3	$2 \times 3^2 = 18$
4	$2 \times 4^2 = 32$
5	$2 \times 5^2 = 50$

1 Electrons fill from the lowest shells first (the shells closest to the nucleus).
2 Electrons fill the shells to a certain number then move on to the next shell. They may not entirely fill the shell.

For the first 20 elements, they fill like this:

- first shell ($n = 1$) with two electrons
- second shell ($n = 2$) with eight electrons
- third shell ($n = 3$) with eight electrons (note that this shell can hold 18 but fills with only eight electrons at this stage)
- fourth shell ($n = 4$) with two electrons.

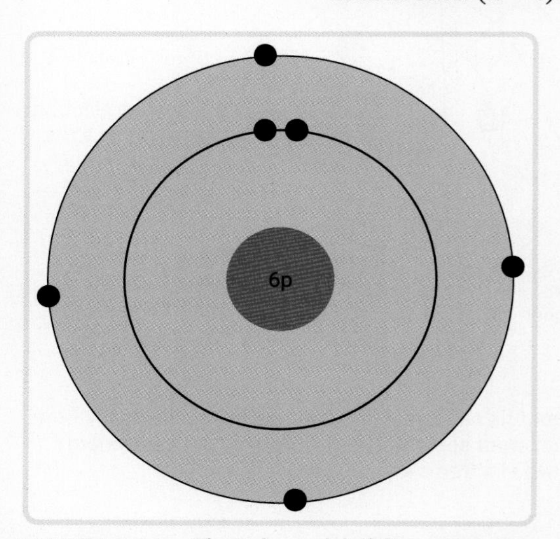

▲ **FIGURE 3.2.4** The Bohr model of the carbon atom

So $2 + 8 + 8 + 2 = 20$, and we have reached the end of the first 20 elements.

Let's look at the specific example of carbon with six electrons (see Figure 3.2.4). Following the rules above:

- two electrons go into the first shell. It is now full and there are four electrons remaining
- four electrons go into the second shell. There are no electrons remaining.

The electron configuration is represented as shown in Figure 3.2.5.

The electron configurations of the first 10 elements in the periodic table are shown in Table 3.2.2.

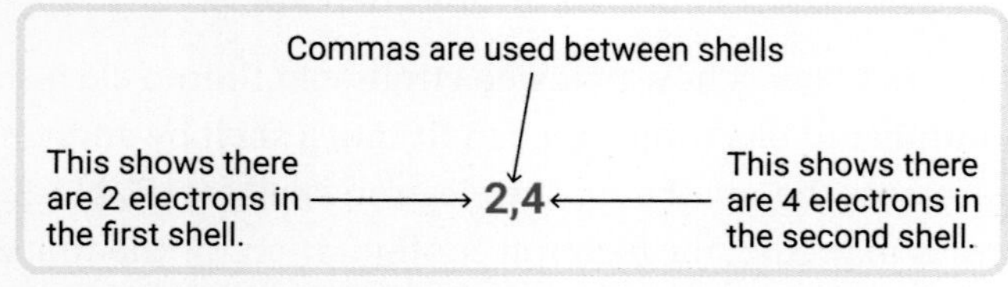

▲ **FIGURE 3.2.5** The electron configuration of carbon

▼ **TABLE 3.2.2** The electron configurations of the first 10 elements

Element	Number of electrons	Electron configuration
Hydrogen	1	1
Helium	2	2
Lithium	3	2,1
Beryllium	4	2,2
Boron	5	2,3
Carbon	6	2,4
Nitrogen	7	2,5
Oxygen	8	2,6
Fluorine	9	2,7
Neon	10	2,8

Configurations written in this way have links to the position of the atom in the periodic table and provide more complex information about the way electrons are arranged in atoms. You will learn more about electron configurations in Year 11.

✩ **ACTIVITY**

Modelling electron configuration

You can do this activity by drawing models or by using plasticine. In this activity, you are going to model each atom's electron configuration until you can extend Table 3.2.2 to include all the first 20 elements. Copy Table 3.2.2 and use your copy to record your answers.

- Use a small circle or small sphere to represent the nucleus.
- Draw a circle around the nucleus to represent the first energy shell.
- Place one electron in the shell (by drawing a dot or using a small plasticine sphere). This represents hydrogen.
- Place a second electron in the first shell. What element is this?
- Draw a second circle with a larger circumference around the first one. What does this represent?
- Add electrons one by one until you have eight electrons in this circle. After each electron is added, identify the element, and write the electron configuration in your table.
- Continue the modelling process with a third electron shell, and then a fourth, until you have built up the elements 11–20 to add to your table.
  - **a** How does this modelling help you understand electron configuration?
  - **b** How is this model inaccurate?

3.2 LEARNING CHECK

1 Write the electron configuration of:
  a fluorine (9 electrons).
  b potassium (19 electrons).

2 Use your table to **identify** the elements represented by:
  a 2,3.
  b 2,8,4.

3 **Describe** the Bohr model of the atom.

4 **Identify** one limitation of the Bohr model.

3.3 Atomic structure and the periodic table

BY THE END OF THIS MODULE, YOU WILL BE ABLE TO:

- ✓ identify key features and sections of the periodic table
- ✓ explain the organisation of the periodic table in terms of atomic structure
- ✓ predict properties of elements based on their position in the periodic table.

Video activities
Introduction to the periodic table
Atomic structure
Noble gases

GET THINKING

Many words in chemistry are common words we use every day that have specific meanings in science. In this module, the word 'group' refers to a particular set of elements on the periodic table. Are there other scientific words you know like this that might cause confusion if used incorrectly in science?

Features of the periodic table

The periodic table represents a range of information about atoms. Figure 3.3.1 shows the current understanding of the elements.

Every square contains an element symbol (e.g. H), the element name (e.g. hydrogen) and the atomic number (e.g. 1). If you look along each row and read it like lines of text in a book, then you will see the elements are arranged in order of increasing atomic number.

groups
the vertical columns on the periodic table

The vertical columns are called **groups**. There are 18 groups in total, and the group numbers are on the top of each column.

Key: atomic number / Symbol / element name
State at room temperature: Solid | Liquid | Gas | Synthetic
Classification: Metal | Non-metal | Metalloid

Period \ Group	1	2	3	4	5	6	7	8	9	10	11	12	13	14	15	16	17	18
1	1 H hydrogen																	2 He helium
2	3 Li lithium	4 Be beryllium											5 B boron	6 C carbon	7 N nitrogen	8 O oxygen	9 F fluorine	10 Ne neon
3	11 Na sodium	12 Mg magnesium											13 Al aluminium	14 Si silicon	15 P phosphorus	16 S sulfur	17 Cl chlorine	18 Ar argon
4	19 K potassium	20 Ca calcium	21 Sc scandium	22 Ti titanium	23 V vanadium	24 Cr chromium	25 Mn manganese	26 Fe iron	27 Co cobalt	28 Ni nickel	29 Cu copper	30 Zn zinc	31 Ga gallium	32 Ge germanium	33 As arsenic	34 Se selenium	35 Br bromine	36 Kr krypton
5	37 Rb rubidium	38 Sr strontium	39 Y yttrium	40 Zr zirconium	41 Nb niobium	42 Mo molybdenum	43 Tc technetium	44 Ru ruthenium	45 Rh rhodium	46 Pd palladium	47 Ag silver	48 Cd cadmium	49 In indium	50 Sn tin	51 Sb antimony	52 Te tellurium	53 I iodine	54 Xe xenon
6	55 Cs caesium	56 Ba barium		72 Hf hafnium	73 Ta tantalum	74 W tungsten	75 Re rhenium	76 Os osmium	77 Ir iridium	78 Pt platinum	79 Au gold	80 Hg mercury	81 Tl thallium	82 Pb lead	83 Bi bismuth	84 Po polonium	85 At astatine	86 Rn radon
7	87 Fr francium	88 Ra radium		104 Rf rutherfordium	105 Db dubnium	106 Sg seaborgium	107 Bh bohrium	108 Hs hassium	109 Mt meitnerium	110 Ds darmstadtium	111 Rg roentgenium	112 Cn copernicium	113 Nh nihonium	114 Fl flerovium	115 Mc moscovium	116 Lv livermorium	117 Ts tennessine	118 Og oganesson

57 La lanthanum	58 Ce cerium	59 Pr praseodymium	60 Nd neodymium	61 Pm promethium	62 Sm samarium	63 Eu europium	64 Gd gadolinium	65 Tb terbium	66 Dy dysprosium	67 Ho holmium	68 Er erbium	69 Tm thulium	70 Yb ytterbium	71 Lu lutetium
89 Ac actinium	90 Th thorium	91 Pa protactinium	92 U uranium	93 Np neptunium	94 Pu plutonium	95 Am americium	96 Cm curium	97 Bk berkelium	98 Cf californium	99 Es einsteinium	100 Fm fermium	101 Md mendelevium	102 No nobelium	103 Lr lawrencium

▲ **FIGURE 3.3.1** The periodic table (current as of February 2023)

The horizontal rows are called **periods**. There are seven periods, and the period numbers are next to each row on the left. These periods correspond to the electron shells described in Module 3.2. The first period has two elements (two electrons) and the second period has eight elements (eight electrons). This matches the expected number of electrons in the electron shells.

periods
the horizontal rows on the periodic table

3.3

What are those extra rows?

There are two extra rows of elements at the bottom of the table. These are not extra periods. They fit into the periodic table in periods 6 and 7. Period 6 has a gap between Ba (56) and Hf (72). The missing elements are the first of the rows at the bottom from La (57) to Lu (71). The elements in this row are called the lanthanoids. The second row fits into period 7 in the same way. The elements in this row are called the actinoids.

Metals, non-metals and metalloids

The periodic table contains many patterns. One simple pattern is seen in Figure 3.3.1 and shows the elements that are metals, non-metals and metalloids. The elements in yellow are metals. Most elements are metals. The light blue elements on the right of the table are the non-metals. The purple strip between them is the metalloids – elements with properties of both metals and non-metals.

Groups of the periodic table

The vertical groups of the periodic table are often given names that represent their common properties. The way the periodic table is organised means that elements in the same group behave in a similar way in chemical reactions. They also have similar physical properties, such as electrical conductivity or hardness. Therefore, we can use the periodic table to predict information about elements based on where they are on the table (see Table 3.3.1).

Hydrogen is the one element that does not really fit these trends. Although it is found in group 1, it does not have the same properties as other group 1 elements.

TABLE 3.3.1 Groups of the periodic table

Group	Name	Common elements	Properties
1	Alkali metals	Sodium, potassium	• Low-density metals • Highly reactive
2	Alkali earth metals	Magnesium, calcium	• Low-density, silvery metals • Mildly reactive
3–12	Transition metals	Copper, nickel, zinc	• High-density metals • Mild–low reactivity
17	Halogens	Fluorine, chlorine	• Toxic substances • Highly reactive
18	Noble gases	Helium, neon, xenon	• Gases • Very unreactive

Electron arrangement and the periodic table

The electron configurations of the atoms involved determine the chemistry of chemical reactions, bonding and how reactive elements are. In Module 3.2 you learned how to determine electron configuration. The electron configuration is directly related to the position of the element on the periodic table (see Figure 3.3.2).

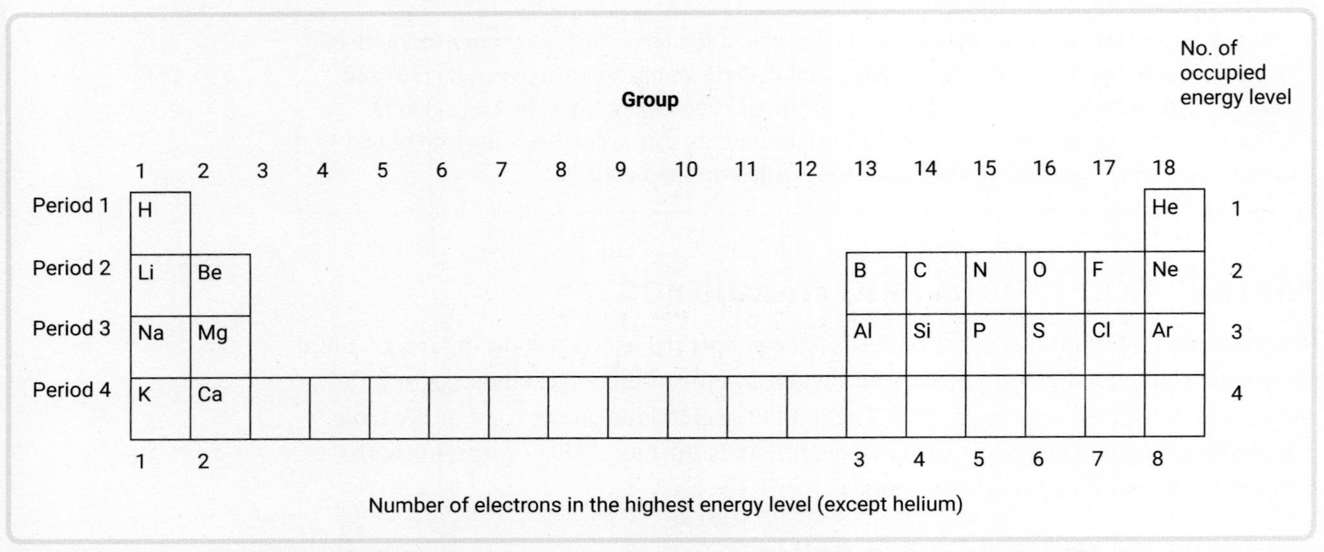

▲ **FIGURE 3.3.2** **A periodic table showing the number of electrons in the highest energy electron shell below each group**

If we look at group 1, we see the following electron configurations:

- H = 1
- Li = 2,1
- Na = 2,8,1
- K = 2,8,8,1.

All elements in group 1 have one electron in their highest energy shell.

You can repeat this process for elements in group 2. They all have two electrons in their highest energy shell. This pattern repeats across groups 1, 2 and 13–18. The number of electrons in the highest shell can be seen on the bottom of the periodic table in Figure 3.3.2. The electrons in the highest energy shell are known as **valence electrons**.

valence electrons electrons in the highest electron shell of an atom

Elements that have the same number of valence electrons are in the same group on the periodic table and behave in similar ways in chemical reactions.

Properties of the elements

Most of the elements on the periodic table are solids. This includes most of the metals – with the exception of mercury, which is a liquid at room temperature. Most of the non-metals are gases; one is a liquid (bromine), and some are solids (such as carbon and phosphorus).

Metals, non-metals and metalloids

The properties of metals, non-metals and metalloids can be attributed to the way their atoms bond together. We will look at the properties of metals in Module 3.6 and the properties of non-metal elements in Module 3.7. Metalloids have properties of both metals and non-metal elements.

Group 1 and group 2 metals

As shown above, group 1 and group 2 metals have one and two valence electrons, respectively. This makes elements in these groups reactive.

Group 1 elements have vigorous or explosive reactions with water and oxygen. Figure 3.3.3 shows the fizzing and sparking that occurs when potassium is added to water. Some of the group 1 elements need to be stored in oil or in containers with the air removed to prevent them from reacting explosively when not being used.

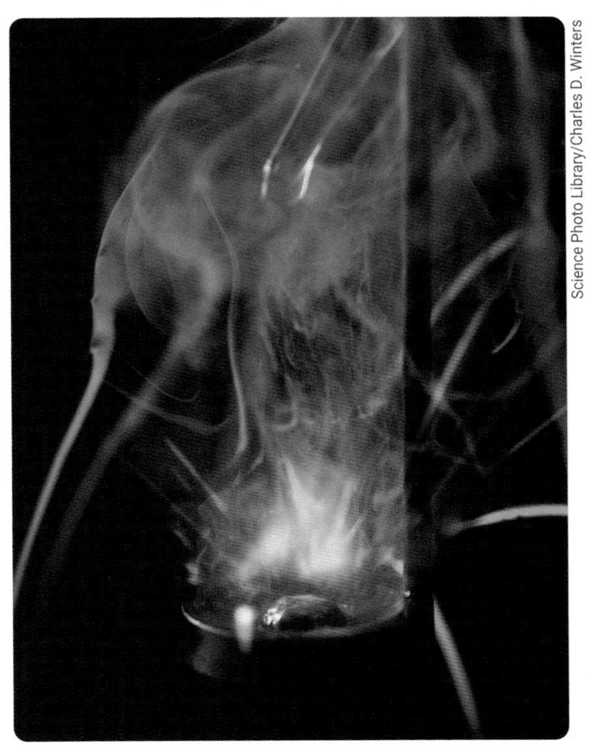

Science Photo Library/Charles D. Winters

▲ **FIGURE 3.3.3** Potassium reacts explosively with water.

Group 2 elements are still quite reactive, but the extra valence electron means they are not as reactive as group 1 elements. This is a good example of how the electron configuration affects chemical properties. The group 2 elements do react with water, but slowly. Some of these elements need heat to start a reaction with water or air.

Reactions of the halogens

The **halogens** are the group 17 elements. They include chlorine, bromine and fluorine. These elements all have seven valence electrons. All halogens are very reactive. All halogens will form acids when dissolved in water.

halogens
the name for the group 17 elements on the periodic table

Noble gases

noble gases
the name for the group 18 elements on the periodic table; they have 'full' valence shells

Group 18, the **noble gases**, all have eight electrons in the outer shell. This makes noble gases very stable and very unreactive, which we will look at in Module 3.4. This unreactive property is where the name 'noble' comes from. Like royalty, who don't mix with the commoners, the noble gases do not mix (or react) with other elements.

☆ ACTIVITY

Creating your own periodic table summary

You will need a blank periodic table like the one shown in Figure 3.3.4. It may have element names and symbols, but no other information should be present. You can copy the one below on a piece of grid paper or find a blank copy on the Internet. You will need some coloured pencils, textas or highlighters.

1 **Label** groups 1–18.
2 **Identify** and label each group that has a name.
3 **Label** rows 1–7.
4 Use three colours to shade the metals, metalloids and non-metals. Create a key so you know which colour represents which type of element.
5 **Label** the number of valence electrons in groups 1, 2 and 13–18.

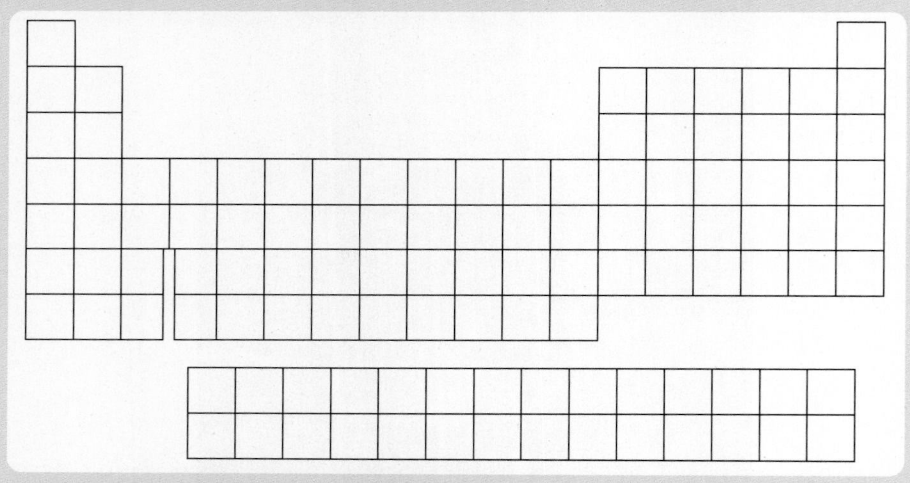

▲ FIGURE 3.3.4 A blank periodic table

3.3 LEARNING CHECK

1 What information can be gained about an element's electron configuration by knowing what period it is in?
2 From what you know about hydrogen, **explain** why it doesn't fit into group 1 in terms of its properties.
3 Group 1 elements are very reactive in water. Group 2 elements are less reactive in water. **Explain** what causes this difference.

3.4 Bonding and stable atoms

BY THE END OF THIS MODULE, YOU WILL BE ABLE TO:

- ✓ explain how atoms achieve a stable electron configuration
- ✓ explain why atoms form bonds and identify the different types of bonding
- ✓ describe how atomic structure relates to the properties of elements.

GET THINKING

Chemistry includes terms such as 'compound', 'mixture', 'element' and 'atom'. In this chapter you will come across a lot of new terms that have very specific meanings. Create a word list in your class notes so you can revise definitions easily.

Video activity
Introduction to chemical bonding

Interactive resource
Drag and drop:
Types of bonding

Why are single atoms rare?

Most elements do not exist as single atoms. In fact, only the noble gases and mercury vapour can exist as single atoms. All other elements are unstable when they are individual atoms. Why is this?

In Module 3.3, you learned that the electron configuration plays a large role in the chemical behaviour of elements. The reason most atoms are unstable is related to electrons.

Stable atoms have a full outer electron shell. This means they have eight electrons in the outer shell. You may recall that the only group on the periodic table that has this arrangement is the noble gases (see Figure 3.3.2). All other elements have between one and seven electrons in the outer shell. This makes them unstable.

Atoms other than noble gases undergo changes to achieve the same stable electron configuration as a noble gas. Figure 3.4.1 shows how atoms can combine to make stable structures.

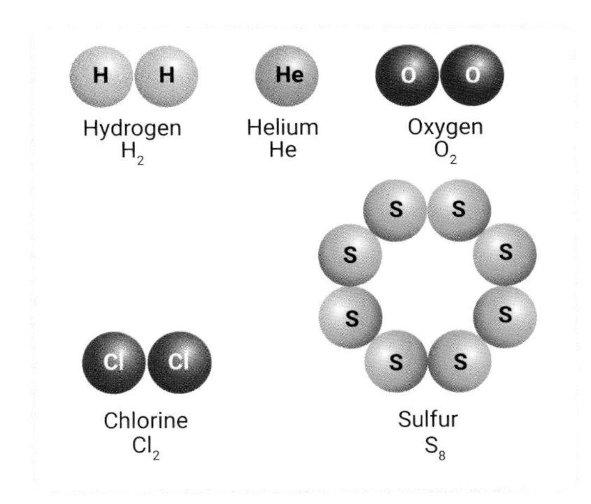

▲ **FIGURE 3.4.1** Only the noble gases and mercury vapour exist as single atoms. Other elements form molecules and compounds to achieve a stable electron configuration.

Gaining or losing electrons to become stable

Figure 3.4.2a shows a sodium atom with electron configuration 2,8,1. Figure 3.4.2b shows the stable configuration of the neon atom, 2,8. You can see that if sodium were to lose its one valence electron, then it would have the same electron configuration as neon. Figure 3.4.2c shows a sodium atom that has lost one electron. This is now a sodium ion. We will look at **ion** formation in detail in Module 3.5.

ion
a charged particle formed when an atom loses or gains valence electrons

The chlorine atom has electron configuration 2,8,7. If it were to gain one more electron, it would have the same configuration as argon, 2,8,8.

Thus, atoms can achieve a stable electron configuration by gaining or losing electrons.

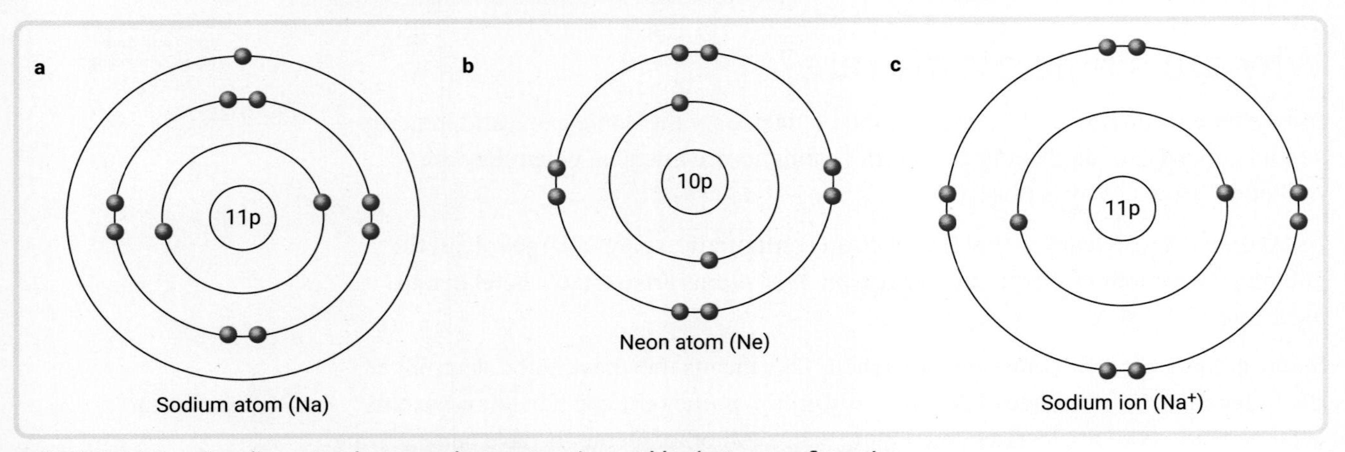

▲ **FIGURE 3.4.2** A sodium atom loses an electron to gain a stable electron configuration.

Sharing electrons to become stable

Some atoms share electrons to achieve a stable electron configuration. Figure 3.4.3 shows two chlorine atoms with electron configuration 2,8,7. Each of the atoms needs only one more electron to achieve a stable 2,8,8 configuration. Each chlorine atom shares one electron with the other chlorine atom. This way they have their own electrons, plus the extra one they need to achieve a stable configuration.

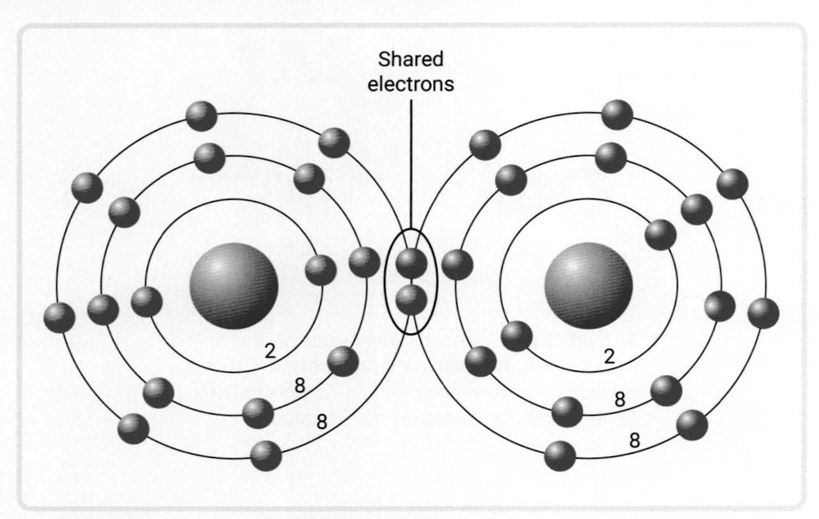

▲ **FIGURE 3.4.3** Two chlorine atoms sharing an electron to achieve a 2,8,8 electron configuration

Types of bonding

Chemical bonding is all about electrons. It occurs when atoms lose or gain electrons, or when they share electrons. Bonding involves atoms joining together, driven by the need to have a stable electron configuration.

chemical bonding
the joining of atoms via the transfer, loss or sharing of electrons to achieve a stable electron configuration

Bonding can occur in elements. This happens when atoms of the same element join to form structures. Examples of element structures can be seen in Figure 3.4.1.

Bonding also creates **compounds** – where two or more different elements join to form substances such as carbon dioxide (CO_2) or water (H_2O).

compound
two or more different elements joined by a chemical bond

Three types of bonding can occur. Each involves a different method of achieving a stable electron configuration. We will cover each type of bonding in more detail in Module 3.5, Module 3.6 and Module 3.7.

▼ **TABLE 3.4.1** The types of bonding

Type of bonding	Method of achieving stable electron configuration	Examples of where this bonding occurs
Metallic bonding	Electrons are lost from metal atoms and become free moving in the metal structure.	Copper, iron, magnesium
Ionic bonding	Electrons are lost from some atoms and gained by others, so they form ions.	Sodium chloride, calcium oxide
Covalent bonding	Electrons are shared between atoms.	Water, oxygen, carbon dioxide

3.4 LEARNING CHECK

1 **Explain** why only one group of the periodic table exists as atoms.
2 **Write** the electron configuration of calcium and suggest how it might achieve a stable electron configuration.
3 Use examples to **describe** two methods by which atoms achieve a stable electron configuration.
4 By describing similarities and differences between the types of bonding, **compare**:
   a metallic bonding and covalent bonding.
   b ionic bonding and metallic bonding.

3.5 Forming ions

BY THE END OF THIS MODULE, YOU WILL BE ABLE TO:

- ✓ explain how atoms form ions by gaining or losing electrons
- ✓ explain why atoms of elements in the same group form ions with the same charge.

Interactive resource
Label: Predicting ions

GET THINKING

Look at Figure 3.5.3. It shows how you can use the periodic table to make conclusions about elements based on their groups rather than learning about each individual element. Why would this make it easier to learn chemistry? If you couldn't make general conclusions like this, what would be your alternative?

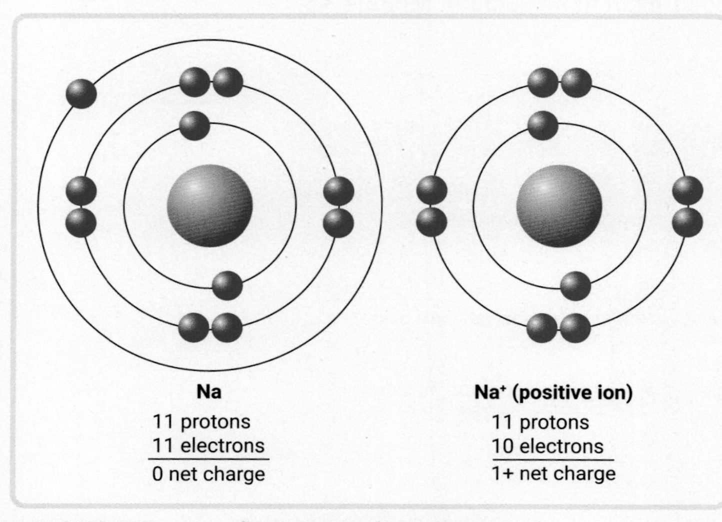

▲ **FIGURE 3.5.1** A sodium atom and a sodium ion

What is an ion?

An ion is a charged particle that forms when an atom loses or gains valence electrons. Atoms are electrically neutral because the number of positive protons equals the number of negative electrons. When the number of electrons is changed, this balance no longer exists, and an ion is formed.

The structure of a sodium atom is shown in Figure 3.5.1. It has 11 protons and 11 electrons, which gives it zero net charge. The one valence electron is making it unstable, so it loses this electron to another atom to form a stable 2,8 electron configuration. The sodium ion has 11 protons, but only 10 electrons. It has a 1+ charge overall because it now has one more positive proton than it does negative electrons.

Positive and negative ions

cation
an ion with a positive charge

anion
an ion with a negative charge

Positive ions are called **cations**. Cations form when atoms lose electrons. This results in the ion having more protons than electrons and a net positive charge.

Negative ions are called **anions**. Anions form when atoms gain electrons. This results in the ion having more electrons than protons and a net negative charge.

So, what determines whether an electron is lost or gained from an atom? It's all about energy. Atoms will gain or lose as few electrons as possible to achieve a stable electron configuration. Each electron lost or gained requires the same amount of energy. Most chemical processes use the path that requires the least energy.

For example, oxygen has the electron configuration 2,6. An oxygen atom can either gain two electrons or lose six electrons. It is energetically easier for the atom to gain two electrons than lose six, so the oxygen atom will form a negative anion.

Atoms with one, two or three valence electrons will lose electrons and form positive cations.

Atoms with five, six or seven valence electrons will gain electrons and form negative anions, as demonstrated in Figure 3.5.2 with the formation of a negative fluoride ion by gaining an electron.

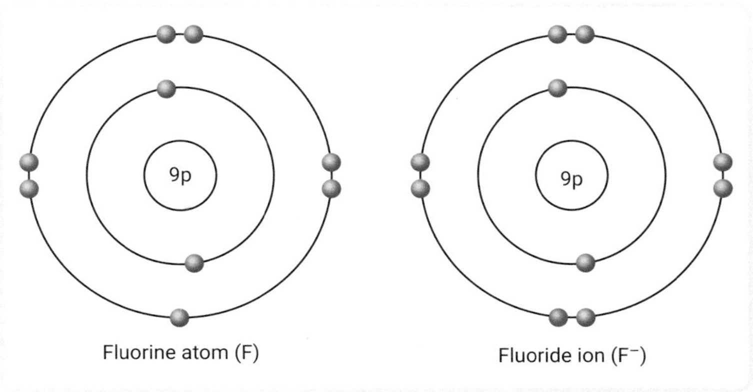

9p

Fluorine atom (F)

9p

Fluoride ion (F^-)

▲ **FIGURE 3.5.2** The formation of a negative fluoride ion by a fluorine atom gaining an electron

Predicting the charge of an ion

The periodic table can be used to predict the charge of an ion for elements in groups 1, 2 and 13–17.

We know that the elements in group 1 all have one valence electron. We also know that elements with one valence electron form a 1+ ion. This means all elements in group 1 have a 1+ charge.

We can use the same logic for group 2 (two valence electrons, charge of 2+) and for group 3 (three valence electrons, charge of 3+). Groups 1, 2 and 3 contain metals. This means that metal ions are always positive.

Most elements in group 4 do not form ions. Elements such as carbon and silicon do not form ions. The non-metal elements in groups 15, 16 and 17 form ions of charge 3–, 2– and 1–, respectively. This information is summarised in Figure 3.5.3.

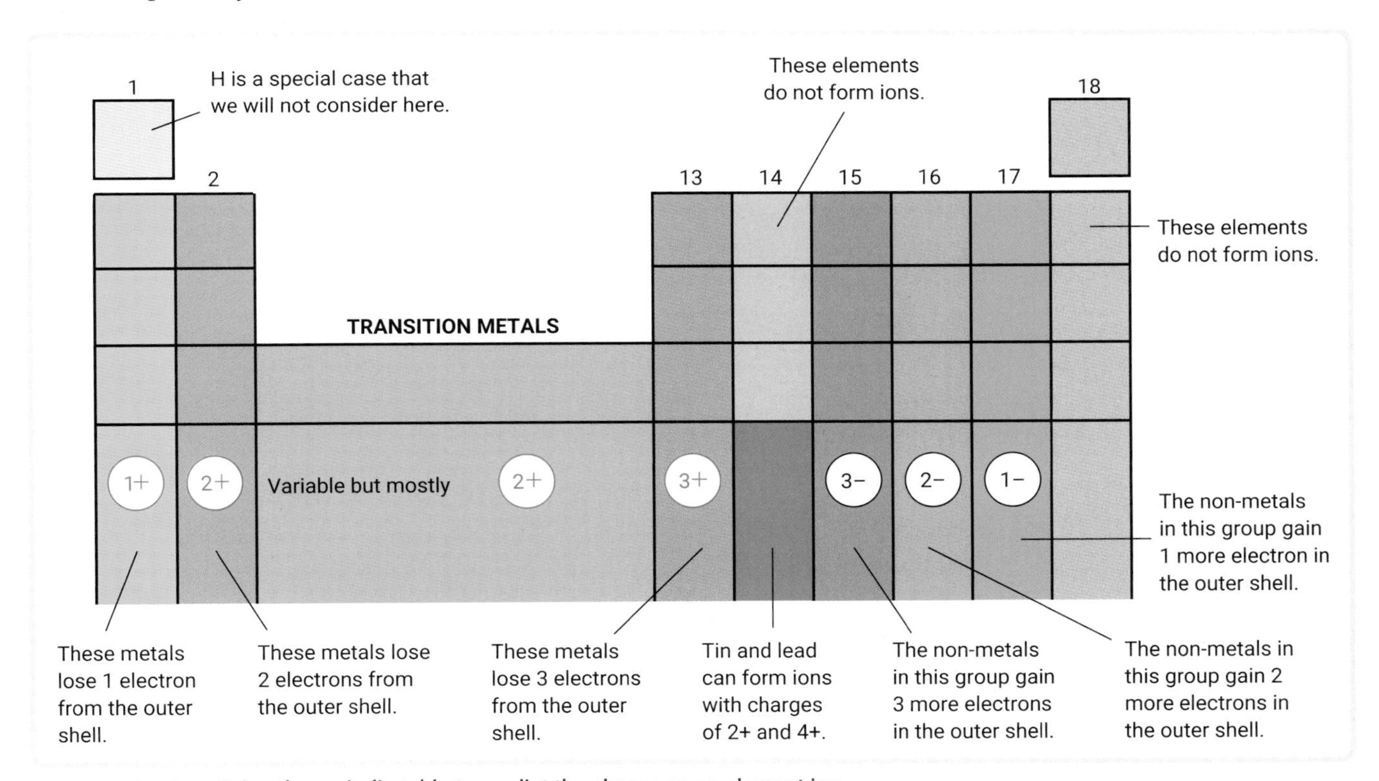

▲ **FIGURE 3.5.3** Using the periodic table to predict the charge on an element ion

Writing formulas and naming ions

Metal ions keep the name of the metal and add 'ion'. Potassium atoms form potassium ions, calcium atoms form calcium ions. To write the formula of the ion, the element symbol is used with the charge in superscript. Potassium ions are K^+, while calcium ions are Ca^{2+}.

Non-metal ions of the elements in groups 15–17 have a change to their name when they form an ion. The end of the name becomes '-ide'. Oxygen atoms become oxide ions, O^{2-}. Chlorine atoms become chloride ions, Cl^-.

Ionic bonding

ionic bonding
bonding between metal and non-metal atoms involving a transfer of electrons

This module is an introduction to **ionic bonding**. For more about ionic compounds and their properties, see digital Module 10.2 in Chapter 10 on Nelson MindTap. Ionic substances include chemicals such as sodium chloride (table salt) and calcium carbonate (found in shells and chalk).

ionic lattice
an organised structure of alternating positive metal ions and negative non-metal ions

Ionic substances form from positive metal ions and negative non-metal ions. Metal ions lose electrons, which transfer to non-metal ions that gain them. The resulting positive and negative ions attract and form a structure called an **ionic lattice** (see Figure 3.5.4).

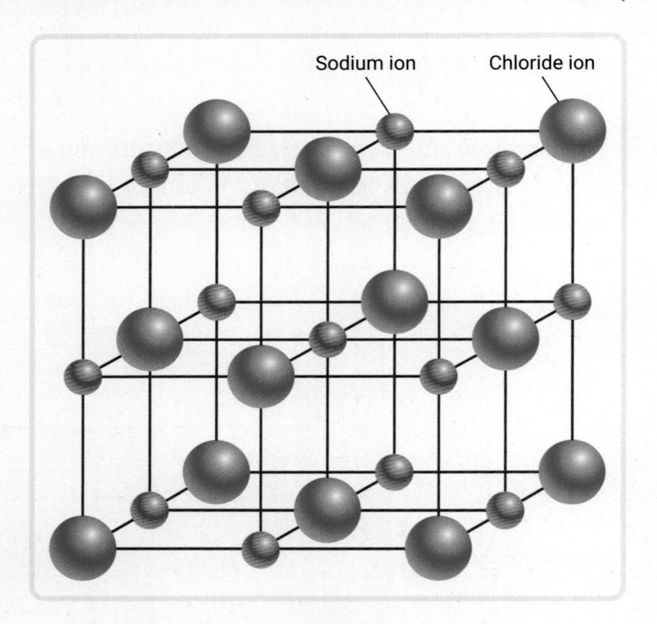

◀ **FIGURE 3.5.4** Sodium chloride is an ionic lattice of alternating positive sodium ions and negative chloride ions.

▲ **FIGURE 3.5.5** Bath salt crystals are popular and are used for relaxation.

Ions in solution

Ionic substances are solids, but many dissolve in water. We use this property when we add sodium chloride to water when cooking, or bath salt crystals to bathwater (see Figure 3.5.5).

When a soluble ionic substance is added to water, the lattice structure breaks down and the positive and negative ions separate and are free to move. We call dissolved ionic substances in water aqueous solutions.

Some ionic substances are insoluble and will keep their lattice structure in water. Calcium carbonate that makes up coral and shells is an ionic substance that does not dissolve readily in water.

Ion quiz

☆ ACTIVITY

3.5

1 Pair up with another student. For this game, you can use a copy of the periodic table that only has the names of the elements and groups written on it.
2 One student is to call out the names of elements at random from groups 1, 2 and 13–17 for 60 seconds (set a timer). The second student is to identify the charge on the ion. Do not move onto the next element until you have the charge correct.
3 Swap roles. The winner of the game is the person who gets the most correct ions in 60 seconds.

3.5 LEARNING CHECK

1 Using Figure 3.5.1 as a model, **justify** the charge on the:
   a sulfide ion.
   b aluminium ion.
   c calcium ion.
   d oxide ion.
2 **Identify** two elements that do not form ions.
3 **Explain** why the magnesium atom does not form a 6− ion.
4 **Describe** how you can use the periodic table to predict the charge on ions in some groups.
5 **Describe** how an ionic lattice forms.
6 Coral is made of calcium carbonate. **Research** and then **explain** how the structure of calcium carbonate relates to its insolubility in water.

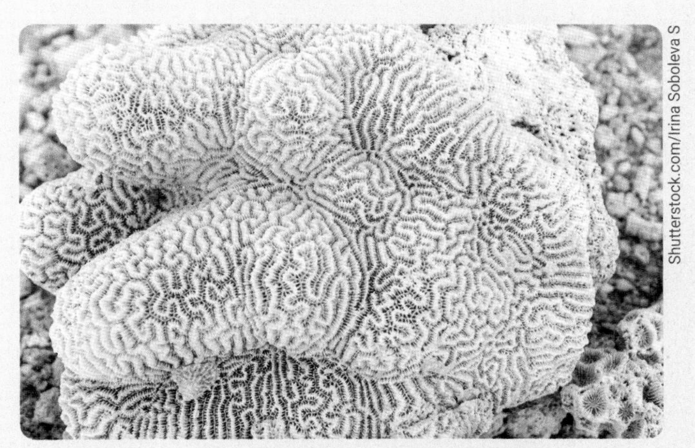

Shutterstock.com/Irina Soboleva S

▲ **FIGURE 3.5.6** Coral is made of calcium carbonate, an ionic substance.

3.6 Bonding in metals

BY THE END OF THIS MODULE, YOU WILL BE ABLE TO:

- ✓ describe how metal atoms bond
- ✓ use the structure of metals to explain their physical properties.

GET THINKING

In chemistry, we often talk about the properties of substances. There are physical properties and chemical properties. What do you think the difference is? Properties can include hardness, boiling point, reactivity, colour and density. Is it easy to identify which properties are physical and which are chemical? Why, or why not?

Interactive resource
Drag and drop:
Metal properties

Quiz
Metal atom bonds

Extra science investigation
Comparing metal properties

Uses of metals

Metallic elements dominate the periodic table, making up nearly 75 per cent of all elements. Metals are widely used in society for construction, musical instruments and electrical wiring. The transport industry relies on metals for the structural components of cars, trains and planes as well as railway lines. In your kitchen you will find metal storage, food cans, cutlery, saucepans and other cooking utensils.

Metals are widely used because of their physical properties. Physical properties relate to observable and/or measurable properties of an object, such as density or boiling point. Chemical properties relate to the reactions substances undergo. The chemical properties of metals will be explored in Chapter 4.

metallic lattice
the organised structure formed with rows of metal cations surrounded by delocalised electrons

delocalised electrons
electrons in a metallic lattice that do not belong to any particular atom

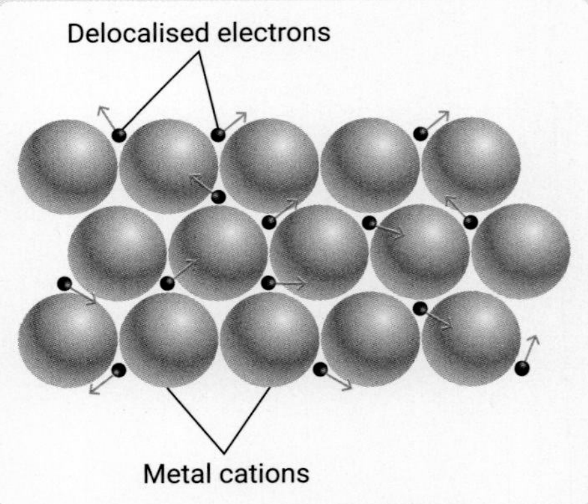

Delocalised electrons

Metal cations

▲ **FIGURE 3.6.1** The internal structure of metals

The structure of metals

Physical properties are determined by the structure of the substance. Metal atoms lose electrons to achieve a stable electron configuration. For example, magnesium has two valence electrons, so loses two electrons to form the Mg^{2+} ion.

The metal cations arrange themselves in an ordered structure called a **metallic lattice** (see Figure 3.6.1). The electrons that were lost by the metal atoms are called **delocalised electrons**. They do not belong to any atom but are free to move around the lattice formed by the metal ions.

The lattice is held together by the force of attraction between the positive metal ions and the negative electrons. This force of attraction is the **metallic bond**.

metallic bond
the force of attraction between metal cations and delocalised electrons

lustrous
shiny when cut or polished

malleable
can be beaten into different shapes

ductile
can be stretched into a wire

Metallic properties

Metals share a common set of physical properties. All properties can be explained by the structure of the metallic lattice.

Metals mostly have a medium–high boiling point and are hard and dense. Metals are **lustrous** (shiny), good conductors of heat and electricity, **malleable** (their shape can be changed) and **ductile** (they can be stretched into a wire), with the exception of mercury, which is the only non-solid metal (see Figure 3.6.2).

Property: hardness and density

The metallic bond is a very strong **electrostatic bond**. This binds the metal cations and delocalised electrons into a very close-packed structure.

The close-packed lattice can absorb a lot of force without distorting, snapping or breaking. This makes metals hard. The close-packed nature of the particles also means that a large amount of mass is present in a given volume. This makes metals denser than many other substances.

▲ **FIGURE 3.6.2** Mercury is the only non-solid metal at room temperature.

Science Photo Library

Property: melting point

To melt a metal, the electrostatic attraction between the cations and electrons needs to be weakened or broken. The strong metallic bond means a large amount of energy is needed to disrupt the force of attraction. This means that metals have medium–high melting points.

Different metals have different melting points. Cadmium has a relatively low melting point for a metal, only 321°C. Aluminium has a melting point of 660°C, whereas nickel has a melting point of 1453°C.

Engineers and designers choose metals for different purposes based on their melting point. Tungsten has a melting point of 3400°C. It is used where the metal used must have a high melting point, such as in industrial heating elements that convert electricity to heat.

▲ **FIGURE 3.6.3** Metals that are cut or polished show lustre.

Science Photo Library/Andrew Lambert Photography

Property: lustre

Metals are described as having lustre when they are polished or freshly cut (see Figure 3.6.3). Metals that have been exposed to air for a long time often develop a coating on the surface, making them dull. If you sand this layer away or cut the metal, it will be shiny.

The shine from metals comes when light reflects off the delocalised electrons moving through the metallic lattice structure, as seen in Figure 3.6.4.

electrostatic bond
the force of attraction between positive and negative particles

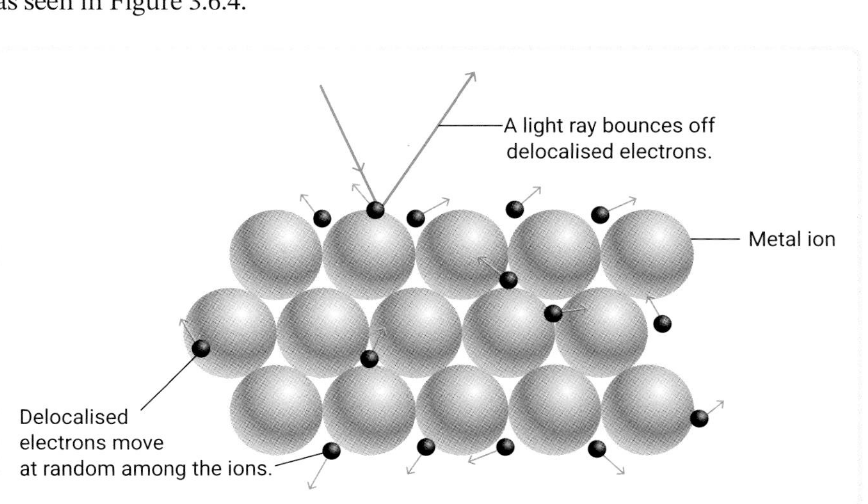

◄ **FIGURE 3.6.4** Metals are shiny because the delocalised electrons reflect light.

Property: conduction of heat and electricity

Metals conduct both heat and electricity well. If you apply electricity or heat to one end of a piece of metal, it will move through to the other end. If you have ever left a metal spoon in a saucepan while it was heating, you would have quickly learned this when you went to pick it up again! This is why wooden spoons are used for stirring, as they do not conduct heat very well.

Metals are used in a lot of electrical appliances to allow electricity to flow. Other materials, such as plastics, are non-conductive, so they stop the flow of electricity. Metals are used for contact points, internal wiring or anywhere you need electricity to flow through a circuit.

Both properties can be explained by the movement of the delocalised electrons. As they move through the lattice structure, the electrons easily carry electrical energy and heat energy through the metal (see Figure 3.6.5).

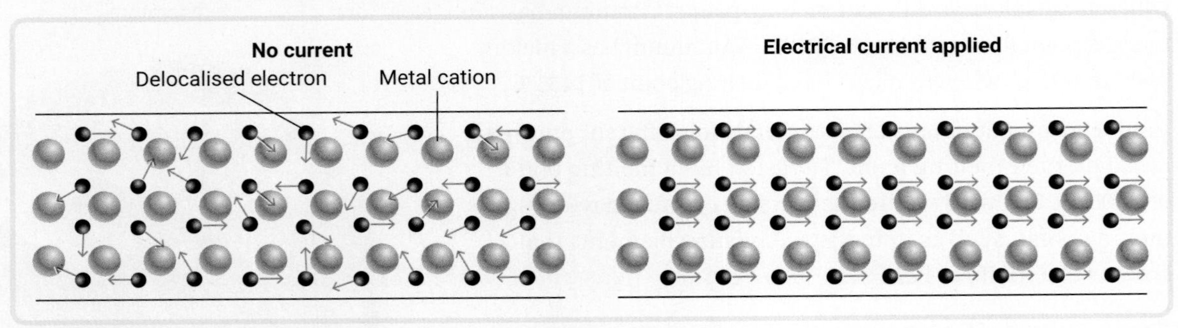

▲ **FIGURE 3.6.5** The movement of electric current through a metal lattice. When a current is applied, the electrons are attracted to the positive terminal.

Not all metals conduct heat and electricity with the same efficiency. One of the best conductors is copper. Copper wiring is most often used for electrical wiring in household circuits and electrical appliances.

Property: malleable and ductile

All metals, except mercury, are malleable and ductile. A malleable metal can be beaten into new shapes. A ductile metal can be stretched or drawn into a wire. These two properties make metals very useful because they can be converted into shapes such as containers or furniture. They can also be used to form wires. Copper wiring carried phone and Internet signals for many years, until replaced by fibre optic cables.

Figure 3.6.6 shows what happens when force is applied to a metallic lattice. Instead of breaking, the metal cations and delocalised electrons simply rearrange themselves into the new shape. This does not damage the lattice in any way and it retains its strength and other properties, just in a different shape!

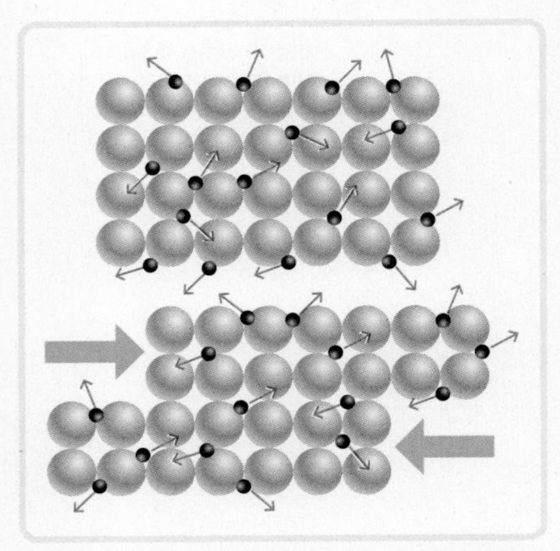

▲ **FIGURE 3.6.6** Metals are malleable and ductile because the cations and delocalised electrons can rearrange themselves to form a different shape.

Modelling the metallic lattice

☆ ACTIVITY

3.6

1 You will need a flat container (such as a takeaway food container) of small polystyrene beads. This is going to act as a model for showing how the metallic lattice gives specific properties.

2 Construct a table like the one shown in Table 3.6.1.

▼ **TABLE 3.6.1** Modelling the metallic lattice

Property	Ways the model is good	Ways the model is poor

3 Using the headings throughout Module 3.6, select three properties and put them into the three rows in your table. For each property you selected, **describe** how the model:

a shows the features of the property.

b could be confusing.

Hint: an example is the property of lustre. The container of polystyrene beads would show the light hitting the lattice structure (good), but it could be confusing, as it does not reflect light the way the delocalised electrons would in a real metal (poor).

3.6 LEARNING CHECK

1 Draw the structure of a metallic lattice, **identifying** the key features of the structure.

2 **Explain** why most metals are dense.

3 **Explain** two reasons many electrical circuit components are made from metal.

▲ **FIGURE 3.6.7** An electrical circuit

3.7 Bonding in non-metals

BY THE END OF THIS MODULE, YOU WILL BE ABLE TO:

- ✓ describe how non-metal atoms bond to form covalent molecules and covalent network structures
- ✓ use the structure of covalent molecules to explain their properties
- ✓ use the structure of covalent network substances to explain their properties.

Interactive resource
Crossword: Bonding

GET THINKING

Pair up with another student and make a list of all the new chemistry words you can remember from this chapter. Have you added them all to your word list? Look for new words in this module to add to the word list you started in Module 3.4.

Sharing electrons

Non-metal atoms occur mostly in groups 14–18 of the periodic table. You already know that group 14 elements such as carbon and silicon do not form ions, so how do they get a stable electron configuration?

Non-metal atoms such as chlorine can bond with other non-metal atoms such as oxygen. If you look at the electron configuration of chlorine (2,8,7) and oxygen (2,6), you will see they both need to gain electrons.

When a non-metal bonds with another non-metal, forming ions is not an option for achieving a stable electron configuration because there is no atom to give them electrons. Instead, non-metal atoms bond with other non-metal atoms by sharing valence electrons.

This means that an electron belongs to two atoms at the same time. Let's look at a simple example. Figure 3.7.1 shows two hydrogen atoms. Hydrogen atoms have one electron in their valence shell. They need to gain one more electron to achieve the stable configuration of the helium atom. Each hydrogen atom shares its one electron with the other hydrogen atom. The shared electrons belong to both atoms, so each has the same configuration as the stable helium.

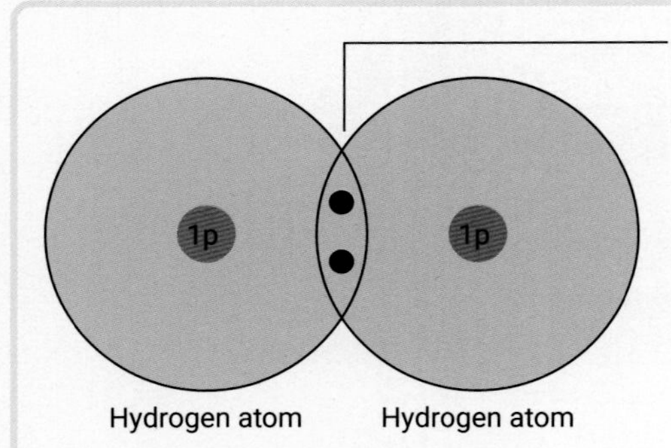

▲ **FIGURE 3.7.1** Hydrogen atoms share electrons to achieve a stable configuration.

Figure 3.7.2 shows the electrostatic force of attraction between the shared pair of negative electrons and the positive nucleus of each atom. This force of attraction is called a **covalent bond**. It is very strong and holds the two atoms together very tightly.

covalent bond
an electrostatic force of attraction between a shared pair of electrons and the nuclei of the atoms sharing the electrons

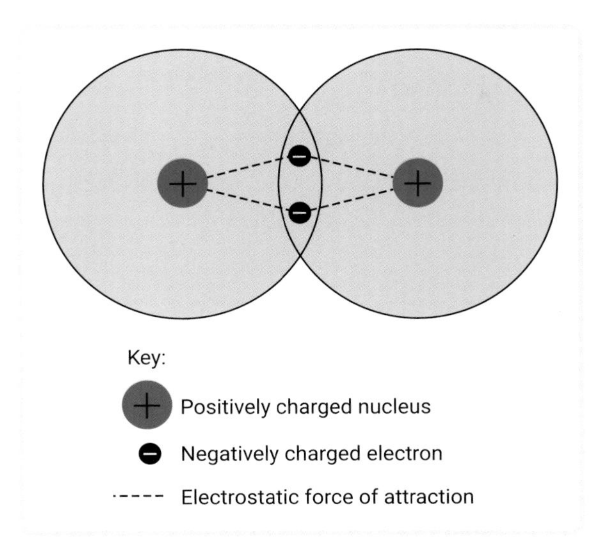

▲ **FIGURE 3.7.2** A covalent bond is an electrostatic force of attraction.

In some cases, atoms need to share more than one of their electrons. Figure 3.7.3 shows two oxygen atoms. Oxygen has the electron configuration 2,6, so it needs two more electrons to form a stable 2,8 configuration. Each oxygen atom shares two of its valence electrons with the other oxygen atom. The two shared pairs form a double covalent bond.

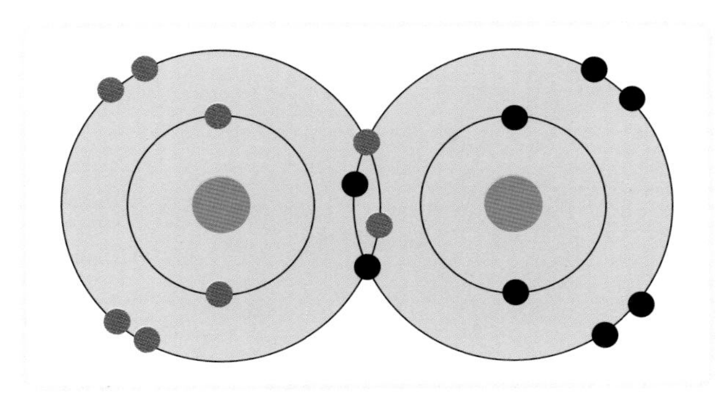

▲ **FIGURE 3.7.3** Oxygen forms a double covalent bond.

There are a number of different ways to represent covalent bonds. Figure 3.7.4 shows (from left to right) an electron dot diagram, a **structural formula** and a physical model. Electron dot diagrams will be explored in Chapter 10 – Chemistry Extension on Nelson MindTap. They show the arrangement of valence electrons in covalent structures, with electrons being represented by dots or crosses.

structural formula
a representation of a chemical structure showing covalent bonds as lines

H : H	H—H	
Electron dot formula The symbol H represents the nucleus of the hydrogen in this case.	**Structural formula** Each pair of electrons in an electron dot formula is represented by a short line.	**Physical model** This shows how atoms have partially merged.

▲ **FIGURE 3.7.4** Ways of representing covalent bonds (shared electron pairs).

A structural formula shows each covalent bond, or shared electron pair, as a single line. For the oxygen example in Figure 3.7.3, a double line would be used because there are two pairs of electrons (O=O).

A physical model is a simple representation of the shape of the resulting atom arrangement.

Forming covalent molecules

covalent molecule a distinct structure formed when two or more non-metal atoms join through covalent bonding

When most non-metals join by covalent bonding, they form a structure known as a **covalent molecule**. A molecule is two or more atoms joined together by covalent bonds to make a distinct structure.

Unlike the lattice structures you have seen for metals and ionic substances, molecules have a fixed number of atoms in them. The hydrogen molecule formed in Figure 3.7.1 only ever has two hydrogen atoms and is represented by H_2 to show the two atoms. The oxygen molecule in Figure 3.7.3 is also made from two atoms of oxygen (O_2). Molecules with the same atoms are elements (such as H_2 or O_2).

Molecules can have different atoms in them, as seen in Figure 3.7.5. Water is a molecule where an oxygen atom forms covalent bonds with each of two hydrogen atoms. The molecule is represented by H_2O. Molecules with different atoms are compounds.

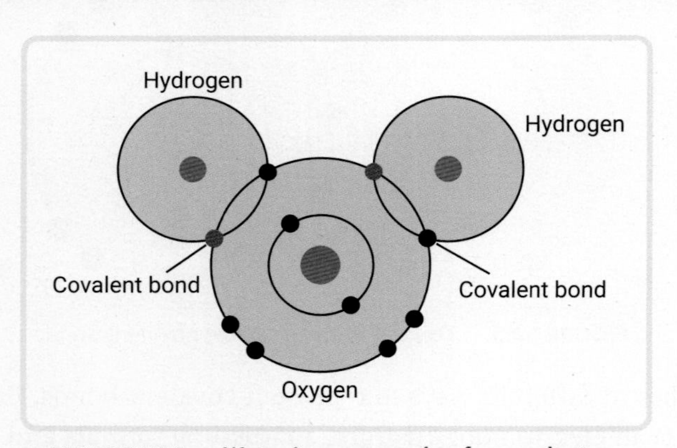

▲ **FIGURE 3.7.5** Water is an example of a covalent compound.

The properties of covalent molecules

Covalent molecular substances have a common set of physical properties related to their structure. When we talk about a covalent molecular substance (e.g., water in a glass), we are talking about a huge number of molecules. As long as the water is still water (or ice or steam), the individual water molecules do not change structure. The molecules in the water are held together by **intermolecular forces**. 'Inter' means between, so intermolecular forces are between molecules. You can see this in Figure 3.7.6.

intermolecular forces forces of attraction between covalent molecules

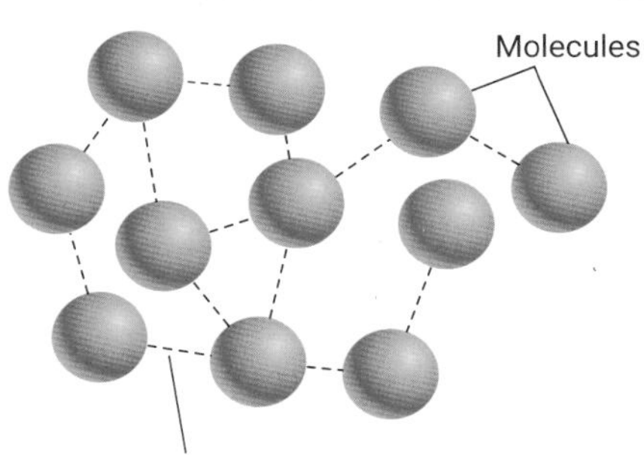

▲ **FIGURE 3.7.6** Intermolecular forces hold molecules together and are responsible for most physical properties.

The properties of covalent molecules are summarised in Table 3.7.1.

▼ **TABLE 3.7.1** The properties of covalent molecules

Property	Explanation
Low melting and boiling points	• Intermolecular forces are very weak. • To melt or boil a substance, you need to weaken or break these intermolecular forces. • As the forces are weak, low energy is needed for this to occur.
Do not conduct electricity	• Covalent molecules are uncharged structures. • There are no free-moving charged particles to conduct electricity.
Usually found as gas, liquid or soft solid	• For a substance to be a liquid, it needs to melt at room temperature. • For a substance to be a gas, it needs to boil at room temperature. • Many substances have melting or boiling points below room temperature, so they are gases or liquids. • Solid substances made by covalent bonding are soft because the intermolecular forces are weak, meaning that not much force is required to distort them.

Forming covalent networks

covalent network
a structure formed when non-metal atoms join in a covalently bonded lattice

The elements carbon and silicon can form a different type of covalent compound called a **covalent network**. Carbon and silicon both have four valence electrons, and this allows a symmetrically shaped lattice to form. Each carbon or silicon atom forms four bonds with other carbon or silicon atoms to form large lattices, as seen in Figure 3.7.7 and Figure 3.7.8.

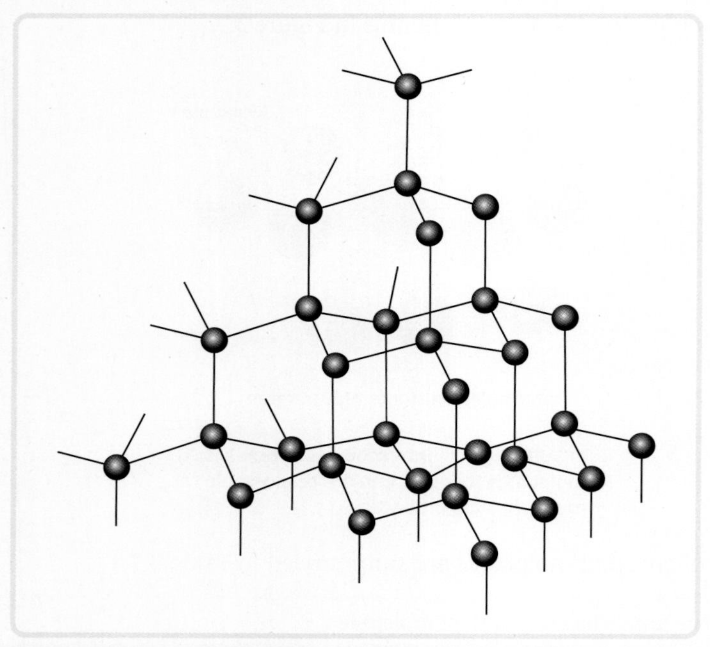

▲ **FIGURE 3.7.7** The structure of diamond; each carbon atom is bonded to four other carbon atoms in a continuous lattice structure.

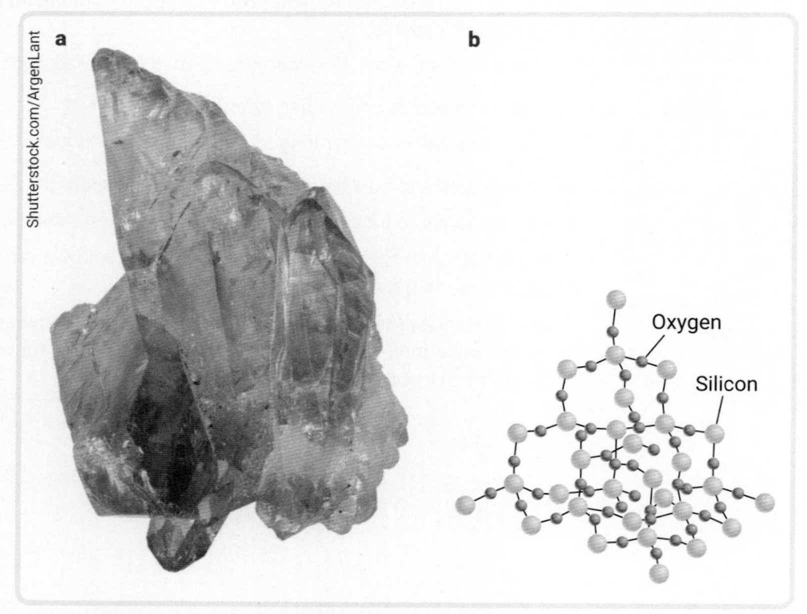

▲ **FIGURE 3.7.8** The mineral quartz (**a**) is formed when silicon bonds with oxygen atoms to form silicon dioxide in a continuous lattice (**b**).

The properties of covalent networks

Covalent networks do not have intermolecular forces because they do not have molecules! The properties of covalent network substances arise from the very strong covalent bonds between atoms. As these bonds are strong, they lead to the following properties.

- Covalent networks have high melting and boiling points because the covalent bonds are hard to weaken or break.
- They are hard and tough substances because the strong covalent bonds can withstand a lot of force.
- They do not conduct electricity because all the valence electrons are involved in covalent bonds and there are no charged particles free to conduct.

Diamond (carbon) is the hardest natural substance and cannot be scratched except by another diamond. Saws and drills in industry are diamond-tipped to enable them to cut through very hard and tough materials (see Figure 3.7.9).

There are a few exceptions to the general properties. One example is graphite, which is composed of carbon atoms in a lattice structure, but with a different arrangement of the atoms than diamond. It has free electrons and can conduct electricity.

▲ **FIGURE 3.7.9** Due to their hardness, diamonds are sometimes used in drills

3.7 LEARNING CHECK

1. Using fluorine atoms bonding together as an example, **explain** why non-metal atoms sometimes have to share electrons.
2. **Explain** why chlorine atoms form a single covalent bond, while oxygen atoms form a double covalent bond.
3. **Predict** how many pairs of electrons would be shared between two nitrogen atoms.
4. Why do most covalent substances not conduct electricity?
5. **Explain** why covalent molecular substances are usually gases or liquids, while covalent network substances are always solids at room temperature.

3.8 Periodic table development

SCIENCE AS A HUMAN ENDEAVOUR

BY THE END OF THIS MODULE, YOU WILL BE ABLE TO:

- ✓ describe how the first periodic table was developed by Mendeleev
- ✓ describe how the periodic table has been improved over time as science has progressed and new elements have been discovered.

Video activity
Mendeleev and the periodic table

Early attempts at organising elements

A number of scientists, including Antoine Lavoisier and Johann Dobereiner, proposed methods of grouping elements into an organised structure, but none managed to organise all the elements that were known.

John Newlands was a British chemist who noticed that there was a pattern of properties repeating about every eight elements. He called this the law of octaves. He arranged elements into groups of eight elements, as seen in Table 3.8.1. You may notice that not all elements are present on his table. The flaw in his proposal was that he left no gaps for undiscovered elements. As not all elements were known at this time, when new elements were discovered, his table was no longer accurate.

▼ TABLE 3.8.1 Newland's law of octaves

Octave 1	H	Li	Be	B	C	N	O
Octave 2	F	Na	Mg	Al	Si	P	S
Octave 3	Cl	K	Ca	Cr	Ti	Mn	Fe
Octave 4	Co, Ni	Cu	Zn	Y	In	As	Se
Octave 5	Br	Rb	Sr	Ce, La	Zr	Di, Mo	Ro, Ru
Octave 6	Pd	Ag	Cd	U	Sn	Sb	I
Octave 7	Te	Cs	Ba, V	Ta	W	Nb	Au
Octave 8	Pt, Ir	Os	Hg	Tl	Pb	Bi	Th

Alamy Stock Photo/Pictorial Press Ltd

▲ **FIGURE 3.8.1** Dmitri Mendeleev

Mendeleev's periodic table

In 1869, Dmitri Mendeleev proposed the periodic table seen in Figure 3.8.2. He arranged the elements in order of atomic weight and used Newland's idea of properties that would repeat after a certain number of elements. He found that the patterns were not perfect and left spaces on his table, proposing that elements not yet known would fill the gaps.

I	II	III	IV	V	VI	VII	VIII		
H 1.01									
Li 6.94	**Be** 9.01	**B** 10.8	**C** 12.0	**N** 14.0	**O** 16.0	**F** 19.0			
Na 23.0	**Mg** 24.3	**Al** 27.0	**Si** 28.1	**P** 31.0	**S** 32.1	**Cl** 35.5			
K 39.1	**Ca** 40.1		**Ti** 47.9	**V** 50.9	**Cr** 52.0	**Mn** 54.9	**Fe** 55.9	**Co** 58.9	**Ni** 58.7
Cu 63.5	**Zn** 65.4			**As** 74.9	**Se** 79.0	**Br** 79.9			
Rb 85.5	**Sr** 87.6	**Y** 88.9	**Zr** 91.2	**Nb** 92.9	**Mo** 95.9		**Ru** 101	**Rh** 103	**Pd** 106
Ag 108	**Cd** 112	**In** 115	**Sn** 119	**Sb** 122	**Te** 128	**I** 127			
Ce 133	**Ba** 137	**La** 139		**Ta** 181	**W** 184		**Os** 194	**Ir** 192	**Pt** 195
Au 197	**Hg** 201	**Ti** 204	**Pb** 207	**Bi** 209					
			Th 232		**U** 238				

▲ **FIGURE 3.8.2** Mendeleev's first periodic table (1869)

Mendeleev predicted the properties of these missing elements. One such element was gallium. He was able to predict its properties accurately based on its position in his periodic table.

Over time, the Mendeleev model was refined and adjusted as more elements were discovered. In the time of Mendeleev, only 63 elements were known. Now we know 118 elements with more elements constantly being created by scientists in laboratories.

3.8 LEARNING CHECK

1 **Research** the work of Johann Dobereiner. **Describe** the 'triads' he observed in elements.

2 **Describe** why Mendeleev's periodic table was a better proposal than Newland's periodic table.

SCIENCE INVESTIGATIONS

3.9 Justifying conclusions

SCIENCE SKILLS IN FOCUS

IN THIS MODULE, YOU WILL FOCUS ON LEARNING AND IMPROVING THESE SKILLS:

- using evidence from experiments and data to form justified conclusions
- describing how to test for physical properties
- describing observations about the physical properties of metals and covalent substances.

Using evidence to form justified conclusions

Scientists develop laws and theories that help us to predict and explain phenomena. Scientists perform experiments, gather data and look for patterns. When they see patterns, they will often propose a new hypothesis. Scientists perform new experiments to confirm or reject the hypothesis. The most important part of this is using the experimental evidence to form a conclusion.

A simple example is if you were to try to dissolve salt in water. Does temperature affect the amount of salt that can dissolve? A student proposed that heating water would increase the amount of salt that would dissolve. Some students conducted an experiment to verify this (see Table 3.9.1).

Using experimental evidence to justify your conclusion is an important skill.

A poor conclusion for this experiment would be 'The mass of salt that dissolves increases as temperature increases'.

This conclusion does not provide any detail about the reasons this decision was made. Conclusions can be a little longer (or may be explained in more detail in a discussion section). A better conclusion that uses the data might be:

'Student 3's data was discarded in this experiment because it was clearly outlier data (it didn't follow the same pattern as the rest of the data). It seemed there was a malfunctioning hot plate, and the water was not heating up. From the data for students 1, 2 and 4, it seems as if increasing the temperature has influenced the mass of salt able to be dissolved. At the lowest temperature (30°C), an average of 36.4 g salt dissolved. At the highest temperature (90°C), an average of 38.8 g salt dissolved. This shows that as the temperature increased, the mass of salt dissolved also increased. This was consistent across all three students. Thus, the hypothesis is supported.'

Conclusions should:

- be more than one sentence long
- refer to the initial aim or hypothesis being tested
- use data from the experiment – the highest and lowest data points usually show the trend clearly
- summarise findings. Do not repeat each piece of data from the experiment
- link your findings to the science – explain why your results occurred.

TABLE 3.9.1 Dissolving salt in water

Temperature (°C)	Maximum mass salt dissolved (g)			
	Student 1	Student 2	Student 3	Student 4
30	36.5	36.2	36.5	36.5
50	37.2	37.4	36.5	37.1
70	37.9	37.8	36.6	37.8
90	38.8	38.8	36.6	38.7

Video
Science skills in a minute: Justifying conclusions
Science skills resource
Science skills in practice: Justifying conclusions

9780170463041

INVESTIGATION 1: PERIODIC TABLE TRENDS

AIM

To use provided data to investigate how position in a period affects the melting point of an element

METHOD

▼ **TABLE 3.9.2** Melting points of the first 20 elements

Element	Atomic number	Melting point (°C)	Element	Atomic number	Melting point (°C)
Hydrogen	1	−259	Sodium	11	98
Helium	2	−272	Magnesium	12	649
Lithium	3	180	Aluminium	13	660
Beryllium	4	1278	Silicon	14	1410
Boron	5	2300	Phosphorus	15	44
Carbon	6	3500	Sulfur	16	115
Nitrogen	7	−209	Chlorine	17	−101
Oxygen	8	−223	Argon	18	−189
Fluorine	9	−220	Potassium	19	63
Neon	10	−248	Calcium	20	839

1 Using Excel or grid paper, plot a graph of position in the period versus the melting point for the elements of periods 2, 3 and 4 shown in Table 3.9.2. To prepare the graph, on the:
- horizontal axis, create a scale with the numbers 1–8. This will represent the position in the period. For example, in period 2, the number 1 represents lithium as it is in group 1. The number 8 will represent neon because it is in group 8.
- vertical axis, create a scale of temperature ranging from −500 to 4000°C.

2 Plot the period 2 elements (lithium to neon). Join each point to the next with a ruled line. Label this line P2.

3 Plot the period 3 elements (sodium to argon). Join each point to the next with a ruled line. Label this line P3.

4 Plot the period 4 elements (potassium and calcium). Join each point to the next with a ruled line. Label this line P4.

EVALUATION

1 Explain why you think the period 1 elements were not included.

2 Describe, using evidence, any similarities between the patterns for period 2 and period 3 elements.

3 Describe, using evidence, any differences between the patterns for period 2 and period 3 elements.

4 Can you make a general conclusion from this data about the melting point versus position in the period? Justify your answer.

5 Can you be confident that the period 4 elements will follow the same trend?

INVESTIGATION 2: PROPERTIES OF METALS

AIM

To investigate the physical properties of metals, including lustre, hardness, electrical conductivity and malleability

MATERIALS

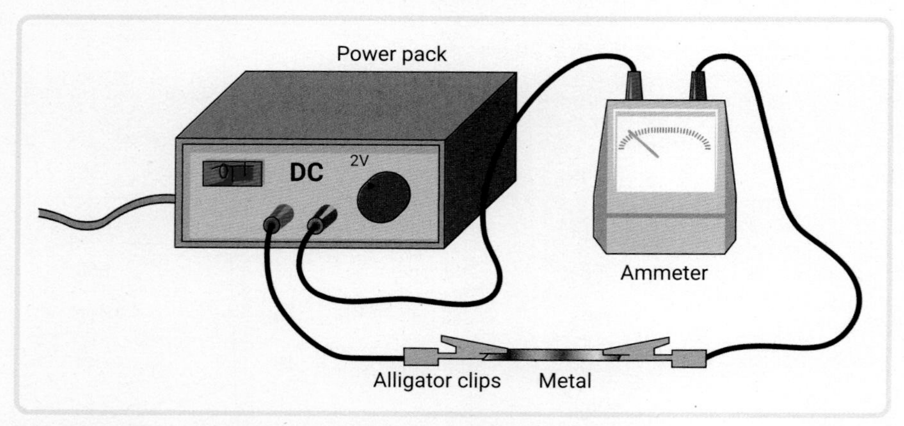

▲ **FIGURE 3.9.1** The equipment needed for testing conductivity

Note: You can substitute the ammeter for a light globe, which will light up if the substance conducts electricity.

- ☑ samples of metals such as magnesium, aluminium, iron, zinc, copper and tin
- ☑ hammer and wooden board (for the malleability test)
- ☑ quartz crystal or iron nail (for the hardness test)
- ☑ sandpaper (for the lustre test)
- ☑ simple electrical circuit, as seen in Figure 3.9.1 (for the electrical conductivity test)

METHOD

You will conduct four tests, repeating each step for all the metal samples. Take photographs or videos and make notes on your observations for each metal.

Lustre test

1. Use sandpaper to clean each metal for about 30–45 seconds.
2. Record the metal colour and whether it is shiny or not.

Hardness test

1. Use an iron nail or quartz crystal (or both) to try to scratch each metal.
2. Record whether each metal could be scratched and whether it was easy or difficult to scratch.

Electrical conductivity test

1. Set up the equipment as shown in Figure 3.9.1, placing each metal in turn into the circuit.
2. Record the ammeter reading or brightness of the light globe. If there is no reading on the ammeter or the globe does not light up, the substance does not conduct electricity.

Malleability test

1. With care for fingers and crush damage, try to hammer the metal into a new shape.
2. Record the level of difficulty it took to change the shape of the metal – easy, medium or difficult.

RESULTS

Create a table(s) to record your results. You may wish to use the format shown in Table 3.9.3 or design your own.

TABLE 3.9.3 Testing the properties of metals

Metal	Lustre observations	Hardness observations	Electrical conductivity observations	Malleability observations
Metal 1 name				
Metal 2 name				
Metal 3 name				
Metal 4 name				

As an extension, you may wish to rank each metal for each observation. For example, if you have four metals, rank their lustre from 1 to 4, with 1 being the shiniest and 4 being the least shiny.

EVALUATION

1 Did any of the metals not conduct electricity? Sometimes if a light globe is used in these experiments, it does not light up for all metals. Suggest a reason for this observation.

2 Rank the metals in order of hardness. Suggest a reason for the variability in results based on the structure of metals.

3 Metal used for structures such as bridges and buildings needs to be hard and not very malleable, as it must withstand massive forces to do its job. Which of the metals you tested would be most suitable for structural uses?

4 Describe problems that arose during this experiment that might affect your confidence when making a conclusion about the properties of metals.

CONCLUSION

Based on this experiment, make a justified conclusion about the properties of metals.

INVESTIGATION 3: PROPERTIES OF COVALENT MOLECULES AND NETWORKS

AIM

To investigate the physical properties of covalent molecules and networks, including hardness, electrical conductivity and melting point

MATERIALS

- ☑ samples of the covalent molecules quartz (covalent network), and ethanol, candle wax and a sugar cube
- ☑ hotplate and stirrer (for the melting point test)
- ☑ electrical conductivity set-up (see Figure 3.9.1)

METHOD

Hardness test

Conduct this test as outlined in Investigation 2.

Electrical conductivity test

1. For the solids sugar, quartz and candle wax, conduct this test as outlined in Investigation 2.
2. Your teacher will demonstrate the electrical conductivity of ethanol by carefully placing the alligator clips from the circuit into the ethanol liquid without letting the clips touch.

Melting point

1. Place a small amount of the candle wax in a beaker and place it on a hotplate. Heat it until it starts to melt. Record the time taken for this to occur.
2. Repeat step 1 for the sugar cube and quartz sample. If no melting is seen after 3 minutes, end the experiment.

RESULTS

Construct a suitable table(s) to record your observations. You can use a table similar to the one suggested in Investigation 2 (Table 3.9.3).

EVALUATION

1. Why do you think no gases were included in this experiment?
2. Why was the melting point test not conducted for ethanol?
3. Explain the vast difference between the quartz sample and the other samples in the hardness and melting point tests.
4. Which of Investigation 2 or Investigation 3 showed more variability in the results?
5. Describe problems that arose during Investigation 3 that might affect your confidence when making a conclusion about the properties of covalent molecules and networks.

CONCLUSION

Based on this experiment, make a justified conclusion about the properties of covalent molecules and networks.

9780170463041

3 REVIEW

REMEMBERING

1 How many electrons fit into the third electron shell?

2 **Identify** two features and one limitation of the Bohr model.

3 Copy Table 3.10.1 and complete it to **summarise** the groups of the periodic table.

4 What is chemical bonding?

5 **Describe** how a covalent bond forms.

▼ **TABLE 3.10.1** The groups of the periodic table

Group	Name (if it has one)	Number of valence electrons	Example of an element in the group
1			
2			
3–12		NA	
14			
15			
16			
17			
18			

UNDERSTANDING

6 What is the electron configuration of:

a hydrogen?

b calcium?

c phosphorus?

7 **Explain** why most elements do not occur as single atoms.

8 **Explain** why ions of the group 2 elements have a 2+ charge.

9 The melting point for most metals varies from around 300°C to over 3000°C. **Describe** why most metals have medium to high melting points and suggest a reason why there is such a range in melting points between different metals.

10 Use diagrams and examples to **explain** the difference between covalent molecules and covalent network structures.

APPLYING

11 Using electron configurations for beryllium, calcium and magnesium, **explain** how many valence electrons are found in group 2 elements.

12 Helium is an unreactive gas. **Identify** one reason why it does fit into group 18, and one reason it doesn't fit into group 18.

13 **Describe** a use of metals where the property of malleability would be required.

14 Carbon dioxide has a low boiling point and is a gas at room temperature. Diamond (carbon) is the hardest natural substance. **Explain** why substances involving the same type of atom (carbon) can have such different properties.

ANALYSING

15 A student claims the electron configuration of an atom is 2,8,10. The teacher points out there is a mistake in the claim.

- **a** **Identify** the element they are using and write the correct electron configuration.
- **b** **Explain** the mistake the student made in writing the configuration.

16 Sodium metal must be stored in oil for it to be safe. Magnesium metal can be stored in air. **Explain** why there is a difference between the two metals.

17 Chlorine atoms can either gain electrons or share electrons to achieve a stable electron configuration. **Identify** whether potassium must gain or share electrons to achieve a stable electron configuration and **explain** why this is so.

EVALUATING

18 Table 3.10.2 shows four metals and some of their properties.

▼ **TABLE 3.10.2** Four metals and some of their properties

Metal	Melting point	Electrical conductivity	Density
A	Medium	Low	Low
B	High	High	Low
C	Low	High	Low
D	Low	Low	High

- **a** **Justify** why metal A is used to build aeroplanes.
- **b** **Predict** and **explain** whether metal B or C would be better for electrical wiring.
- **c** Metal D is used for construction in high-rise buildings. What other properties would be important for this metal to have?

19 A student says that they can determine the charge on the ions of any element from the periodic table. **Evaluate** their claim by providing the pros and cons of this statement.

20 Choose ONE of the items listed below and **explain** whether the materials used to build it are likely to be ionic, metallic, covalent molecular or covalent network. You must refer to the bonding and physical properties in your response.

surfboard, tennis racquet (frame and strings), chopping knife (blade and handle), plastic bag

CREATING

21 **Create** a mind map that links the following keywords.

atom, ion, cation, anion, periodic table, group, proton, electron, ionic bonding, covalent bonding, metallic bonding, covalent molecule, covalent network

Join related words and include annotations to show why you are joining them.

22 Using the information found in Module 3.8, come up with a different way to organise the elements into a visual form other than our current periodic table. Consider other ways the elements might be grouped and how you might organise this. **Draw** your final 'new' periodic table and **explain** your organisation.

BIG SCIENCE CHALLENGE PROJECT #3

1 Connect what you've learned

For each of the items in the photographs, explain the properties that each object has and how they relate to its use.

2 Check your thinking

When cars were first built, they were usually made almost entirely from metal and rubber for the body and glass for the windows. Now, cars are almost entirely made from engineered materials such as plastics, plexiglass and other synthetic materials.

Why are synthetic materials used more than natural materials such as metal and wood?

Think of another example of an object that is now made from different materials than it was when it was originally produced. Explain why the change in materials mostly likely occurred. If you are stuck for ideas, consider how sports equipment such as cricket bats, clothing, running shoes or umpiring technology has changed. Or how computers or televisions have changed with new materials available.

3 Get into action

Select something that you are interested in, such as a sport you watch or play, or a hobby or pastime. Pick an object associated with that interest. For example, if you like surfing you might choose a surfboard. If you like playing computer games, you might choose a console.

Do some research to see what these objects are commonly made from. How do their properties relate to their composition and structure? Are there any new materials being used or developed in this field?

4 Communicate

Create a presentation to show the development of materials in a common object such as a car, boat or mobile phone.

Chemical reactions

BIG SCIENCE CHALLENGE 

Getty Images/Pool

▲ **FIGURE 4.0.1** A mining disaster in New Zealand in 2010

Figure 4.0.1 shows an explosion that occurred at Pike River mine in New Zealand in 2010. Coal mines can explode for many reasons. The two most common are a build up of methane gas and a build up of coal dust. Methane is highly flammable, so a spark can set off an explosion. Coal is difficult to ignite in solid form. When the coal is being drilled, coal dust forms and remains suspended in the air throughout the mine. This ignites very easily when exposed to a flame. So why is solid coal hard to ignite when coal dust is explosive with a single spark? The answer lies in a series of chemical reactions and an understanding of rates of reaction.

- **What are the potential consequences of explosions and fires like this?**
- **What chemical reactions could be occurring to cause this damage?**
- **How could coal pieces and coal dust be the same chemical, yet so different in their explosive potential?**
- **How could an understanding of chemistry prevent these disasters?**

#4 SCIENCE CHALLENGE ACCEPTED!

At the end of this chapter, you can complete the Big Science Challenge Project #4. You can apply the knowledge and skills you learn in this chapter to complete the project.

Assessments
- Prior knowledge quiz
- Chapter review questions
- End-of-chapter test
- Portfolio assessment task: Science investigation

Videos
- Science skills in a minute: Qualitative v. quantitative data **(4.13)**
- Video activities: Energy change of reactions **(4.1)**; Oxidation reactions **(4.2)**; Reactivity series **(4.4)**; Acids and alkalis: Part 1 **(4.6)**; Acids and alkalis: Part 2 **(4.7)**; Plastic-eating microbes **(4.12)**

Science skills resources
- Science skills in practice: Qualitative v. quantitative data **(4.13)**
- Extra science investigations: Reactions between acids and metals **(4.6)**; Comparing reaction rates **(4.8)**; Effect of temperature and concentration on reactions **(4.10)**

Interactive resources
- Drag and drop: Decomposition or synthesis reaction? **(4.3)**; Displacement v. precipitation reaction **(4.5)**
- Label: Collision theory **(4.9)**
- Simulation: Balancing chemical equations **(4.1)**

Nelson MindTap

To access these resources and many more, visit:
cengage.com.au/nelsonmindtap

4.1 Review of chemical equations

BY THE END OF THIS MODULE, YOU WILL BE ABLE TO:

- ✓ identify reactants and products in a word or balanced equation
- ✓ write word and balanced equations to represent a chemical reaction.

Video activity
Energy changes in reactions

Interactive resource
Simulation: Balancing chemical equations

GET THINKING

Examine the equations in this module. Compare them to the equations you normally see in maths. How are chemical equations similar to maths equations? How are they different?

What is a chemical equation?

chemical equations
word equations or balanced equations that are used to represent the substances in chemical reactions

reactants
chemical substances that, when added together, react to form products in a chemical reaction

products
chemical substances that form in chemical reactions

Chemical equations are used to represent the **reactants** and **products** in a chemical reaction. Chemical reactions occur if a chemical change occurs when substances are added together. This means a new substance has formed. Chemical changes can be indicated by observations including colour changes, gas formation (bubbles), a solid forming or heat/light energy being released or absorbed.

Chemical equations can be represented by word equations:

$$\text{hydrogen} + \text{oxygen} \rightarrow \text{water}$$

They can also be written as balanced equations, using chemical formulas:

$$2H_2 + O_2 \rightarrow 2H_2O$$

Both types of equation show reactants on the left side. These are the chemicals required to start the reaction. Products that form in the reaction are found on the right. An arrow separates the reactants and products. Chemical equations never include an equals sign (=).

Balanced equations

A balanced equation provides detailed information about the atoms in a chemical reaction. Consider Figure 4.1.1, showing the reaction between methane (CH_4) and oxygen (O_2) to form carbon dioxide (CO_2) and water (H_2O).

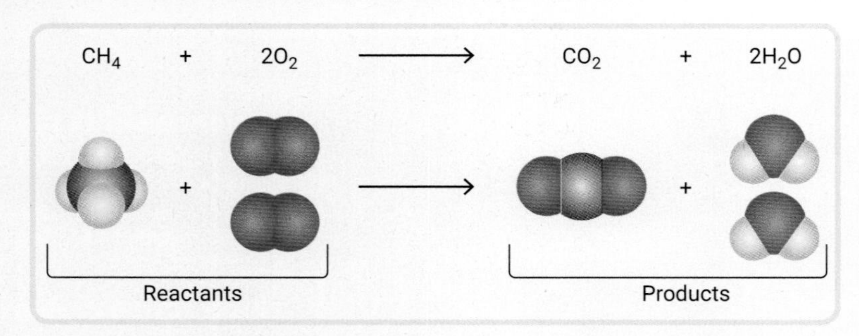

▶ **FIGURE 4.1.1** The reaction between methane and oxygen

The first thing you can see from the equation is the type and number of atoms in each reactant and product. For example, CH_4 is methane (the gas used for cooking and heating), and each molecule contains one carbon atom and four hydrogen atoms.

law of conservation of mass
the total mass of reactants and products in a chemical reaction is equal

The equation also shows how the **law of conservation of mass** is being followed in chemical reactions. The law of conservation of mass states that the number and type of atoms are the same in the reactants and products.

Balanced equations use coefficients – numbers in front of the formulas – to ensure that the atoms on each side of the equation are balanced. In Figure 4.1.1, one carbon atom, four hydrogen atoms and four oxygen atoms occur on each side of the arrow. Therefore, this reaction is balanced and the law of conservation of mass is being followed.

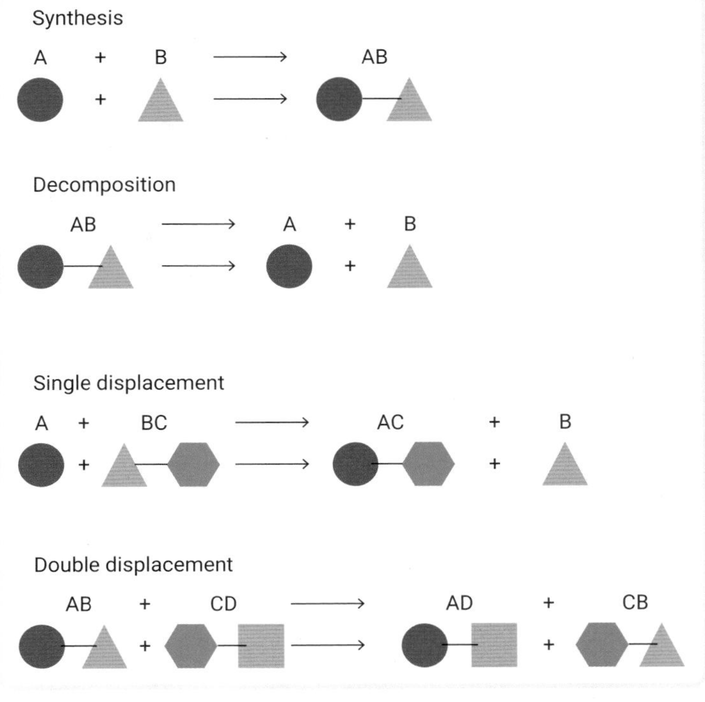

▲ **FIGURE 4.1.2** Types of chemical reactions

Types of reactions

In this chapter, you will be exploring different types of chemical reactions: synthesis, decomposition, single displacement and double displacement reactions.

1 **Synthesis reactions**: two or more elements or compounds combine to form one or more complex products. Synthesis reactions, also called combination or addition reactions, will be explored in Module 4.2.
2 **Decomposition reactions**: a single chemical breaks down into two or more simpler substances. Decomposition reactions will be explored in Module 4.3.
3 Single **displacement reactions**: to displace something means to replace or take the place of another object. In these reactions, one atom replaces an atom in a compound. In Figure 4.1.2, you can see that atom A (blue circle) replaces atom B (pink triangle) in the reactant BC. In Modules 4.4 and 4.6, you will explore metal displacement reactions and the reaction between acids and metals.
4 Double displacement reactions occur when both reactants have two parts that 'swap'. As both substances replace part of the other substance, two displacements occur. Figure 4.1.2 shows AB splitting, with A replacing C in CD. The two products AD and CB are the 'swaps' from the reactants AB and CD. Precipitation reactions and neutralisation reactions will be explored in Modules 4.5 and 4.7.

synthesis reactions
reactions in which two or more elements and compounds combine to form a more complex substance

decomposition reactions
reactions involving the breakdown of compounds into simpler elements or compounds

displacement reactions
reactions involving the replacement of atoms in compounds

4.1 LEARNING CHECK

1 The following equation represents the reaction between hydrochloric acid and calcium carbonate:

$$2HCl + CaCO_3 \rightarrow CaCl_2 + CO_2 + H_2O$$

a The products of this reaction are calcium chloride, carbon dioxide and water. **Write** a word equation for this reaction.
b Show that the law of conservation of mass applies to this reaction.

2 **Identify** the four types of chemical reactions.
3 **Write** a one-sentence summary to describe each type of chemical reaction.

4.2 Synthesis reactions

BY THE END OF THIS MODULE, YOU WILL BE ABLE TO:

✓ describe at an atomic level what occurs during a synthesis reaction
✓ predict the products and write equations to represent synthesis reactions.

Video activity
Oxidation reactions

Quiz
Synthesis reactions

Extra science investigation
Extraction of iron oxide

GET THINKING

The word 'synthesise' can mean to create something from lots of pieces of information or ideas. How do you think this definition relates to chemical synthesis?

What is a synthesis reaction?

In a synthesis reaction, two or more chemicals combine to form a more complex product. Atoms from both the reactants can be found in the product. Figure 4.2.1 shows a general representation of synthesis reactions. A and B are the reactants; AB is the product when they combine.

Predicting the products of simple synthesis reactions

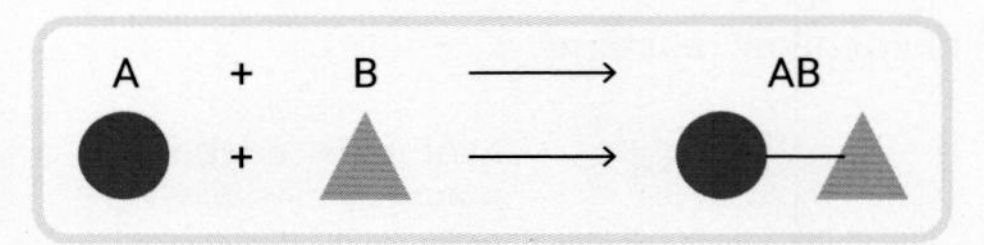

▲ **FIGURE 4.2.1** A synthesis reaction

Many simple synthesis reactions involve the combination of elements such as magnesium, hydrogen, aluminium or oxygen.

One type of synthesis reaction where you can predict the products involves the addition of a metal element to a non-metal element (see Table 4.2.1). The product is a combination of the two elements called an **ionic salt**. An ionic salt has a metal ion and a non-metal ion and is named by:

- keeping the metal name
- changing the non-metal element ending to '-ide'.

For example, magnesium and oxygen combine to form magnesium oxide.

The general equation for this is:

$$\text{metal} + \text{non-metal} \rightarrow \text{ionic salt}$$

ionic salt
a chemical containing a metal ion and a non-metal ion

corrosion
a chemical process that often happens to metals exposed to oxygen

▼ **TABLE 4.2.1** Synthesis reactions involving metal and non-metal elements

Element 1	Element 2	Product	Word and balanced equation
Potassium	Oxygen	Potassium oxide	potassium + oxygen → potassium oxide $4K + O_2 \rightarrow 2K_2O$
Iron	Sulfur	Iron sulfide	iron + sulfur → iron sulfide $Fe + S \rightarrow FeS$
Calcium	Chlorine	Calcium chloride	calcium + chlorine → calcium chloride $Ca + Cl_2 \rightarrow CaCl_2$

▲ **FIGURE 4.2.2** A rusted iron nail. The iron oxide forms from a synthesis reaction

One common example of this type of synthesis reaction is **corrosion** or rusting. Iron metal reacts with oxygen in the air to form iron oxide, which we call rust. This is the red, flaky material you find on iron, as shown in Figure 4.2.2.

Examples of more complex synthesis reactions

There are many other types of synthesis reactions. Medicines are generally complex chemicals as shown in Figure 4.2.3. They are formed by reacting many simpler chemicals, usually in a series of chemical reactions. The structures of some medicines formed in synthesis reactions are:

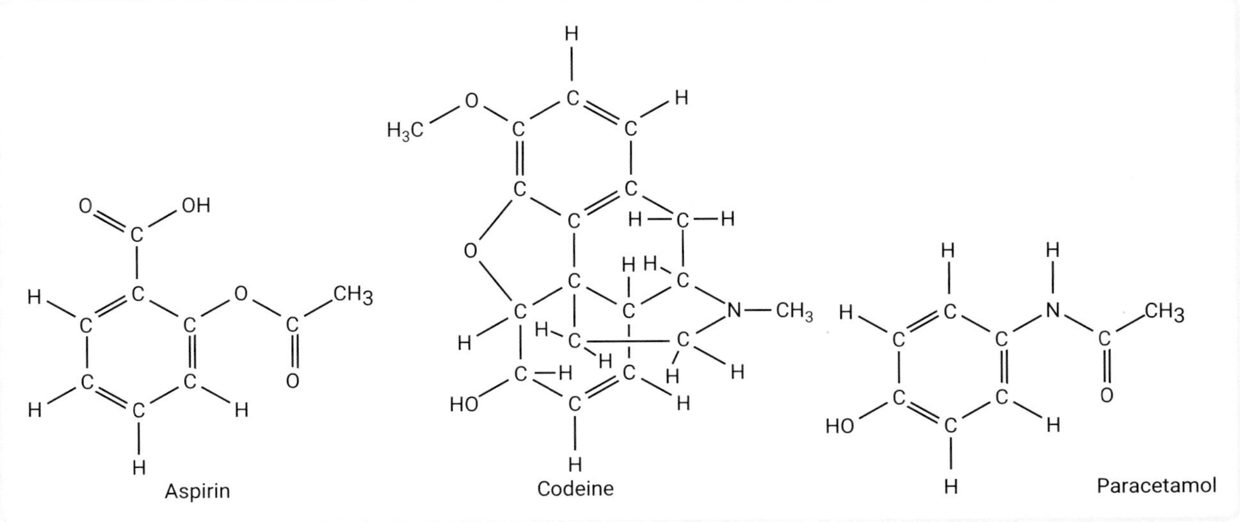

Aspirin Codeine Paracetamol

▲ **FIGURE 4.2.3** Examples of medicines formed from synthesis reactions

Polymers are large chemicals made by synthesising small chemical structures called monomers. 'Mono' means 'one' and 'poly' means 'many', so when you add many monomers together, you get a polymer (see Figure 4.2.4). Polymers include plastics, foam and fibres. Some examples are:

- plastic carry bags – polythene
- non-stick coating on frypans – Teflon
- soft-drink bottles – polyethylene terephthalate
- clothes – polyester and acrylic.

polymer
a large chemical made in a synthesis reaction from repeating, simpler chemicals called monomers

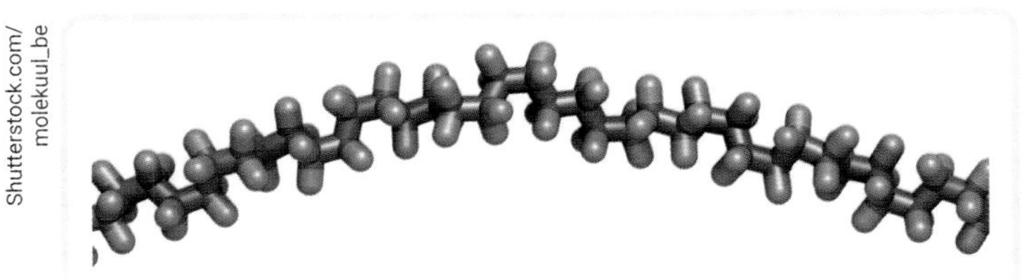

Shutterstock.com/molekuul_be

Monomer

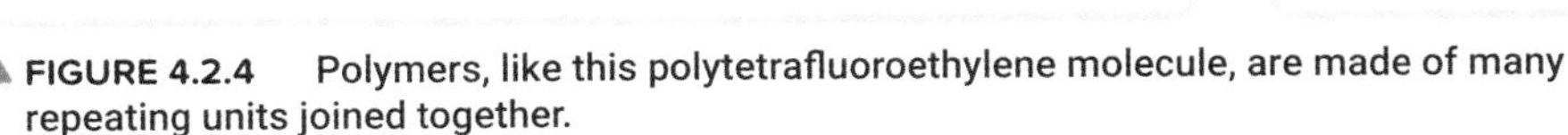

▲ **FIGURE 4.2.4** Polymers, like this polytetrafluoroethylene molecule, are made of many repeating units joined together.

☆ ACTIVITY

Simple synthesis reactions

Materials

- 5 cm strip magnesium and a small sample of steel wool
- crucible
- heating apparatus
- metal tongs

Method

1. Set up the equipment as shown in Figure 4.2.5.
2. Place the magnesium strip in the crucible and heat it.
3. Using metal tongs, hold the steel wool in the Bunsen burner flame for about 60 seconds. It will glow red while the reaction is occurring.
4. Place it on a heatproof mat to cool and observe the product.

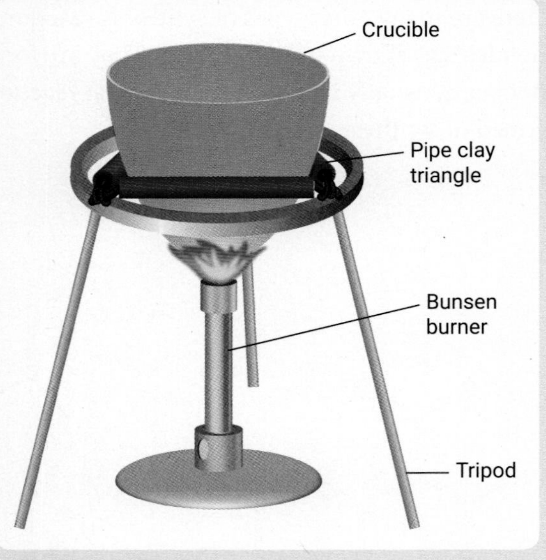

▲ FIGURE 4.2.5

Evaluation

1. **Explain** why these are both synthesis reactions. Identify the other element involved in both reactions.
2. **Write** word equations for both reactions, **predicting** the name of the ionic salt product.

Warning

When the magnesium ignites, it will release a very intense white light. DO NOT look directly at the light. Observe the products only when the reaction is finished.

4.2 LEARNING CHECK

1. **Predict** the name of the products when the following are combined.
   a. Aluminium + oxygen
   b. Calcium + phosphorus
   c. Lithium + fluorine

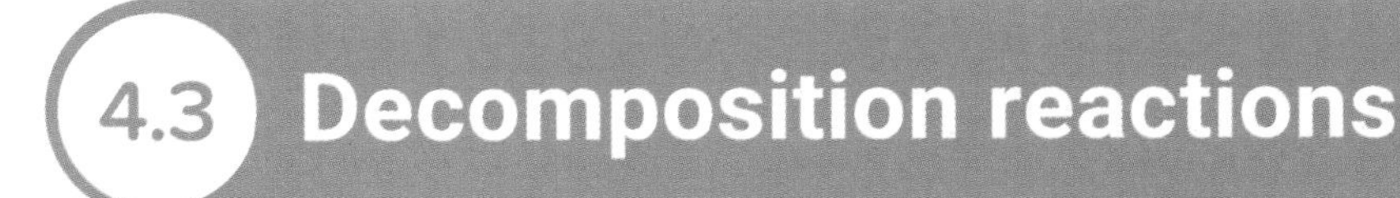

4.3 Decomposition reactions

BY THE END OF THIS MODULE, YOU WILL BE ABLE TO:

- ✓ describe at an atomic level what occurs during a decomposition reaction
- ✓ predict the products and write equations to represent decomposition reactions.

GET THINKING

Decomposition reactions can be initiated by three different types of energy. This gives three types of decomposition: thermal, photo and electrolytic decomposition. Identify at least one other everyday use of each of the three words 'thermal', 'electric' and 'photo' that are related to energy.

Interactive resource
Drag and drop: Decomposition or synthesis reaction?

What is decomposition?

Decomposition is the breaking down of chemical substances into two or more smaller products. A simple example of this is the decomposition of water. Water (H_2O) can be broken down into hydrogen (H_2) and oxygen (O_2). This can be seen in the equations:

decomposition
the breaking down of chemical substances into two or more smaller products

water → hydrogen + oxygen

$$2H_2O \rightarrow 2H_2 + O_2$$

Figure 4.3.1 shows the breakdown of a chemical containing A and B (AB) into each of A and B individually as products.

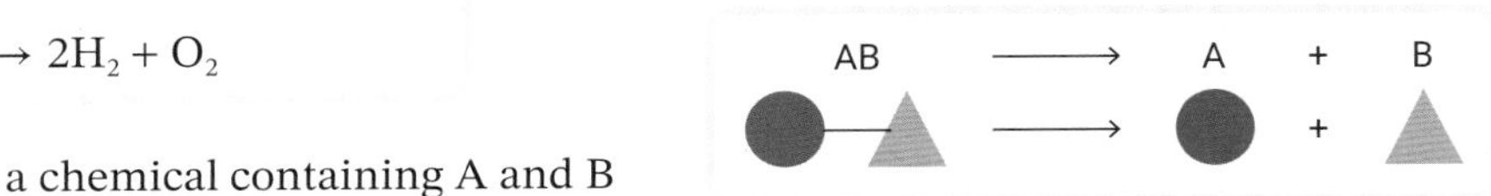

▲ FIGURE 4.3.1 A decomposition reaction

Types of decomposition

Some decomposition reactions can happen at room temperature, but most require energy input to start. This gives three types of decomposition (see Table 4.3.1).

▼ TABLE 4.3.1 The three types of decomposition

Type of decomposition	Example
Thermal – heat is applied	Heating of copper carbonate to form copper oxide and carbon dioxide $CuCO_3 \rightarrow CuO + CO_2$
Electrolytic – electricity is applied	Applying electricity to water to form hydrogen and oxygen $2H_2O \rightarrow 2H_2 + O_2$
Photo – light is applied	Decomposition of silver chloride in sunlight into silver and chlorine gas $2AgCl \rightarrow 2Ag + Cl_2$

Predicting products of simple decomposition reactions

Ionic salts consisting of a metal and non-metal can be decomposed into their elements. You can predict the products by looking at the name of the product. Sodium chloride will decompose into sodium and chlorine. Potassium oxide will decompose into potassium and oxygen.

Metal extraction

ore
a compound containing a metal and other elements that is mined from the ground and processed to extract the metal

The process of decomposition is used to make useful chemicals. One example of this is the extraction of metals from **ores**. Figure 4.3.2 shows the ores of copper, aluminium and iron and the metal that is extracted. Ores are a combination of metal and other elements such as oxygen, sulfur or hydrogen. Most metals are too reactive to exist in their elemental form. Gold is one of the few metals that can be found in nature in its pure form. Most other metals need to be extracted chemically before they can be used.

▲ **FIGURE 4.3.2** Some metals and their ores

▼ **TABLE 4.3.2** Some metals extracted using high temperatures or electricity

Metal	Ore and formula
Iron	Haematite Fe_2O_3 Magnetite Fe_3O_4
Lead	Galena PbS
Calcium	Dolomite $CaMg(CO_3)_2$
Aluminium	Bauxite $Al_2O_3.2H_2O$
Tin	Cassiterite SnO_2

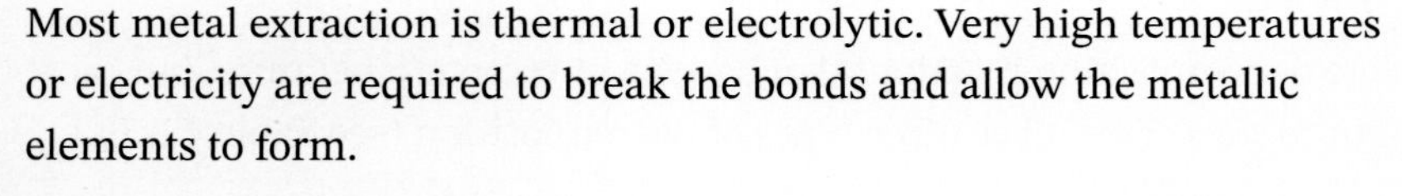

Most metal extraction is thermal or electrolytic. Very high temperatures or electricity are required to break the bonds and allow the metallic elements to form.

Some examples of metals that are produced in this way are shown in Table 4.3.2.

Other examples of decomposition

Airbags in cars contain a chemical called sodium azide (NaN_3). When a car is involved in an accident of sufficient force, an electric circuit is turned on. The electricity causes the decomposition of the solid sodium azide:

$$2NaN_3 \rightarrow 2Na + 3N_2$$

The nitrogen gas formed rapidly inflates the airbag, providing protection for the people in the car as shown in Figure 4.3.3.

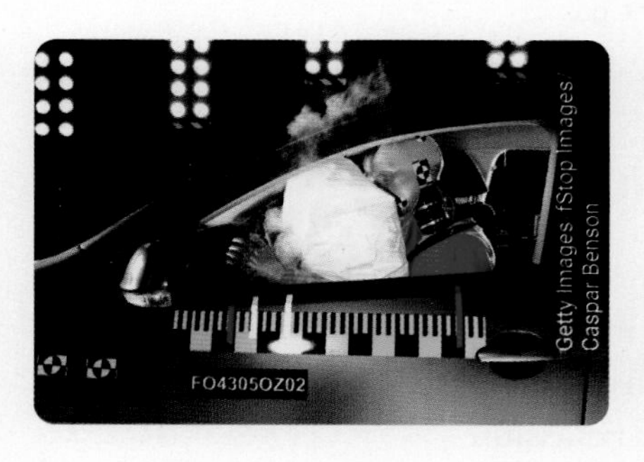

◀ **FIGURE 4.3.3** Airbags use a decomposition reaction to inflate.

☆ ACTIVITY 4.3

Electrolytic decomposition of water

The apparatus in Figure 4.3.4 is a Hoffman voltameter. It allows you to pass an electric current through a liquid. In this experiment, water will be decomposed into two different gases. The water has approximately 1 mL sulfuric acid added to help conduct the electricity.

Set up the apparatus as shown in Figure 4.3.4 and set the power pack to 12 volts. Ensure you are using DC electricity and that all wires are connected securely and safely.

When you turn on the electricity, you should see gas forming in the two side cylinders. One of the cylinders will have double the volume of gas of the other one. Each cylinder contains a different product gas.

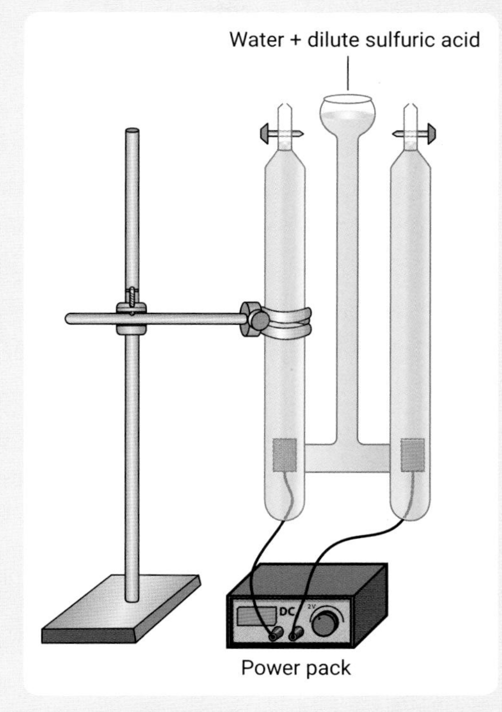

▲ **FIGURE 4.3.4** A Hoffman voltameter

- **a** **Predict** the products of the decomposition of water and write a word equation for this reaction.
- **b** Examine the balanced equation and **suggest** why one gas has twice the volume of the other. Then, **identify** which gas is in which cylinder.
- **c** **Recall** (or **research**) how you could test to confirm the identity of the two product gases. **Write** a method to show how you could test the gases. (Your teacher may demonstrate this for you.)

4.3 LEARNING CHECK

1. **Write** word equations for the decomposition of:
   - **a** potassium sulfide.
   - **b** calcium oxide.
   - **c** magnesium bromide.
   - **d** aluminium fluoride.
2. Using Table 4.3.2, **predict** the two products from the decomposition of:
   - **a** galena.
   - **b** magnetite.
   - **c** cassiterite.
3. Suggest a reason why energy is needed to start most decomposition reactions.
4. Hydrogen peroxide (H_2O_2) does not need much energy to undergo decomposition. It must be stored in a dark, non-transparent bottle. If stored in glass or transparent plastic, it decomposes into water and oxygen. **Explain** why hydrogen peroxide is stored in a dark bottle.

4.4 Metal displacement reactions

BY THE END OF THIS MODULE, YOU WILL BE ABLE TO:

✓ describe at an atomic level what occurs during a metal displacement reaction
✓ predict the products and write equations to represent metal displacement reactions.

Video activity
Reactivity series

Quiz
Displacement reactions

GET THINKING

'Displacement' is a commonly used word in science but is also used in many other situations. What does it mean to displace someone or something? Do a quick online search to find examples of how displacement can be used in areas outside science.

Single displacement reactions

Displacement reactions can be 'single' or 'double'. Both types involve atoms replacing other atoms in a chemical compound. A single displacement reaction is illustrated in Figure 4.4.1. When A reacts with BC, it displaces B. This means it replaces B in the reactant so that the end products are AC and B.

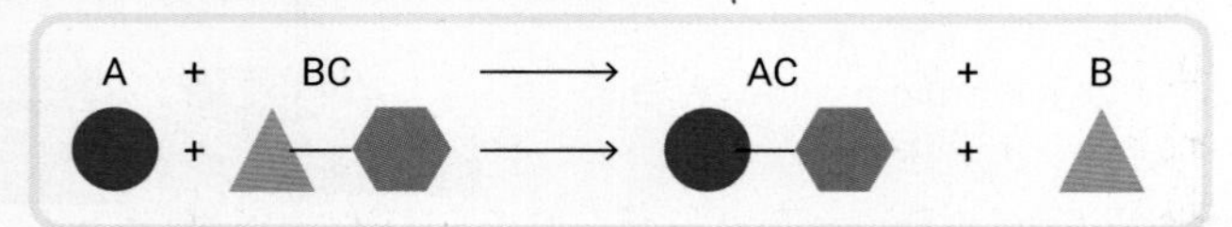

▲ **FIGURE 4.4.1** A single displacement reaction

Figure 4.4.2 shows this in the reaction of aluminium metal with iron oxide. The aluminium displaces the iron from the iron oxide, forming aluminium oxide and iron. The iron has been displaced, so it ends up as an element on its own in the products.

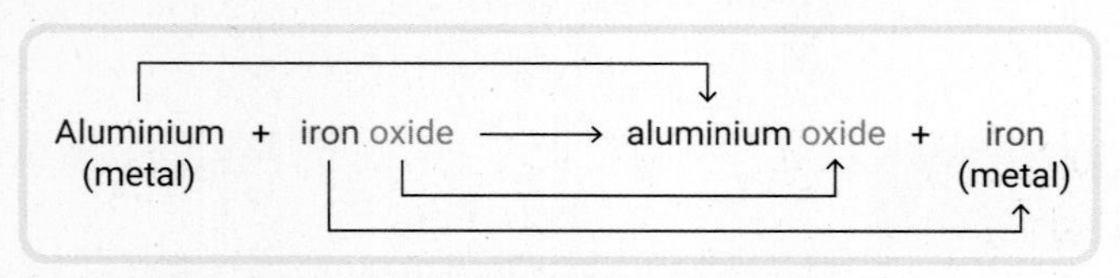

▲ **FIGURE 4.4.2** The metal displacement reaction of aluminium metal with iron oxide

metal displacement reaction
a chemical reaction where a more reactive metal displaces a less reactive metal from a solution

This is an example of a **metal displacement reaction**. To explore this further, you need to understand the different forms in which metals can exist.

Metals and metal ion solutions

You may recall that almost all metal elements are solids. In chemistry, they are represented by their element symbol: magnesium solid is Mg, aluminium solid is Al and copper solid is Cu.

Metals can also exist in ion form in solution. When an ionic salt is added to water and it is soluble, the ions separate. This means metal ions are present in solution. Metal ions are positively charged ions represented by the chemical symbol and charge. Magnesium ions are Mg^{2+}, aluminium ions are Al^{3+} and copper ions are Cu^{2+}.

Remember, you can predict the charge on the ions of metals in groups 1 and 2 from their position on the periodic table.

Reactivity of metals

As mentioned in Module 4.3, many metals are reactive and so are rarely found as pure metal elements. Chemical reactivity generally refers to how easily substances react with other chemicals. Metals show a wide range of reactivity.

Group 1 and 2 metals are generally the most reactive metals. Figure 4.4.3 shows a lithium battery, which would react explosively with water. Group 1 metals have vigorous reactions with other elements. Several of them are so reactive, they are explosive in air and must be kept in oil or in a vacuum to prevent explosions while they are being stored.

Shutterstock.com/wk1003mike

▲ **FIGURE 4.4.3** Water should never be used to put out a lithium battery fire.

Other metals, such as platinum, copper, silver and gold, are far less reactive. Gold is one of the few metals that occurs in its pure form in nature. This is due to its low reactivity with other elements. Gold, silver and platinum are often used for jewellery because of their lack of reactivity. The skin contains acids, and your jewellery comes into contact with sweat, cleaning materials and other chemicals every day. This is why jewellery needs to be made from non-reactive metals.

Activity series of metals

Scientists rank the reactivity of metals in the **activity series**. Figure 4.4.4 shows potassium and calcium as two of the most reactive metals. Copper, silver and gold are among the least reactive metals.

activity series
a list of metals ranked by their chemical reactivity

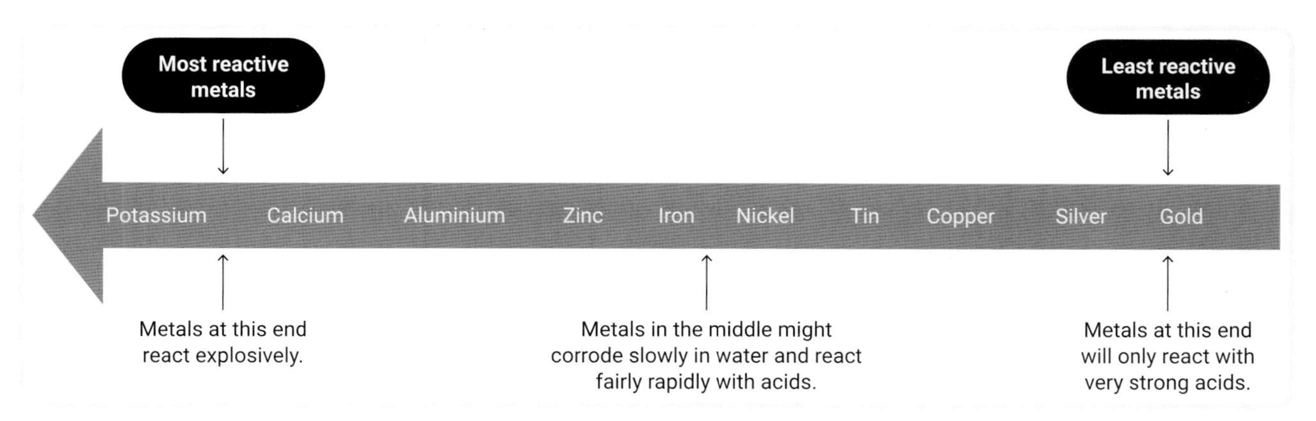

▲ **FIGURE 4.4.4** An activity series of metals

This ranking can be used to predict how metals and metal ion solutions will react when they come into contact. Figure 4.4.5 shows a piece of copper metal that has been placed into a silver nitrate solution containing silver ions. What can you see?

The following observations are made.

- A solid metal (silver) has formed on the piece of copper metal.
- The solution has turned blue. Chemists know this is due to the formation of copper ions in solution.

Charles D. Winters/Science Source

▲ **FIGURE 4.4.5** Copper metal reacts with silver nitrate.

These observations allow conclusions to be made. First, the formation of silver solid means that silver ions have formed silver metal. Second, the formation of copper ions means the copper metal has formed copper ions. This lets us write the word equation:

copper + silver nitrate → copper nitrate + silver
(silver ions) (copper ions)

The copper has displaced the silver from the silver nitrate solution. This is a metal displacement reaction. As only one substance has been displaced, it is a single displacement reaction.

You can see this clearly from the balanced equation:

$$Cu(s) + 2AgNO_3(aq) \rightarrow Cu(NO_3)_2(aq) + 2Ag(s)$$

The copper (Cu) has displaced the silver (Ag) in the reactant compound. The silver nitrate ($AgNO_3$) reactant becomes copper nitrate ($Cu(NO_3)_2$) in the products.

You may recall the use of state symbols in equations from Year 9. In balanced equations such as metal displacement, it is important to show what is solid (s) and what is a solution (aq). Other state symbols show gases (g) and pure liquids (l). You should include state symbols in balanced equations if you know the state of the chemicals.

Predicting the products of metal displacement reactions

Let's relate this back to the activity series. Copper is more reactive than silver. There is a rule about metal activity that states that 'the more reactive metal will displace the less reactive metal from a solution'. In this case, the more reactive metal (copper) has displaced the less reactive metal (silver) from the silver nitrate solution.

What would happen if silver were added to copper nitrate? The more reactive metal (copper) is already in solution (copper nitrate) so it will not be displaced, and no reaction will occur.

Example 1

If zinc metal is added to copper nitrate, will a reaction occur?

- Zinc is the more reactive metal.
- Copper is the less reactive metal and is in solution (copper nitrate).
- The more reactive metal will displace the less reactive metal from the solution.
- Zinc will displace copper from the solution.

Yes – a reaction will occur.

The word equation is:

zinc + copper nitrate → copper + zinc nitrate

The balanced equation is:

$$Zn(s) + Cu(NO_3)_2(aq) \rightarrow Cu(s) + Zn(NO_3)_2(aq)$$

Example 2

If aluminium metal is added to iron nitrate, will a reaction occur?

- Aluminium is the more reactive metal.
- Iron is the less reactive metal and is in solution (iron nitrate).
- The more reactive metal will displace the less reactive metal from the solution.
- Aluminium will displace iron from the solution.

Yes – a reaction will occur.

The word equation is:

aluminium + iron nitrate → iron + aluminium nitrate

The balanced equation is:

$$2Al(s) + 3Fe(NO_3)_2(aq) \rightarrow 3Fe(s) + 2Al(NO_3)_3(aq)$$

4.4 LEARNING CHECK

1 Ionic salts include sodium chloride, magnesium fluoride and aluminium bromide. **Write** the formula of the metal ions present in these metal salts.

2 **Draw** a diagram to show what is meant by a single displacement reaction.

3 A student was given a range of metals to conduct experiments. They were given a jar with a silvery metal stored in oil, like the example shown in Figure 4.4.6. **Predict** which group in the periodic table this metal would belong to. **Justify** your answer.

4 Some people get black marks on their skin from wearing silver jewellery. The black substance is silver oxide, formed when silver reacts with chemicals on the skin. Very few people have this reaction when wearing gold jewellery. By referring to the activity series of metals, **explain** this observation.

Alamy Stock Photo/Science Photo Library

▲ **FIGURE 4.4.6** Some metals are stored in oil.

5 In which of the following will reactions occur? **Justify** your answer.

Calcium metal in tin nitrate solution OR tin metal in calcium nitrate solution

6 **Predict** whether the following combination of metals and ionic salts will react, and **write** word equations for any reactions that occur.

a Aluminium metal and silver nitrate solution

b Iron metal and calcium nitrate solution

c Copper metal and zinc nitrate solution

d Tin metal and copper nitrate solution

e Zinc metal and silver nitrate solution

4.5 Precipitation reactions

BY THE END OF THIS MODULE, YOU WILL BE ABLE TO:

- ✓ describe at an atomic level what occurs during a precipitation reaction
- ✓ predict the products and write equations to represent precipitation reactions.

Interactive resource
Drag and drop: Displacement v precipitation reactions

GET THINKING

In Chapter 3 you learned about the formation of positive and negative ions. Go back and review this work so you know why metal ions are always positive and non-metal ions are always negative.

Double displacement reactions

double displacement reaction
a reaction in which atoms of the reactants displace each other from their compounds

In a **double displacement reaction**, the atoms of the reactants displace each other from their compounds. Figure 4.5.1 shows reactants AB and CD. When they react, the pairs of atoms 'swap' or replace each other. This forms products AD and CB.

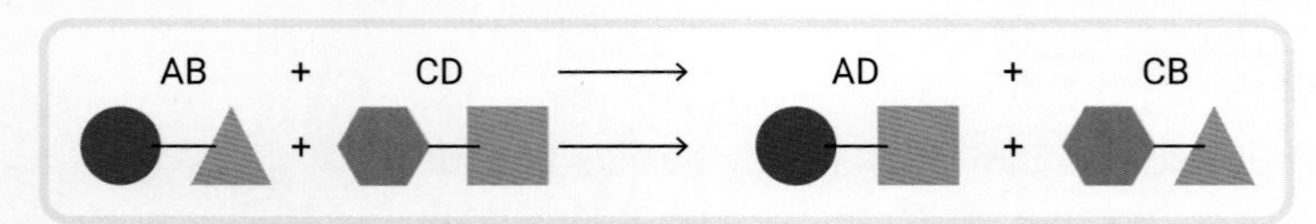

▲ **FIGURE 4.5.1** A double displacement reaction

Figure 4.5.2 shows an example with the reactants lead nitrate and potassium iodide. You can follow the colours to see that the lead and potassium displace each other to form two new products.

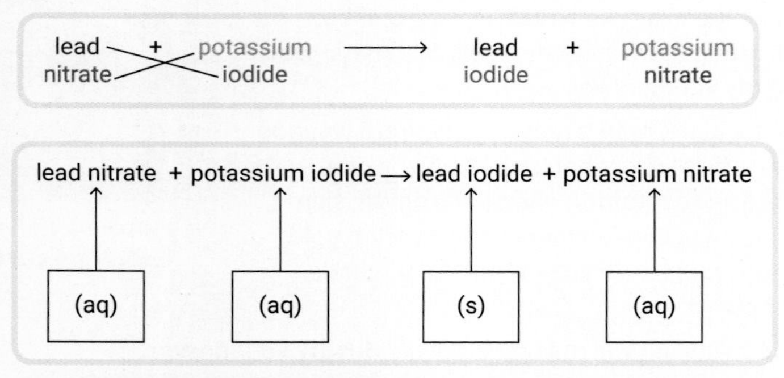

▲ **FIGURE 4.5.2** The reaction between lead nitrate and potassium iodide is an example of a double displacement reaction.

Salts, solutions and precipitates

precipitate
a solid formed from certain combinations of positive and negative ions

A common double displacement reaction forms a **precipitate**. In precipitation reactions, ionic salt solutions react together to form at least one solid ionic compound.

In Chapter 3 we looked at how metals form positive ions and non-metals form negative ions. Sometimes those ions attract to form an ionic lattice. Ionic substances are often called ionic salts. They are made from a metal and a non-metal.

Figure 4.5.2 shows four ionic salts. Potassium iodide has a metal ion (potassium) and a non-metal ion (iodine). Lead nitrate has a metal ion (lead) and a non-metal ion (containing nitrogen and oxygen).

polyatomic ion
an ion that contains more than one non-metal element and has an overall charge

The nitrate ion (NO_3^-) is a negative ion with more than one type of non-metal atom. This type of ion is called a **polyatomic ion**.

Other polyatomic ions are the hydroxide ion (OH^-), sulfate ion (SO_4^{2-}) and carbonate ion (CO_3^{2-}). You can learn more about polyatomic ions in the digital-only Chapter 10 on Nelson MindTap.

Some ionic salts are soluble. When soluble ionic salts are dissolved in water, the ions separate into positive and negative ions. You can see lead nitrate and potassium iodide depicted in solution with separate ions in Figure 4.5.3.

When the two solutions are added, four ions are present: lead ions, potassium ions, nitrate ions and iodide ions.

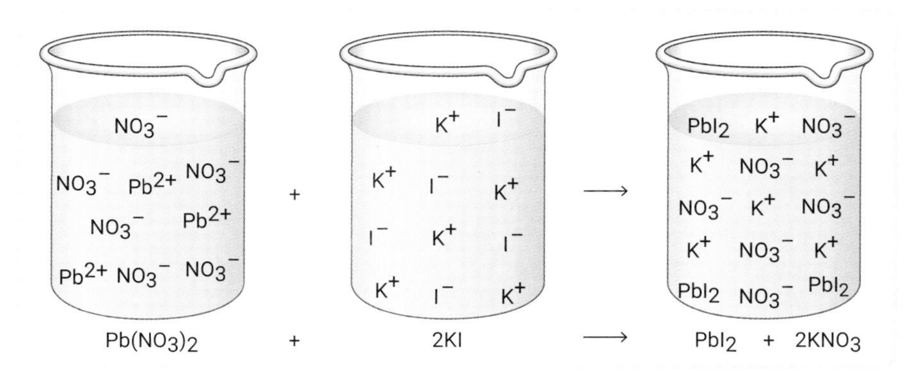

FIGURE 4.5.3 Lead nitrate and potassium iodide in solution with separate ions

When some pairs of ions are in solution together, they can form an insoluble solid. This solid is called a precipitate. In the example in Figure 4.5.4, the combination of lead ions and iodide ions forms an insoluble lead iodide precipitate. Lead iodide is a bright yellow solid.

FIGURE 4.5.4 When lead ions and iodide ions combine, they form a bright yellow precipitate, lead iodide.

Predicting products of precipitation reactions

To predict whether a precipitate will form, we need to know which combinations of ions make an insoluble precipitate. Table 4.5.1 shows the combinations of ions that will form a precipitate.

TABLE 4.5.1 Ion combinations that form precipitates when added to a solution

Positive ions	Negative ions			
	Hydroxide OH^-	**Nitrate NO_3^-**	**Sulfate SO_4^{2-}**	**Chloride Cl^-**
Magnesium Mg^{2+}	Yes	No	No	No
Copper Cu^{2+}	Yes	No	No	No
Strontium Sr^{2+}	No	No	Yes	No
Zinc Zn^{2+}	Yes	No	No	No
Lead Pb^{2+}	Yes	No	Yes	Yes
Calcium Ca^{2+}	Yes	No	Yes	No
Sodium Na^+	No	No	No	No
Silver Ag^+	Yes	No	Yes	Yes
Potassium K^+	No	No	No	No

Note. Yes = precipitate forms.

Using Table 4.5.1, you can see that if silver ions and hydroxide ions are added, then a precipitate will form. To determine whether a precipitate will form:

1 write the word equation
2 check the products against Table 4.5.1
3 identify which product is a precipitate
4 if neither product is a precipitate, then none will form in the reaction.

For example, if calcium nitrate is added to sodium sulfate, the word equation is:

calcium nitrate + sodium sulfate → calcium sulfate + sodium nitrate

Checking the table, you can see that sodium nitrate does not form a precipitate, but calcium sulfate does. So, this reaction will form a precipitate.

4.5 LEARNING CHECK

1 For the following combinations of ionic salts, **write** the word equation for their addition using Table 4.5.1 to determine if any insoluble products are present and circling any precipitate that forms.
   a Magnesium nitrate and sodium hydroxide
   b Zinc hydroxide and potassium chloride
   c Sodium sulfate and lead nitrate
2 What combination of ionic salts could you add to form the following precipitates? The other product must be soluble. **Write** a word equation for the reaction. The first one is partially completed for you.
   a Calcium sulfate—reactant 1 could be calcium nitrate, and reactant 2 could be sodium sulfate. The soluble product is sodium nitrate. (Hint: sodium ions, potassium ions and nitrate ions are always soluble, so they would make a soluble product when combined). Now **write** the word equation.
   b Silver hydroxide
   c Lead chloride

4.6 Metal and acid reactions

BY THE END OF THIS MODULE, YOU WILL BE ABLE TO:

- ✓ describe at an atomic level what occurs during a reaction between acids and metals
- ✓ predict the products and write equations to represent a reaction between acids and metals.

GET THINKING

You had an introduction to acids in Year 9. What do all acids have in common? Can you identify some common acids you might find in your house? What are some uses for acids in your home or everyday life?

Video activity
Acids and alkalis: part 1

Extra science investigation
Reactions between acids and metals

What is an acid?

Acids are chemicals that contain hydrogen atoms that can be removed when they react. All acids contain hydrogen. The physical properties of acids were studied in Year 9 and include a sour taste, good electrical conductivity and solubility in water. Some common acids are listed in Table 4.6.1.

acid
a chemical that can donate hydrogen ions and undergo chemical reactions with metals and bases

This module will look at one of the chemical reactions that acids undergo.

When acids are in solution, they form hydrogen ions (H^+) and the negative ion, as shown in Table 4.6.1. When acids react in chemical reactions, they are always in solution, so they always form these ions.

▼ **TABLE 4.6.1** Some common acids and their formulas

Acid name	Chemical formula	Ions that form in solution
Hydrochloric acid	HCl	H^+ and Cl^- (chloride ion)
Sulfuric acid	H_2SO_4	H^+ and SO_4^{2-} (sulfate ion)
Nitric acid	HNO_3	H^+ and NO_3^- (nitrate ion)
Carbonic acid	H_2CO_3	H^+ and CO_3^{2-} (carbonate ion)

What happens when acid and metals react?

Acids react with some metals in a single displacement reaction. The products formed when an acid and a metal react are an ionic salt and hydrogen gas (H_2). A general equation for all acid–metal reactions is:

$$\text{acid} + \text{metal} \rightarrow \text{ionic salt} + \text{hydrogen}$$

You can use the acid and metal names to predict the name of the ionic salt that will form. As shown in Figure 4.6.1, the ionic salt takes the name of the metal and adds the negative ion from the acid.

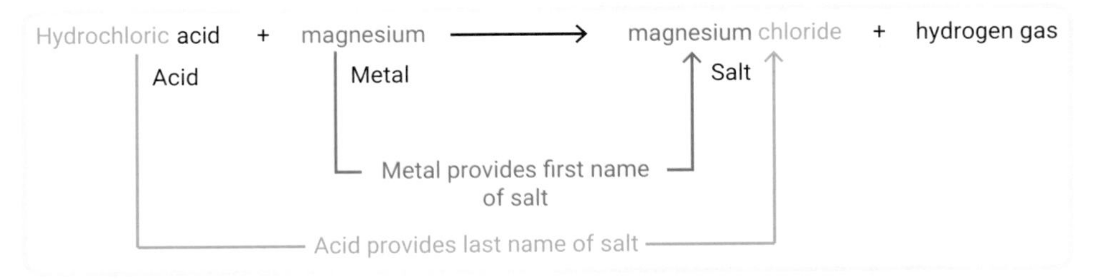

▲ **FIGURE 4.6.1** Using acid and metal names to predict the name of ionic salts that form in acid and metal reactions

Alamy Stock Photo/Science Photo Library

▲ **FIGURE 4.6.2** Magnesium reacting with hydrochloric acid. The bubbles of hydrogen gas appear white.

If zinc were added to sulfuric acid, the ionic salt would be zinc sulfate. Aluminium and nitric acid would form aluminium nitrate. These reactions can be represented by the following word equations:

zinc + sulfuric acid → zinc sulfate + hydrogen

aluminium + nitric acid → aluminium nitrate + hydrogen

When these reactions occur, bubbles of gas form. This is the hydrogen gas formed in the reaction shown in Figure 4.6.1. The amount of bubbling can indicate the reactivity of the metal. As seen in Figure 4.6.2, the bubbling when magnesium is added to hydrochloric acid is vigorous. A lot of bubbles form very quickly. Less reactive metals would form fewer bubbles.

Acid rain

In many places, **acid rain** falls due to high levels of acidic pollutants in the air. The acid can react with metals to form ionic salts. Metal structures can be weakened in this way (see Figure 4.6.3).

acid rain
rain that is acidic due to chemicals dissolved in water in the atmosphere

Getty Images/EyeEm/Lars Dogondke

▲ **FIGURE 4.6.3** Acid rain damage to a metal structure

Metals can be protected from acid rain by different methods that prevent the acid reacting with the metal. Most of these methods also protect against corrosion by providing a barrier to oxygen and water. Some common methods are shown in Table 4.6.2.

▼ **TABLE 4.6.2** Protecting metals from acid rain and corrosion

Method	How it works	Examples
Painting	• Covering the metal with paint means acid rain attacks the paint instead of the metal. • The paint needs to be reapplied often, as it wears away. • The paint also helps protect the metal against regular corrosion.	Bridges and fences are often painted regularly as a form of protection.
Galvanising	• The metal is covered in a layer of zinc so that any acid rain is prevented from contacting the metal underneath. • Over time, the zinc can be eaten away, and the metal may need to be recoated.	Nails and roofing materials are commonly galvanised substances.
Alloy metals	• The original metal (e.g. iron) is combined with another substance such as nickel or silicon, giving the metal protection so that it does not react with the acid rain.	Alloys of iron with nickel (e.g. steel) are commonly used to make metal containers and pipes that transport and store acids.

☆ ACTIVITY

Protecting metal from acid

Individually, or in pairs, design a scientific method to test the effectiveness of one of the protection methods listed in Table 4.6.2. Consider the following points in your scientific method.

- What material will you use? For example, you could paint a piece of metal. Alternatively, galvanised nails are an easy way to find some galvanised metal, or some nuts and bolts for high-stress jobs are made from stainless steel with nickel.
- How will you know if it provides protection for the original, unprotected metal? How will you compare the two?
- What type and concentration of acid will you use? Consult your teacher regarding what is most appropriate.

If time and resources allow, you might be able to test some of your methods. Check with your teacher to organise this.

4.6 LEARNING CHECK

1 What do all acids have in common in their chemical structure?

2 **Identify** the ions present in a solution of:
  a nitric acid.
  b carbonic acid.

3 **Explain** why the reaction of an acid and a metal is a single displacement reaction.

4 **Write** an equation for the reaction between the following acids and metals.
  a Calcium and hydrochloric acid
  b Magnesium and sulfuric acid
  c Lithium and carbonic acid

5 **Predict** the metal and acid you would need to form the following ionic salts and, for each salt, write an equation for its formation.
  a Zinc nitrate
  b Lead sulfate
  c Aluminium chloride

4.7 Acid and metal hydroxide reactions

BY THE END OF THIS MODULE, YOU WILL BE ABLE TO:

✓ describe at an atomic level what occurs during a reaction between acids and metal hydroxides
✓ predict the products and write equations to represent a reaction between acids and metal hydroxides.

Video activity
Acids and alkalis: part 2

Interactive resource
Crossword: Types of reactions

Extra science investigation
Observing chemical reactions

GET THINKING

Several of the modules in this chapter have referred to 'general equations'. Find the general equation in the previous module and in this one. What do you think the purpose of a general equation is? How can it be used to write word equations to represent specific chemical reactions?

Neutralisation reactions

When an acid and a **base** react together, it is called a **neutralisation** reaction. The acid and the base form an ionic salt and water. You may recall that a solution with a **pH** between 0 and 6 is acidic, a pH of 7 is neutral and a pH between 7 and 14 is basic.

base
a group of chemicals that react to neutralise acids; may include metal hydroxides, metal carbonates or metal oxides

neutralisation
the chemical reaction between an acid and a base to produce a salt and water

pH
a measure of the acid or base levels in a solution, measured between 0 and 14

The general equation for a neutralisation reaction is:

$$\text{acid} + \text{base} \rightarrow \text{salt} + \text{water}$$

Farmers and gardeners use neutralisation reactions. Soils can be acidic or basic, depending on the rocks or minerals present and the source of the soil. Some plants grow best in acidic soils; others grow best in basic soils. Farmers and gardeners often add bases or acids to the soil to change its pH so they can grow crops or flowers successfully (see Figure 4.7.1).

Swimming pools that are too acidic can make your eyes burn and itch. Pools that are not acidic enough grow algae and turn green. To properly maintain the pH of a swimming pool, acids or bases are added until the right pH is reached (see Figure 4.7.2).

Getty Images Plus/iStock/Tevarak

▲ **FIGURE 4.7.1** Bases are used to neutralise soils that are too acidic.

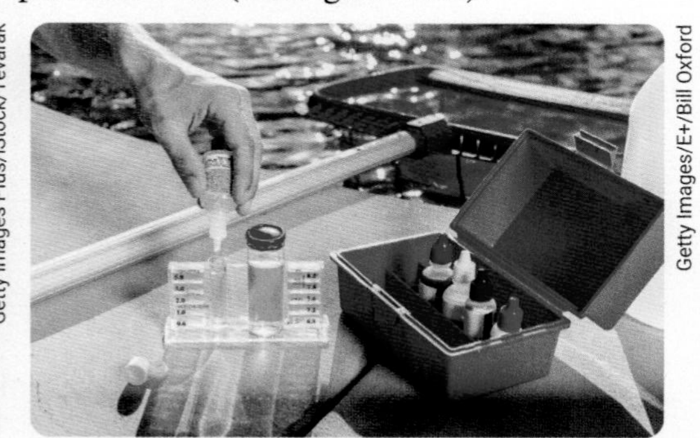

Getty Images/E+/Bill Oxford

▲ **FIGURE 4.7.2** Acids and bases are used to control the pH of swimming pools.

What is a metal hydroxide?

metal hydroxide
an ionic salt containing a metal ion and a hydroxide ion

A **metal hydroxide** contains a metal ion and a hydroxide ion (the polyatomic ion OH^-). Examples of metal hydroxides are magnesium hydroxide, $Mg(OH)_2$, and sodium hydroxide, $NaOH$.

Predicting the products of neutralisation reactions

The ionic salt that forms in a neutralisation reaction is a combination of the metal from the metal hydroxide and the negative ion from the acid (see the previous module for the list). Figure 4.7.3 shows two neutralisation reactions involving metal hydroxides.

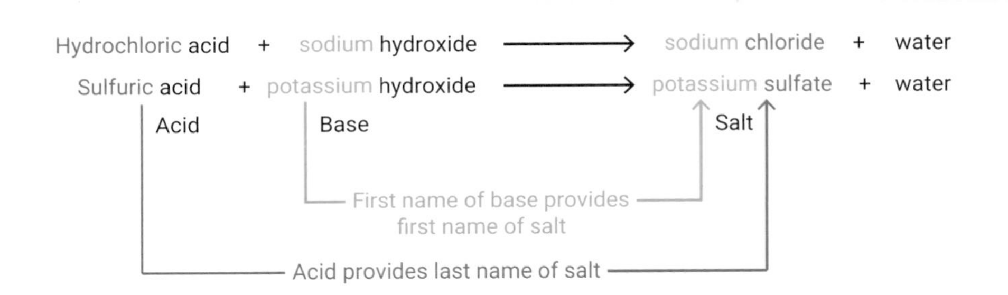

FIGURE 4.7.3 You can use the names of the acids and bases to predict the products of neutralisation reactions.

ACTIVITY

Demonstrating pH change in a neutralisation reaction

- Set up two test tubes.
  - Test tube 1: 5 mL of 0.1 mol L^{-1} hydrochloric acid and 2 drops of universal indicator
  - Test tube 2: 5 mL of 0.1 mol L^{-1} sodium hydroxide and 2 drops of universal indicator
- Carefully pour the contents of Test tube 2 into Test tube 1 a few drops at a time, until it is all transferred.
  - **a** Using Figure 4.7.4 as a guide, **describe** the colour changes of the reactants and products.
  - **b** How do the colours show that neutralisation has occurred?
  - **c** **Write** a word equation to represent this reaction.

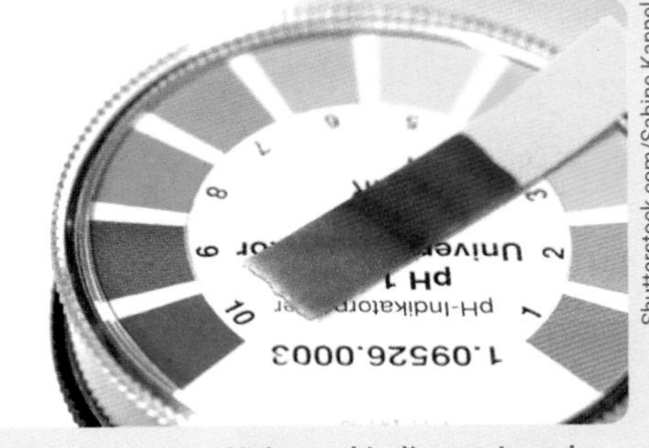

Shutterstock.com/Sabine Kappel

FIGURE 4.7.4 Universal indicator is red, pink or yellow in acids. It turns green in a neutral solution and blue or purple in basic solutions.

4.7 LEARNING CHECK

1. **Write** word equations to represent the reactions between the following acids and metal hydroxides:
   - **a** Hydrochloric acid and calcium hydroxide
   - **b** Sulfuric acid and aluminium hydroxide
   - **c** Carbonic acid and zinc hydroxide
   - **d** Nitric acid and sodium hydroxide
2. Two common uses of neutralisation reactions were given in this module. **Research** another common use of neutralisation reactions in everyday life.

4.8 Rate of reaction

BY THE END OF THIS MODULE, YOU WILL BE ABLE TO:

- ✓ describe examples of chemical reactions with high and low rates of reaction
- ✓ describe methods of observing and measuring the rate of a reaction.

Quiz
Rate of reaction

Extra science investigation
Comparing reaction rates

GET THINKING

Science experiments can use both qualitative observations and quantitative measurements. What is the definition of 'qualitative' and 'quantitative'? What words do they sound like? Which of these two words applies when measuring rate of reaction?

Fast and slow reactions

rate of reaction
a measurement of how fast a reaction is proceeding

In chemistry, the **rate of reaction** is used to describe how fast or how slow reactions take place. When we apply this to chemical reactions, we examine how much reactant is used up in a period of time. Alternatively, you can examine how much product forms in a period of time.

Fireworks involve chemical reactions that are very fast (see Figure 4.8.1). Many chemical reactions are completed after a few seconds or minutes.

Rusting (or corrosion) of iron is an example of a synthesis reaction between iron and oxygen. It is a very slow reaction, with rust sometimes taking years to form on an iron structure (see Figure 4.8.2).

Shutterstock.com/bajars

▲ **FIGURE 4.8.1** Fireworks are an example of a fast chemical reaction.

Shutterstock.com/Gilmanshin

▲ **FIGURE 4.8.2** The synthesis reaction involving iron and oxygen (corrosion or rusting) is a very slow chemical reaction.

9780170463041

Rate of reaction graphs

The progress of a chemical reaction can be measured and plotted onto a graph.

Consider the progress of a reaction involving reactant A forming B (A → B). As A is used up, the amount of A decreases. As B forms, the amount of B increases. Figure 4.8.3 shows two graphs representing the reaction A → B. Graph **a** shows the change in B over time. Graph **b** shows the change in the amount of A over time.

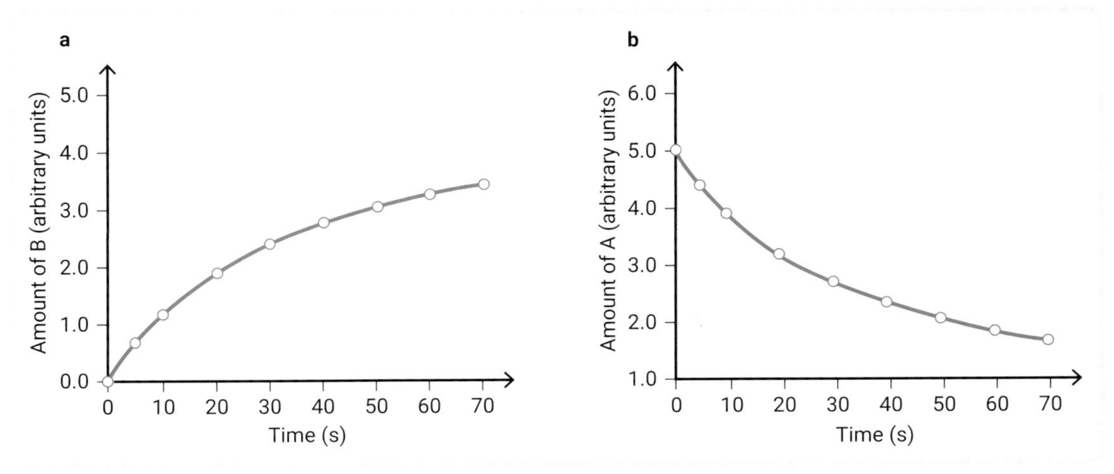

▲ **FIGURE 4.8.3** Rate of reaction graphs showing the change in the amount of (a) B and (b) A

How can the rate of reaction be measured?

The rate of reaction can be calculated quantitatively using the formula:

$$\text{Rate of reaction} = \frac{\text{amount of reactant used}}{\text{time}}$$

$$\text{Rate of reaction} = \frac{\text{amount of product formed}}{\text{time}}$$

For example, magnesium reacts with hydrochloric acid to form magnesium chloride and hydrogen gas:

magnesium + hydrochloric acid → magnesium chloride + hydrogen

1.3 g of magnesium reacted with hydrochloric acid. The time required for all the magnesium to react was 3 minutes.

$$\text{Rate} = \frac{\text{mass used}}{\text{time}} = \frac{1.3\,\text{g}}{3\,\text{min}} = 0.043\ \text{g/min}$$

The unit is g/min or grams per minute. This means that 0.043 grams of magnesium was used in each minute.

The rate of reaction can also be calculated using the slope or gradient of a graph. You will probably recall from your maths studies that the gradient of a graph can be calculated by using the $\frac{\text{rise}}{\text{run}}$ formula. As the graph has the amount of product on the vertical axis and time on the horizontal axis, the gradient is $\frac{\text{amount of product}}{\text{time}}$, which is the same as the formula for the rate of reaction. Thus, the gradient is the rate of reaction.

Figure 4.8.4 shows you how to use a graph to calculate the rate of formation of a product in a chemical reaction.

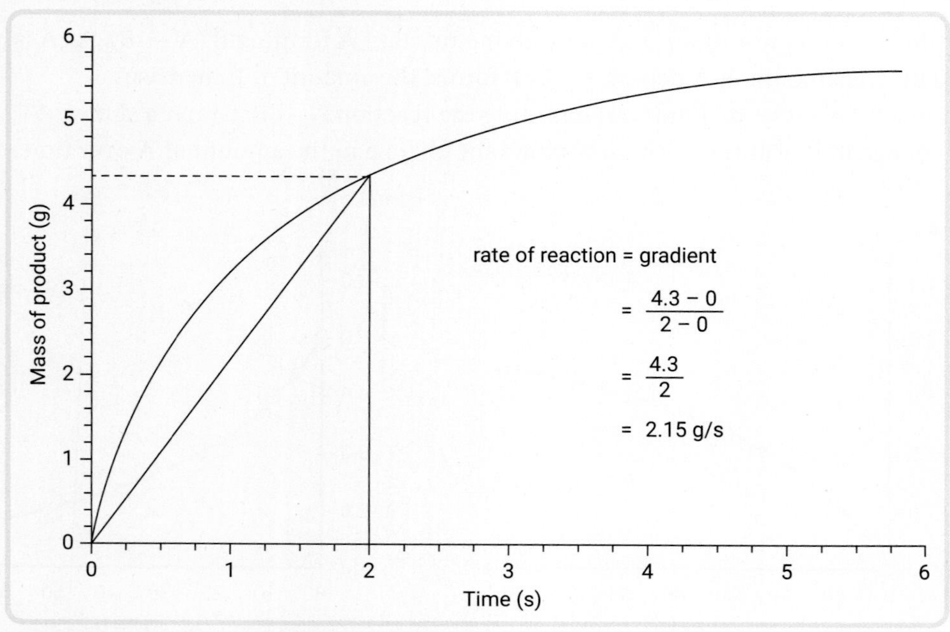

▲ **FIGURE 4.8.4** Using a graph to calculate the rate of reaction

☆ **ACTIVITY**

Measuring the rate of reaction

You will need

- electronic balance
- 20 mL of 2.0 mol/L hydrochloric acid
- small piece (approximately 1 cm strip) of magnesium metal
- 100 mL beaker
- stopwatch

Method

- Measure and record the mass of the magnesium strip.
- Measure 20 mL of the hydrochloric acid and place it into a 100 mL beaker.
- Place the beaker on the electronic balance.
- Add the magnesium piece and at the same time zero the electronic balance. At the same time have another student start a stopwatch.
- Record the time taken for the magnesium piece to completely react so there is no visible sign of it remaining.
- Record the reading on the electronic balance at this time.
  - a **Write** an equation for the reaction that is occurring in the beaker.
  - b **Explain** why the final mass on the electronic balance was negative. What does the negative mass represent?
  - c **Calculate** the rate of reaction of the magnesium metal, using the data.
  - d **Calculate** the rate of formation of the product in step **b**, using the data.

4.8 LEARNING CHECK

1 Figure 4.8.5 shows the formation of carbon dioxide in a chemical reaction.

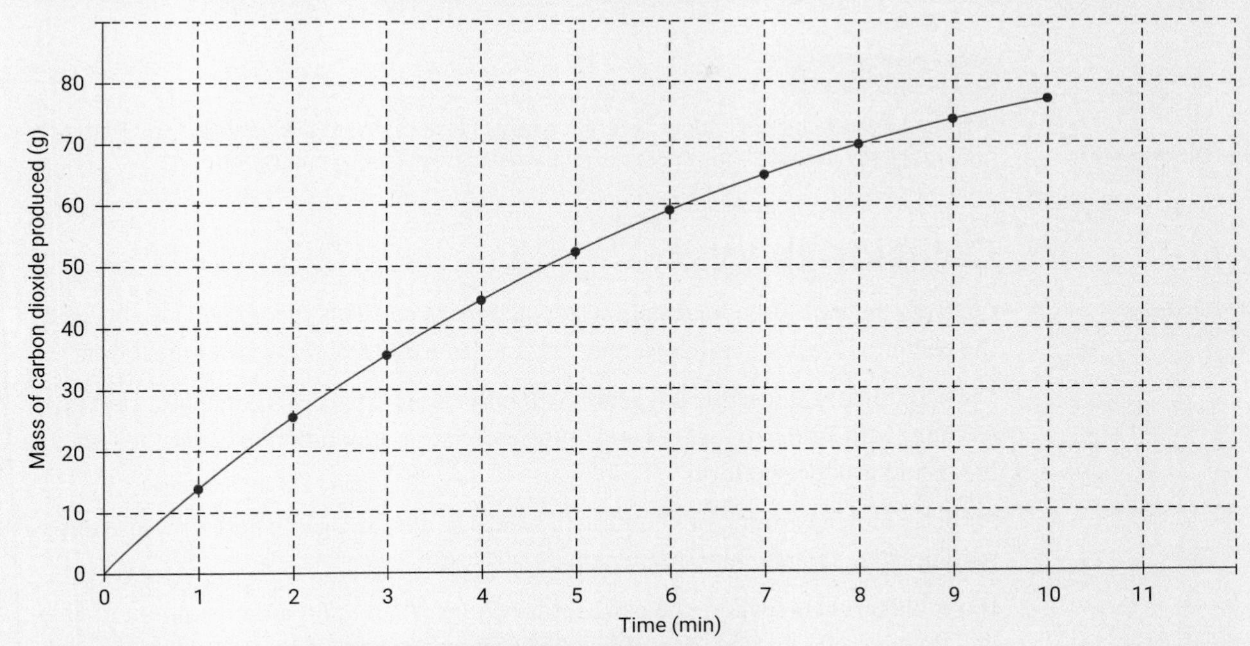

▲ **FIGURE 4.8.5** The formation of carbon dioxide in a chemical reaction

   a **Calculate** the rate of formation of the carbon dioxide in the first 8 minutes of the reaction.
   b How does the graph show that carbon dioxide is a product and not a reactant?

2 **Justify** whether the following involve fast or slow rates of reaction:
   a cooking food using a high temperature
   b combusting fuel in a car engine
   c the breaking down of rocks as they react with chemicals in the environment.

3 In a chemical reaction, 2.5 grams of a chemical takes a time of 83 seconds to fully react. **Calculate** the rate of reaction for this chemical in grams per second.

4 If a chemical is produced in a reaction at a rate of 10 grams per minute, **calculate** the mass of chemical that would form if the reaction were conducted for 6 minutes.

9780170463041

4.9 Collision theory

BY THE END OF THIS MODULE, YOU WILL BE ABLE TO:

✓ explain the role of collision theory and activation energy in rates of reaction.

Interactive resource
Label: Collision theory

GET THINKING

What do you remember about different types of energy? What is kinetic energy? Particles need energy to react. Why do you think particles need kinetic energy to react?

Collision theory

collision theory
the theory that a reaction will only occur if particles come into contact with sufficient energy and at the correct orientation to cause a successful collision

Collision theory is used to explain why reactions occur. This theory will also be used in the next module to explain how and why we can make reactions faster or slower.

Collision theory states that a reaction will only occur if particles come into contact (collide) with sufficient energy and at the correction orientation to cause the bonds to break and products to form.

Particles moving around or vibrating in solids, liquids and gases have kinetic energy. All particles demonstrate kinetic energy.

successful collision
a collision that results in products forming

If a collision occurs where the particles have more than the required amount of energy and the correction orientation, it is called a **successful collision**. Collisions where the particles do not have enough energy to form products are called unsuccessful collisions. In an unsuccessful collision, the reactants bounce off each other and do not undergo any change. Both types of collisions are shown in Figure 4.9.1.

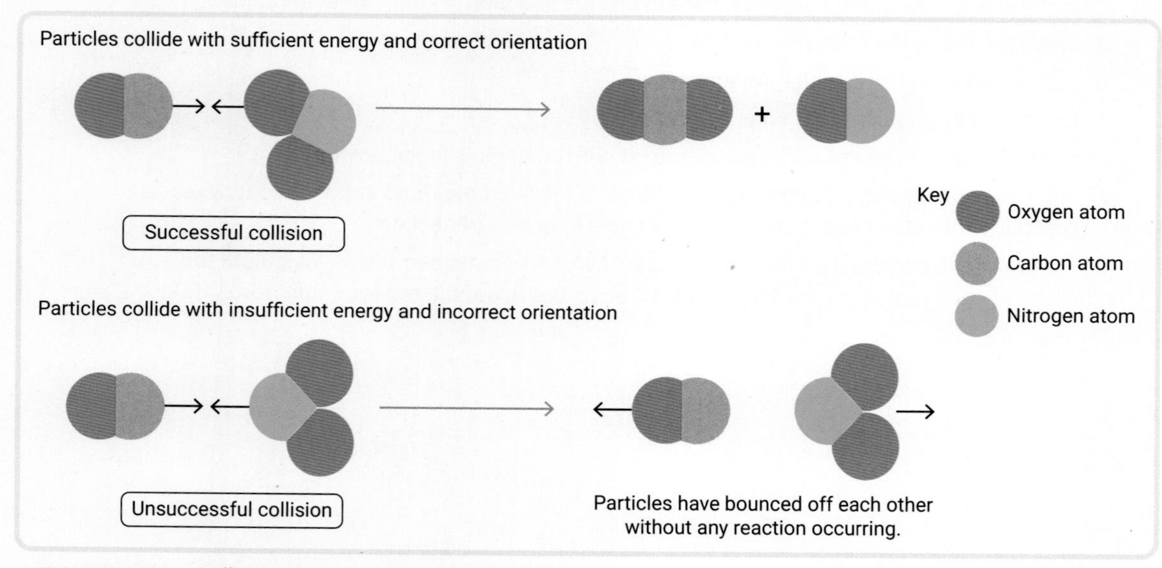

▲ FIGURE 4.9.1 Collision theory

Activation energy

activation energy
the minimum energy particles need to have for a successful collision to occur

Activation energy is the term used to describe the minimum amount of energy required for a successful collision to occur and products to form. Each chemical reaction has a particular activation energy that is unique to that reaction.

9780170463041

Some reactions have a very low activation energy. Sodium metal and chlorine gas react explosively at room temperature (see Figure 4.9.2). Not a lot of energy is required to start the reaction because it has a low activation energy. Even at room temperature, the particles have sufficient kinetic energy for a successful collision.

The reaction between magnesium and oxygen is much harder to start (see Figure 4.9.3). In Module 4.2 you heated magnesium with oxygen in a synthesis reaction. The fact you had to heat it with a Bunsen burner indicates the particles needed a great deal of kinetic energy to provide the activation energy for the reaction.

Alamy Stock Photo/Science Photo Library

▲ **FIGURE 4.9.2** Sodium metal and chlorine gas reacting

Alamy Stock Photo/sciencephotos

▲ **FIGURE 4.9.3** Magnesium ribbon in contact with oxygen gas

☆ ACTIVITY

Designing a model to show collision theory

Your teacher can provide you with a range of suitable materials for this modelling activity. This might include tennis balls, plasticine or cardboard, or you may choose to create a stop-motion animation, short video or other electronic presentation.

Your task is to create a series of photos or electronic presentation that:

- illustrates what is required for a successful collision to occur
- shows the before and after of a successful and an unsuccessful collision
- illustrates the idea of activation energy.

4.9 LEARNING CHECK

1. **State** the collision theory.
2. **Describe** the conditions for a successful collision to occur.
3. **Suggest** a way that an unsuccessful collision could become a successful collision.
4. If a reaction occurs at room temperature, does it have a high or low activation energy? **Justify** your answer.

4.10 Factors affecting the rate of reaction

BY THE END OF THIS MODULE, YOU WILL BE ABLE TO:

✓ explain how altering the concentration of reactants, surface area, temperature and catalysts affect the rate of a chemical reaction.

Interactive resource
Drag and drop: Factors that speed up reactions

Extra science investigation
Effect of temperature and concentration on reactions

GET THINKING

Speeding up chemical reactions is a key part of the chemical industry. Production of medicines, food and materials is more profitable if the chemical reaction is faster. Using the idea of collision theory from the previous module, brainstorm ways that chemical reactions might be made faster.

How can we speed up reactions?

Speeding up (or slowing down) chemical reactions is based on the two ideas of collision theory.

1 Particles must come into contact to have a collision.

2 Particles must have enough energy to provide enough activation energy.

There are four ways of speeding up reactions:

- increasing the concentration of reactants
- increasing the surface area of reactants
- increasing the temperature of reactants
- adding a **catalyst**.

catalyst
a substance that increases the rate of a chemical reaction by lowering the activation energy of the reaction without itself being changed

Concentration

When you increase the concentration of a solution or a gas, you have more particles in the same space. You can see this effect in Figure 4.10.1. In a crowded room, you are more likely to collide with someone. More collisions occur between reactants when there is a higher concentration, so more product is able to form in a given time.

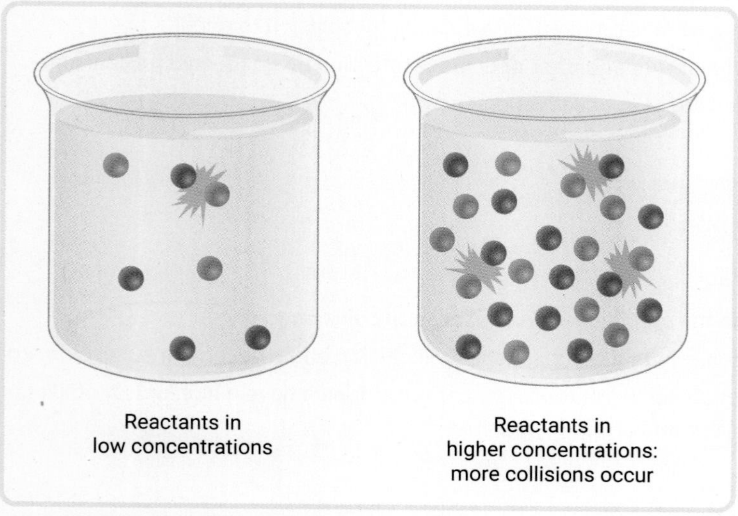

▲ **FIGURE 4.10.1** Higher concentrations result in more particle collisions.

Surface area

Increasing the surface area of a chemical involves chopping or breaking it into smaller pieces. Figure 4.10.2 shows how increasing the surface area makes more of the material accessible to another reactant. This also increases the number of collisions. As a result, the rate of reaction is increased.

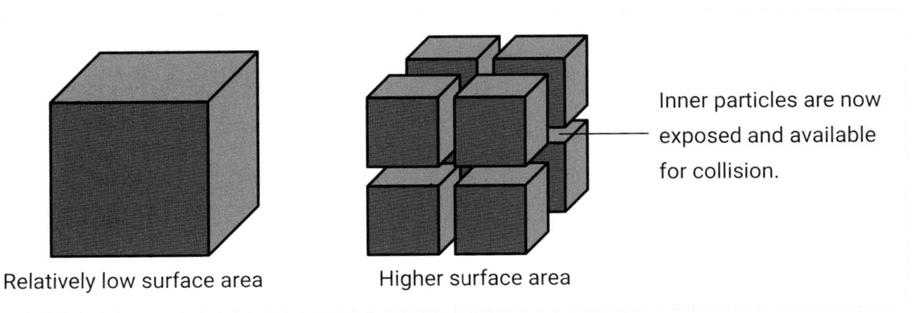

▲ **FIGURE 4.10.2** Increasing the surface area results in more particle collisions.

☆ **ACTIVITY 1**

The surface area of sugar

This activity does not show a chemical reaction, but it does show the effect of surface area through a model.

Add a solid sugar cube to 50 mL of water and stir it to dissolve. Repeat with the same amount of sugar grains (larger surface area).

- **a** **Explain** why the grains of sugar dissolve more quickly than the sugar cube.
- **b** **Assess** this model as a method of demonstrating surface area. Describe at least one way it is a good model and one way it is a poor model. Make an overall justified judgement on the usefulness of the model.

Shutterstock.com/Cozine

▲ **FIGURE 4.10.3** Sugar cubes and grains of sugar dissolve at different rates.

Temperature

When a reactant is heated, all the particles gain more kinetic energy. This increases the rate of reaction in two ways.

1 As the particles gain energy, there are now more particles with kinetic energy above the activation energy for the reaction, as shown in Figure 4.10.4. This results in more successful collisions.

2 The particles are moving faster and so will collide more often. As a result, more successful collisions occur.

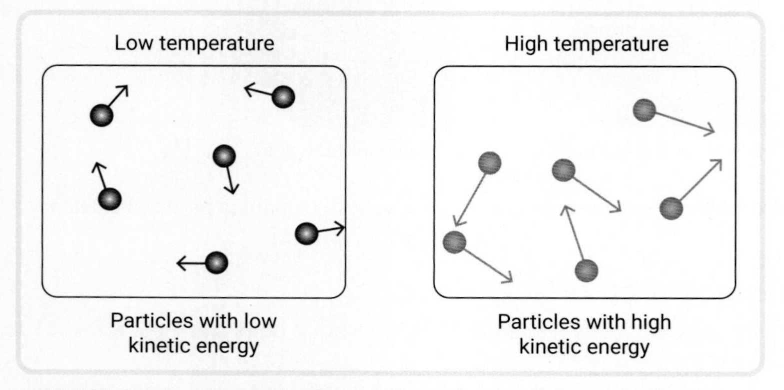

▲ **FIGURE 4.10.4** Increasing temperature gives particles more kinetic energy (represented by the longer and different coloured arrows).

Catalysts

A catalyst is a chemical that speeds up the rate of reaction but is not used up or changed in a chemical reaction. A catalyst reduces the activation energy, meaning more particles will have enough energy to react. As a result, more particles exceed the activation energy and there are more successful collisions. The rate of reaction is increased. The conical flasks shown in Figure 4.10.5 have increasing amounts of catalyst added. The increased rate of reaction can be seen by the formation of a greater volume of bubbles.

▲ **FIGURE 4.10.5** Increasing amounts of catalyst have been added to these conical flasks from left to right.

Examining the effect of a manganese dioxide catalyst on hydrogen peroxide

☆ ACTIVITY 2

4.10

Your teacher will most likely demonstrate this for you. It is a very quick reaction, so you might want to film it so that you can watch it back later.

Add 75 mL of 100 volume hydrogen peroxide solution to a 250 mL conical flask, as shown in Figure 4.10.6. Observe it for a moment. While it may not look like it, there is a reaction going on. The hydrogen peroxide is decomposing into water and oxygen.

- **a** Why does it appear as though no reaction is occurring?
- **b** Add 0.5 g powdered manganese dioxide and record all observations.
- **c** If you were to collect the manganese dioxide at the end, what mass would you expect to collect? **Justify** your answer.
- **d** **Explain** how the manganese dioxide increased the rate of reaction.
- **e** **Compare** how a catalyst and heating can increase the rate of a chemical reaction.

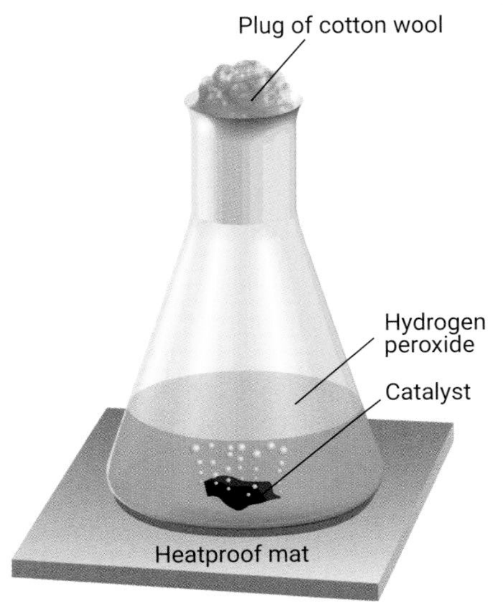

▲ **FIGURE 4.10.6** Equipment set up for the decomposition of hydrogen peroxide

4.10 LEARNING CHECK

1. **Identify** the four ways of increasing the rate of reaction.
2. **a** **Describe** what happens to the number of particles when the concentration of a solution is increased.
   b **Explain** how this increases the rate of reaction.
3. **Identify** what happens to the kinetic energy of particles when a chemical is heated.
4. **Explain** how heating a chemical reaction increases the rate of reaction.
5. Heating a reaction and adding a catalyst both increase the rate of reaction. **Explain** how these two different methods each work to produce the same result.

FIRST NATIONS SCIENCE CONTEXTS

4.11 First Nations Australians use of chemical reactions to produce a range of products

IN THIS MODULE, YOU WILL:

✓ investigate some of the chemical reactions used by First Nations Australians to produce products for consumption.

Fermentation of plant products

First Nations Australians have long used chemical reactions to process plant products for consumption. Fermentation is a chemical process that converts carbohydrates to cellular energy using micro-organisms. The products of the reaction are alcohol or organic acids.

The Noongar People of Western Australia are recognised as producers of a fermented liquid called *mangaitch*, produced from the nectar of a species of banksia tree. The banksia flowers containing nectar were soaked in water and then fermented in a bark vessel taking advantage of native yeasts. In Tasmania, First Nations Peoples tapped *Eucalyptus gunnii* trees and allowed the sweet sap to pool in tree hollows. Native yeasts fermented the sap to produce an alcoholic beverage called *wayalinah* (see Figure 4.11.1).

Dr Joanna Sundstrom, University of Adelaide

▲ **FIGURE 4.11.1** A scientist from the University of Adelaide and Australian Wine Research Institute collects sap from the Tasmanian cider gum, *Eucalyptus gunnii*.

It has been suggested that Noongar People also fermented toxic *Macrozamia* seeds, a type of cycad plant. The seeds, including their fleshy outer covering, were soaked and then buried underground, creating an anaerobic environment for the fermentation process (see Figure 4.11.2). This produces an extremely nutritious, fat-rich food source that is safe for consumption. Uniquely, in this region, some groups were recorded to consume the red outer flesh of the cycad, rather than the seed.

iStock.com/ GerhardSaueracker

▲ **FIGURE 4.11.2** *Macrozamia* cone with flesh-covered seeds

Detoxification of cycads

In other parts of Australia, poisonous cycads were detoxified by many different First Nations Peoples using chemical processes other than fermentation to produce an edible product.

9780170463041

Soaking the cycad seeds in water hydrolyses the toxin, cycasin, so that the water-soluble chemical bonds are broken. This allows the toxic component to be washed away. Under normal conditions, this reaction proceeds very slowly. However, historical accounts show how the First Nations Peoples of the rainforest region in North Queensland increased the rate of the reaction by thinly slicing the seeds to increase the surface area, so more of the seeds were exposed to the action of hydrolysis. This also ensured that all of the cycad seed material was exposed to the action of hydrolysis. The reaction was allowed to proceed in special baskets for sufficient time to ensure complete detoxification. This process resulted in an abundant food resource that was safe to eat. Figure 4.11.3 shows a specialised colander-like basket used by First Nations Peoples of North Queensland to detoxify cycads.

Joe Sambono

▲ **FIGURE 4.11.3** A *Jawun* basket made by the Jirrbal Peoples of Tully, North Queensland.

☆ **ACTIVITY**

Investigating how processing food products causes chemical changes

First Nations Australians have extensive knowledge of the chemical reactions used to process toxic food products and make them edible.

It is not safe to use toxic cycads to investigate chemical reactions in this activity. Instead, you will use pineapples to model how processing foods can cause chemical changes. Pineapples contain an enzyme, bromelain, that digests collagen proteins. This is also a hydrolysis reaction.

Materials

- fresh pineapple
- frozen pineapple
- canned pineapple
- powdered gelatine
- water
- kettle
- ice water bath
- 4 test tubes or cups
- scales
- measuring cylinder

Method

1 Cut equal-sized pieces of fresh, canned and frozen pineapple. Ensure these pieces are small enough to fit into the test tubes (or cups).
2 Weigh each piece and record the weight.
3 Prepare the gelatine according to the instructions on the packet, using the kettle to boil the water.
4 Use the measuring cylinder to pour equal amounts of hot gelatine solution into each test tube.
5 Place each type of pineapple into separate test tubes, leaving one without fruit as a control. Label the test tubes.
6 Mix the test tubes gently and place them into the ice water bath.
7 Check the control tube every few minutes. When the control tube has solidified, remove all the tubes from the ice bath and observe the consistency of the gelatine. Record your observations.

Evaluation

1 What did you observe? Why do you think this happened?
2 How does processing pineapple through freezing and canning cause chemical changes to the enzyme bromelain?
3 How does this experiment reflect the chemical practices employed by First Nations Australians in processing foods?

SCIENCE AS A HUMAN ENDEAVOUR

4.12 How can we solve the plastic problem?

BY THE END OF THIS MODULE, YOU WILL BE ABLE TO:

- ✓ explain why we need to develop methods for disposal and breakdown of plastics
- ✓ explain how microbes may be able to be used to control plastic waste.

Shutterstock.com/Benny Marty

▲ **FIGURE 4.12.1** The Great Pacific Garbage Patch is a collection of waste plastic in the Pacific Ocean covering an area approximately seven times the size of the state of Victoria.

How big is the plastic problem?

In 2020, more than 350 million tonnes of plastics were produced around the world. Despite efforts to recycle plastics, less than 20% of waste plastic is recycled. Each year, more than 200 million tonnes of plastic waste is generated. Some of this goes into the oceans, but most ends up in landfills, where it can take hundreds or thousands of years to break down.

What can microbes do to help?

Video activity
Plastic-eating microbes

In 2016, Japanese scientists visiting a plastics recycling facility observed that a bacteria (*Ideonella sakaiensis*) was breaking down PET plastics. PET, or polyethylene terephthalate, is a long-chain polymer made of many smaller monomers.

Scientists discovered that the bacteria produces a type of enzyme called PETase that is able to break the long PET chains into the monomers it was produced from. Enzymes are proteins that cause specific reactions to occur. The monomers were then further broken down by the enzymes to release energy and help the bacteria grow.

Studies are currently underway to improve the efficiency of the bacteria, so they can be used as a method of helping to break down PET plastics. Other scientists are working on using other bacteria to break down different types of plastics.

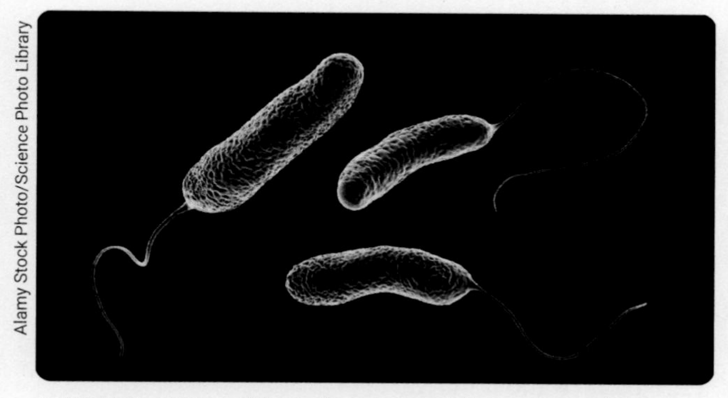

Alamy Stock Photo/Science Photo Library

▲ **FIGURE 4.12.2** *Ideonella sakaiensis* bacteria produce enzymes that break down PET into its monomers.

4.12 LEARNING CHECK

1. What does PET stand for?
2. **Conduct** research to describe impacts of plastics in our oceans.
3. **Describe** the process used by the bacteria *Ideonella sakaiensis* to break down PET.
4. **Explain** how use of bacteria could help solve our waste plastic problem.

SCIENCE INVESTIGATIONS

4.13

Qualitative observations versus quantitative measurements

SCIENCE SKILLS IN FOCUS

IN THIS MODULE, YOU WILL FOCUS ON LEARNING AND IMPROVING THESE SKILLS:

- gathering and organising qualitative observations and quantitative measurements about chemical reactions
- describing how making observations and taking measurements can provide information about the products formed in chemical reactions
- writing word equations for chemical reactions observed in experiments.

Qualitative observations are any observations you can make with your senses during an experiment. This could be sight, smell, touch or hearing. These might include:

- colours or colour change
- heat being released or absorbed
- bubbles forming
- odours being produced.

Quantitative measurements are any experimental results you can collect with a numerical value (i.e. they can be measured). This might include quantities such as mass or time, which are both very useful.

The rate of reaction is an example where quantitative measurements are important. Collecting data about mass and time allows you to calculate the rate of reaction in an experiment.

There are also many experiments where it is impossible to collect quantitative data. For example, the experiments in this chapter involve qualitative observations alone. Examples of this include precipitation and metal displacement reactions. In these experiments, there is nothing to measure. You are only determining whether a reaction occurred or not. This can be seen through a simple observation such as a solid precipitate forming.

There are potential problems with both qualitative and quantitative measurements.

Qualitative observations are usually affected by subjectivity. Consider a colour change occurring. If you ask several people to observe a colour change, there will often be different opinions. When colours are 'in between' clear colours, such as a colour between yellow and orange, the judgement is subjective. For example, this colour might be seen as yellow by one person but orange by another person. This may lead to an incorrect positive identification of a reaction, or misidentification of a product if the observations are subjective.

Quantitative measurements can be inaccurate if equipment is not read or used correctly. To avoid problems, it is important that you know how to use the equipment before you start an experiment.

EXPERIMENT 1: DECOMPOSITION OF COPPER CARBONATE AND TEST FOR CARBON DIOXIDE

AIM

To investigate the decomposition of copper carbonate

MATERIALS

- ☑ side arm test tube with rubber stopper
- ☑ rubber or plastic hose
- ☑ Bunsen burner heating set-up
- ☑ large test tube half full of limewater (calcium hydroxide)
- ☑ matches
- ☑ copper carbonate powder
- ☑ spatula

Video
Science skills in a minute: Quantitative v. qualitative data

Science skills resource
Science skills in practice: Quantitative v. qualitative data

METHOD

1. Place two heaped spatulas of copper carbonate powder into a side arm test tube and clamp the test tube to a retort stand.
2. Set up the equipment as shown in Figure 4.13.1, ensuring the hose from the side arm test tube is submerged in the limewater (calcium hydroxide).

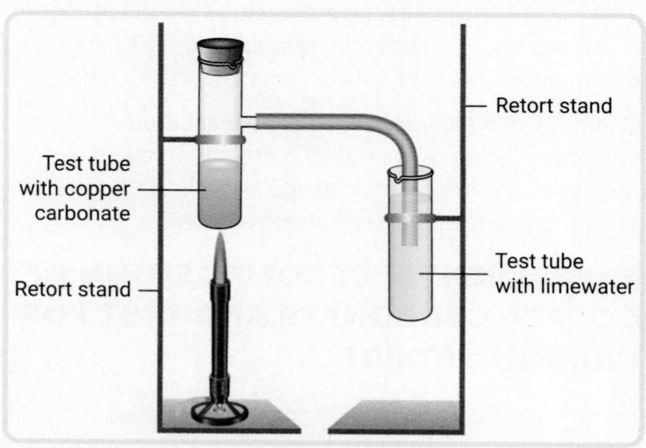

▲ FIGURE 4.13.1

3. Light the Bunsen burner and gently heat the copper carbonate until it entirely changes colour.

RESULTS

Record your observations by taking photographs or making notes.

EVALUATION

1. What did you observe that indicated a gas formed in this reaction?
2. Write a word equation for this reaction.
3. The limewater test detects the presence of carbon dioxide. Describe the observations you made and state whether they indicate if carbon dioxide was present.
4. Explain whether the observations made in this experiment are qualitative or quantitative.
5. Identify two observations you made in this experiment that confirmed a chemical reaction took place.

EXPERIMENT 2: METAL DISPLACEMENT REACTIONS

AIM

To investigate metal displacement reactions involving magnesium, zinc, copper and iron

MATERIALS

- ☑ test tubes and test-tube rack
- ☑ samples of magnesium, zinc, copper and iron
- ☑ sandpaper to clean the metals
- ☑ solutions of magnesium nitrate, zinc nitrate, copper nitrate and iron nitrate

METHOD

1. Set up four test tubes in a test-tube rack.
2. Place the zinc, magnesium, copper and iron into the four different test tubes, one metal per test tube.
3. Cover each metal with about 2–3 cm of magnesium nitrate and record observations.
4. Repeat steps 2 and 3 using all the metals and the zinc nitrate solution.
5. Repeat again using copper nitrate, and then again with iron nitrate.

RESULTS

Create an appropriate table to record the results of each metal with each solution. Your results should include observations and whether a reaction occurred.

EVALUATION

1. Explain, using a word equation from the results, whether this experiment is demonstrating single or double displacement reactions.
2. Write word equations for any reactions that occurred in this experiment.
3. Use the results to rank the metals from most to least reactive. Justify your rankings.
4. Describe any issues you had in deciding whether a reaction had occurred.

9780170463041

EXPERIMENT 3: PRECIPITATION REACTIONS

AIM

To observe the results of precipitation reactions and write equations for the reactions that occurred

MATERIALS

- ☑ dropper bottles of copper nitrate, zinc nitrate, silver nitrate, lead nitrate, calcium nitrate and magnesium nitrate (Group A)
- ☑ dropper bottles of potassium iodide, sodium hydroxide, sodium carbonate, sodium chloride and sodium sulfate (Group B)
- ☑ laminated black sheet of cardboard

METHOD

1 On the laminated cardboard, create a grid of large squares with five columns and six rows.

2 Above each vertical column, write one of the chemicals from Group B.

3 To the left of each horizontal row, write one of the chemicals from Group A.

4 Add 3 drops of each chemical to the relevant grid boxes. When you add the second chemical, ensure you add the drops directly on top of the first chemical so they mix.

RESULTS

Take a photograph of your results.

Create a table recording whether a precipitate formed. If a precipitate formed, record its colour.

EVALUATION

1 Explain whether precipitation reactions are single or double displacement reactions.

2 Write word equations for all reactions that formed a precipitate.

EXPERIMENT 4: REACTIONS INVOLVING ACIDS

AIM

To investigate the reaction between metals and acids

MATERIALS

- ☑ samples of a range of metals. This could include copper, zinc, magnesium and aluminium
- ☑ 2 mol/L hydrochloric acid (HCl)
- ☑ test tubes and test-tube rack
- ☑ 10 mL measuring cylinder
- ☑ matches
- ☑ cotton wool

METHOD

1 Place a small piece of each metal in the bottom of a test tube.

2 Add 10 mL of hydrochloric acid to each test tube.

3 Place a small plug of cotton wool loosely in the top of each test tube.

4 Record your observations. Focus on the level of bubbling that is occurring in each test tube.

5 To test for the gas produced, light a match, remove the cotton wool and hold the match to the top of the test tube. A 'pop' sound indicates that hydrogen gas has been produced.

RESULTS

Create a table to record your observations for each metal.

EVALUATION

1 Write a word equation for each metal's reaction with the hydrochloric acid.

2 Use the equation you wrote in Question 1 to explain why the 'pop' test in step 5 of the method was conducted.

3 Did all the metals react with the same intensity?

4 How do the results of this experiment show you the reactivity of metals? Rank the metals you used in order of their reactivity.

5 The observations you made in this experiment are qualitative. Describe any difficulties you had in ranking the metals based on your observations alone.

6 Suggest an improvement to the method that would allow you to make quantitative measurements.

EXPERIMENT 5: THE RATE OF REACTION BETWEEN MAGNESIUM AND HYDROCHLORIC ACID

AIM

To investigate how changing the concentration of hydrochloric acid affects the rate of reaction between hydrochloric acid and magnesium metal

MATERIALS

- ☑ 9 × 3 cm strips of magnesium (Mg)
- ☑ 2 mol/L hydrochloric acid (HCl)
- ☑ 3 × 100 mL conical flasks
- ☑ electronic balance
- ☑ 20 mL measuring cylinder
- ☑ distilled water

METHOD

1 Measure 20 mL of hydrochloric acid using a measuring cylinder and place it into one of the conical flasks.

2 Measure 15 mL of hydrochloric acid and make up to a total of 20 mL with distilled water. Place this in the second conical flask.

3 Measure 10 mL of hydrochloric acid and make up to a total of 20 mL with distilled water. Place this in the third conical flask.

4 Place the first conical flask on the electronic balance. Add a 3 cm strip of magnesium and immediately zero the electronic balance.

5 After 60 seconds, note the mass reading. Since gas is being formed and escaping, the value will be negative. Ignore the negative sign and record the value of the mass.

6 Repeat step 5 with the second and third conical flasks.

7 If time and resources allow, repeat steps 1–6 twice more and average your results. Alternatively, collect data from other groups in your class and calculate the average.

RESULTS

1 Create a suitable table to record your data. For each trial (repetition) of steps 1–6 of the method, include the volume of hydrochloric acid used in the conical flask, the mass lost and the average mass lost.

2 The mass lost is the mass of hydrogen gas formed in 60 seconds. Use this information to calculate the average rate of reaction for each of the three concentrations. Show all workings.

EVALUATION

1 Write a word equation to represent the reaction that occurred.

2 Use your answer to Question 1 to explain why mass is lost in this reaction.

3 Suggest a different method of measuring the rate of reaction that could be used instead of the loss of mass.

4 Write a conclusion about the effect of the concentration of hydrochloric acid on the rate of this reaction. Use data from your experiment to support your conclusion.

5 Describe two errors that could have occurred during this experiment and explain how they could have affected the results.

4 REVIEW

REMEMBERING

1 **Write** the general word equation for:
   a an acid and a metal.
   b an acid and a metal hydroxide.

2 A piece of magnesium reacted with sulfuric acid. Bubbles of gas formed. **Identify** the gas that was forming the bubbles.

3 **List** four methods of increasing the rate of reaction.

4 **Describe** the structure of an ionic salt and **describe** how it is named.

5 **Describe** two examples of useful synthesis reactions.

6 **Describe** the three ways that energy can be supplied in a decomposition reaction.

UNDERSTANDING

7 The graph below shows the volume of oxygen gas produced during the decomposition of hydrogen peroxide.
   a **Calculate** the rate of reaction after 4 minutes and after 14 minutes.
   b **Suggest** a reason why there is no change in the volume of oxygen after 16 minutes.

8 **Describe** how a catalyst increases the rate of reaction.

9 **Explain** how an understanding of acids, bases and pH can assist a gardener grow flowers more successfully.

10 **Describe** how you could confirm through an experiment that the product(s) of an acid and metal hydroxide reaction was neutral.

11 For the following pairs of ionic salts, **write** an equation and **predict** whether a precipitate will form.
   a Lead nitrate and sodium hydroxide
   b Magnesium nitrate and potassium chloride
   c Calcium chloride and magnesium sulfate

12 Using an example, **describe** one way First Nations Australians have used their knowledge of chemical reactions in food preparation.

APPLYING

13 Some hydrochloric acid has been spilled in a laboratory. The teacher states that it would be risky to wipe it up directly and suggests adding some sodium carbonate to the spill to make it safer to wipe up. Use a word equation to help **explain** why this suggestion makes it safer to wipe up the acid.

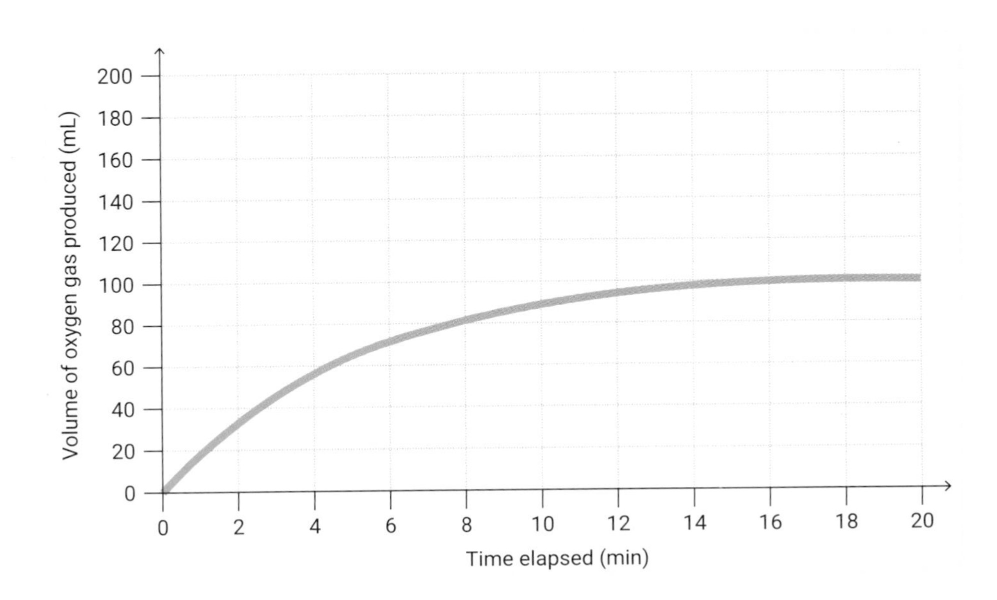

14 **Identify** whether the following reactions are examples of synthesis (S), decomposition (D), precipitation (P), metal/acid reaction (M) or neutralisation (N), and for each reaction (except e), write a word equation to show the products formed.

- **a** Lead nitrate and sodium chloride
- **b** Iron reacting with oxygen
- **c** Heating copper carbonate and forming black copper oxide powder and carbon dioxide gas
- **d** Magnesium reacting with hydrochloric acid
- **e** Adding together monomers to make polystyrene
- **f** Potassium sulfate reacting with silver nitrate
- **g** Zinc metal reacting with sulfuric acid

15 Consider the experiment shown below. Rank the five beakers in order of the expected rate of reaction from fastest to slowest. **Justify** your ranking.

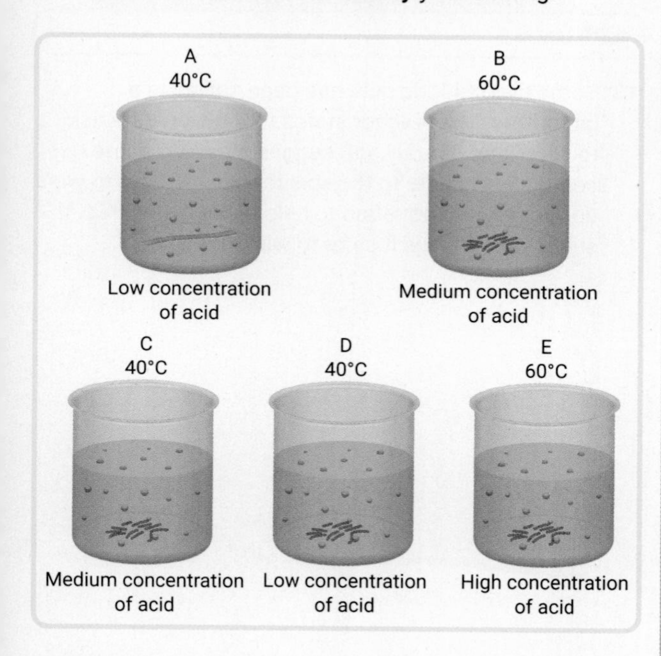

ANALYSING

16 The human stomach contains hydrochloric acid. Sometimes excess acid moves upwards, causing a burning sensation. Antacids containing chemicals such as magnesium hydroxide can remove the excess acid. **Write** an equation to show how antacids work.

17 When silver nitrate is mixed with potassium chloride, a white solid is formed.

- **a** What type of reaction is this?
- **b** **Write** the word equation for this chemical reaction.
- **c** Which of the products is the white solid?

18 A student conducted an experiment where 5 g of calcium metal was added to hydrochloric acid at different temperatures. The time taken for all the calcium to react was recorded in the following table.

- **a** Copy and **complete** the table by calculating the rate of reaction at each temperature.
- **b** **Explain** these results using collision theory.

Student experiment results for calcium metal and hydrochloric acid reaction

Temperature (°C)	Time taken for calcium to react (min)	Rate of reaction (g/min)
40	9.1	
50	7.3	
60	4.6	
70	2.1	

EVALUATING

19 'Chemical reactions have a positive impact on society.' **Discuss** this statement.

CREATING

20 **Create** a summary poster with general equations and specific examples to show the four types of reactions studied in this chapter: include everyday examples of each type, find photographs to illustrate them or draw your own models to represent the reactions.

BIG SCIENCE CHALLENGE PROJECT

1 Connect what you've learned

At the start of the chapter, you were asked why solid coal is hard to ignite but coal dust can explode with a spark. Use what you have learned in this chapter to provide a chemical explanation for this phenomenon.

2 Check your thinking

Australia has a number of coal mines. Coal is a major export for Australia and provides fuel for coal-fired power stations.

Research ways that miners in coal mines can be kept safe from a build-up of coal dust.

Research the cause of a coal mine disaster. Suggest ways it could have been prevented. Examples of disasters are those at the Westray mine in Canada in 1992, the Pike River mine in New Zealand and the Upper Big Branch mine in the United States, both in 2010.

Getty Images/Pool

3 Get into action

Chemical disasters are the accidental release of chemicals that can potentially be harmful to people or the environment. Three examples of major disasters that happened overseas are the Bhopal disaster in India, the ammonium nitrate explosion in Beirut, and the Deepwater Horizon oil spill disaster in the United States.

Choose one of these disasters (or identify another that interests you) and research the details of the event. This could include the cause of the disaster, the consequences for people or the environment or how it was cleaned up.

4 Communicate

Think about the chemicals that you use around your home. This might include gas for cooking or heating, cleaning chemicals, garden chemicals or other items in your shed or garage. Create a safety poster for your family that shows the safe use of some of these chemicals.

5 The universe

9780170463041

BIG SCIENCE CHALLENGE 

Ángel R. López-Sánchez (AAO-MQ)

▲ FIGURE 5.0.1 The Milky Way over Siding Spring Observatory

Humans have always been naturally curious about the night sky. Like so many before us, we look up into the sky at night and wonder about what we can see. Have you ever tried to count how many stars you can see? It was estimated that, once, we could see 5000 stars with the naked eye on a clear cloudless night. However, that is no longer the case. Unless you get well away from populated areas, it is getting harder to see the stars with the amount of light pollution in our cities at night.

- **Have you ever noticed the difference between the night sky in the city and the night sky in the bush or outback?**
- **What could be some of the other impacts of light pollution?**

 **SCIENCE CHALLENGE ACCEPTED!**

At the end of this chapter, you can complete the Big Science Challenge Project #5. You can apply the knowledge and skills you learn in this chapter to complete the project.

Assessments

- Prior knowledge quiz
- Chapter review questions
- End-of-chapter test
- Portfolio assessment task: Project

Videos

- Science skills in a minute: Validating scientific work **(5.10)**
- Video activities: The birth of our solar system **(5.2)**; Life cycle of a star **(5.3)**; Big Bang theory **(5.5)**; Big Bang evidence **(5.6)**; Dark matter **(5.7)**; Gravitational waves **(5.7)**; SKA **(5.9)**

Science skills resources

- Science skills in practice: Validating scientific work **(5.10)**
- Extra science investigations: Modelling separating galaxies **(5.2)**; Calculating parallaxes **(5.2)**; Expanding universe **(5.6)**

Interactive resources

- Label: Organisation of the universe **(5.1)**; Types of galaxies **(5.2)**
- Crossword: Starlight **(5.4)**
- Simulation: Blackbody spectrum **(5.4)**
- Quiz: Big Bang evidence **(5.6)**

Nelson MindTap

To access these resources and many more, visit: **cengage.com.au/nelsonmindtap**

5.1 The universe

BY THE END OF THIS MODULE, YOU WILL BE ABLE TO:

✓ describe the major parts of the universe using appropriate scientific terminology and units
✓ outline the way cosmologists categorise the parts of the universe.

GET THINKING

As you work through this chapter, try to remember the ideas you had about the universe when you were younger. Where did these ideas come from? Have any of your ideas changed over time? How have they changed? Share your answers with a partner.

supercluster
a large group made up of clusters of galaxies

galaxy
a collection of gas, dust and billions of stars and their solar systems, all held together by gravity

astronomical
a term to relate ideas or objects to astronomy

dark matter
matter that cannot be seen but is known to exist because its effect on objects can be observed

planetoid
a small body that is like a planet but does not meet specific criteria, such as minimum diameter

comet
an object made of rock and ice that orbits the Sun at regular intervals

asteroid
a rocky object smaller than a planet that orbits the Sun

Interactive resource
Label: Organisation of the universe

Every human culture has had its own ideas about the universe, where it came from and what can be found in it. Scientific theories about the formation of the universe have changed over time due to advances in technology, such as the invention of the telescope and supercomputing, and through international collaborations between research organisations.

What's in the universe?

The universe contains everything that exists in space and time, including all matter and energy. It consists of the smallest of subatomic particles to huge **superclusters** of **galaxies** that would take billions of years to cross. The universe contains many trillions of **astronomical** bodies such as stars, planets and galaxies. As it is so vast and contains so much, including things like **dark matter** that we can't see, it can be hard for us to comprehend the universe's size and make-up.

Think about the size of oceans and continents on our own planet, Earth, and all the life, matter and energy they contain. Now consider all the parts that make up our own solar system: the Sun, our Moon, the other planets and their moons, **planetoids**, **comets** and **asteroids**. Now think about how the Sun is just one of the billions of stars in our galaxy, the Milky Way. Finally, try to visualise the billions of galaxies astronomers have estimated exist in the universe, with each containing billions of stars (see Figure 5.1.1).

▶ **FIGURE 5.1.1** Looking through the centre of our galaxy, the Milky Way, as viewed from La Silla Observatory in Chile. Our galaxy is just one of the estimated 200 billion galaxies in the observable universe.

ESO/B. Tafreshi (twanight.org)

The organisation of the universe

The biggest things in the universe are giant structures called filaments and voids. Filaments are the parts of the universe that contain concentrated **clusters** of galaxies. Voids are the huge spaces between filaments that are relatively empty of galaxies. Astronomers estimate that voids make up between 80 and 90 per cent of the universe. Our own galaxy, the Milky Way, sits within a group of galaxies in one of these voids.

cluster
a group of stars or galaxies

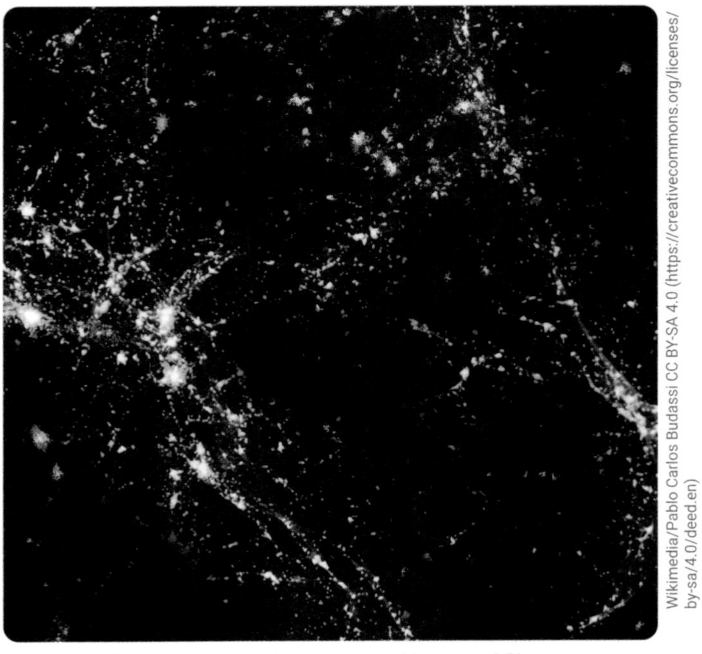

▲ **FIGURE 5.1.2** A computer-generated image of filaments, which are the biggest rotating structures in the universe

The computer-generated image in Figure 5.1.2 shows voids in between a web of filaments. The bright yellow and white region represents galaxy clusters. These structures are so big that light would take millions of years to travel their length.

cosmologist
a scientist who studies the universe

To better understand the size and complexity of the parts of the universe, **cosmologists** have developed a hierarchy to show how the structures in the universe relate to each other.

Figure 5.1.3 shows one interpretation of how cosmologists categorise the structures in the universe. The figure starts with the biggest-sized structures, filaments and voids, at the base and finishes with planets at the top. In between these are many other structures that are either groupings of similar structures that are held together by the force of gravity, or individual bodies, such as stars and planets.

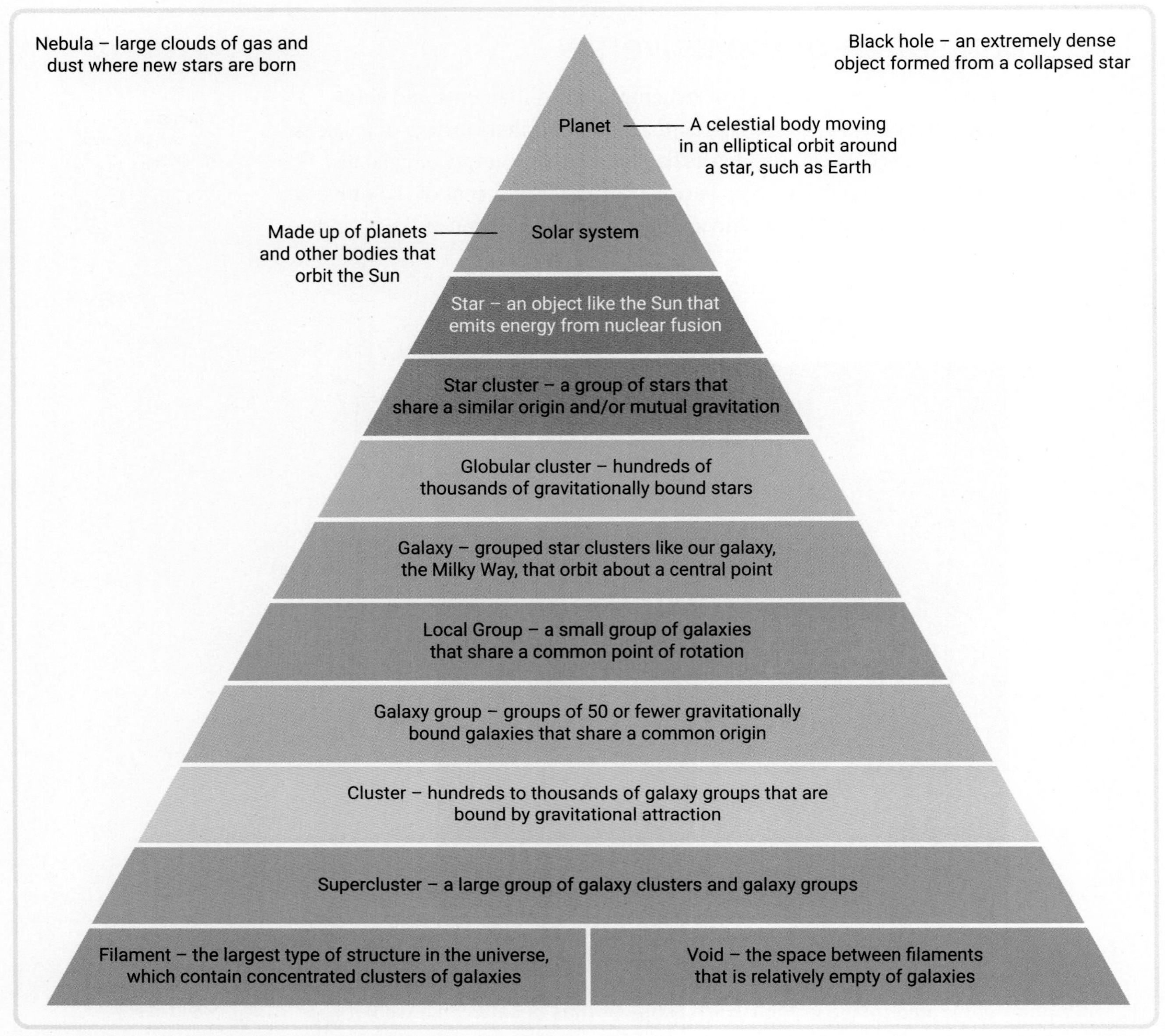

▲ **FIGURE 5.1.3** The hierarchy of the structures in the universe

5.1 LEARNING CHECK

1 **Define** 'cluster'.
2 **List** five astronomical objects found in the solar system.
3 **Name** four types of structures found in the universe.
4 **Explain** how our galaxy, the Milky Way, fits into the hierarchy of the structures in the universe.

5.2 Galaxies

BY THE END OF THIS MODULE, YOU WILL BE ABLE TO:

- ✓ describe the different types of galaxies
- ✓ describe how galaxies form
- ✓ explain why astronomical units are used to compare distances in the solar system and why distances in space are usually measured in light-years.

GET THINKING

When we look at the stars, we mostly see individual points of light. However, there is a lot more to the story of the stars! Most stars belong to much larger collections of stars. You probably know that our galaxy is called the Milky Way. What shape is our galaxy? Why do galaxies form in these different shapes? Why are galaxies different sizes? Share what you already know about the Milky Way with a partner.

Video activity
The birth of our solar system

Interactive resource
Label: Types of galaxies

Extra science investigations
Modelling separating galaxies

Calculating parallaxes

Types of galaxies

We describe and classify types of galaxies by their shape. The main types are spiral, elliptical or, when there is no definitive shape, irregular. Other shapes are barred spiral, peculiar and lenticular. Figure 5.2.1 gives examples of the different types of galaxies.

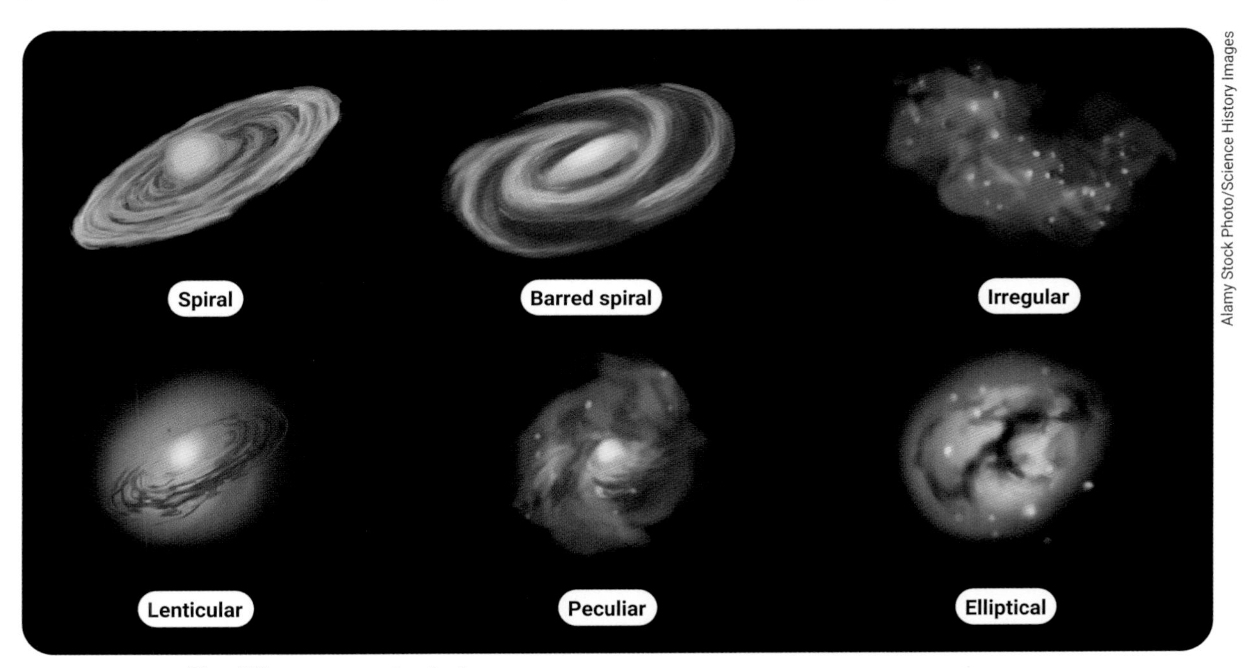

▲ **FIGURE 5.2.1** The different types of galaxies

Our galaxy, the Milky Way, is a barred spiral galaxy made up of billions of stars. The Sun sits on the inner edge of one of the spiral arms. Figure 5.2.2 shows an artist's impression of the Milky Way.

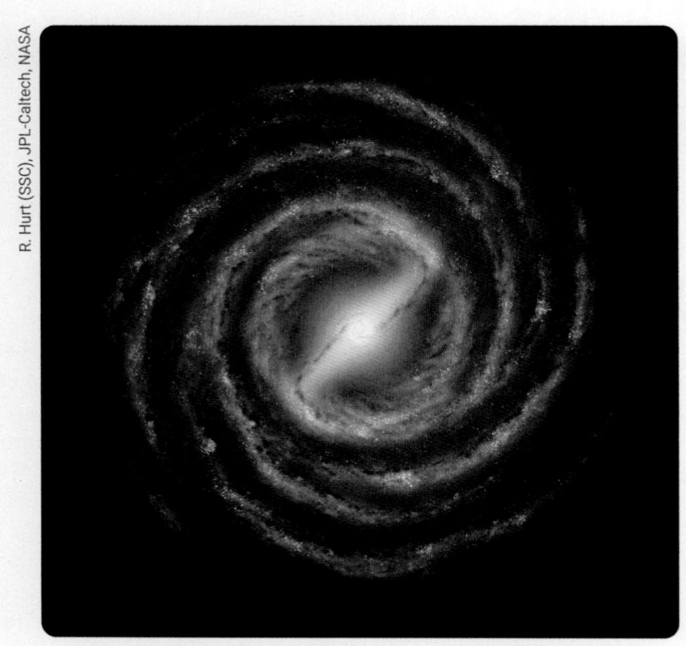

R. Hurt (SSC), JPL-Caltech, NASA

▲ **FIGURE 5.2.2** An artist's impression of the Milky Way

The Milky Way is part of a collection of 40 galaxies that we call the Local Group.

How do galaxies form?

Scientists from space agencies such as the National Aeronautics and Space Administration (NASA) and the European Space Agency (ESA) use supercomputers to simulate how galaxies formed in the early universe. Their modelling shows that galaxies start out as clouds of stars, dust, gas and dark matter swirling through space and being held together by the force of gravity. As the clouds swirl, they encounter other nearby clouds. Gravity forces these clouds closer together until they collide, creating an even larger cloud of stars, dust, gas and dark matter. Over time, collisions continue to occur between other clouds of stars, which results in the formation of a galaxy.

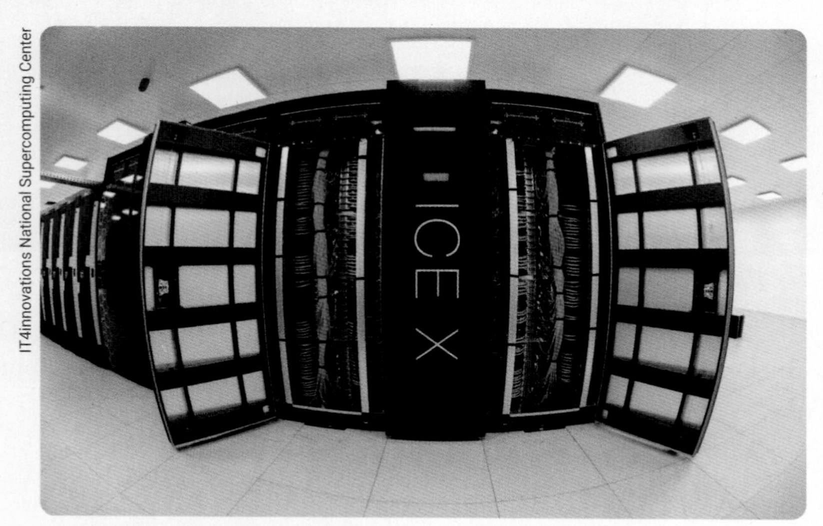

IT4innovations National Supercomputing Center

▲ **FIGURE 5.2.3** IT4Innovation's Salomon supercomputer in the Czech Republic, one of ESA's supercomputers

Computer modelling shows that spiral galaxies tend to form when collisions happen between clouds of stars, and the galaxies push material out to the edges of the forming galaxy, creating the spiral arms. Scientists use specialised telescopes such as the James Webb Space Telescope (NASA, ESA and the Canadian Space Agency) and the Hubble Space Telescope (NASA and ESA) to capture images of colliding galaxies that support the scientific theories of astronomers (see Figure 5.2.4).

The Hubble Space Telescope has captured many images of large, young galaxies, showing their distinctive shapes. In July 2022, the James Web Space Telescope captured images of the oldest, most distant galaxies astronomers have ever seen. These images show that these galaxies are smaller than newer galaxies and are clumpy in appearance (see Figure 5.2.5).

NASA, ESA, CSA, STScI

▲ **FIGURE 5.2.4** A cluster of five galaxies, known as the Stephan Quintet, as seen through the James Webb Telescope. It shows huge shock waves as one of the galaxies, NGC 7318B, smashes through the cluster

ESO/European Southern Observatory's Wide Field Imager

NASA, ESA, CSA, STScI

▲ **FIGURE 5.2.5** **(a)** A young spiral galaxy and **(b)** the galaxy cluster SMACS 0723, showing thousands of galaxies deep in the distant universe. Photos taken in 2022 with the James Webb Space Telescope.

The images being produced by the James Webb Space Telescope will help astronomers compare the more distant, oldest galaxies with younger spiral and elliptical galaxies. This comparison will provide more information about how galaxies formed, particularly those that formed in the very early history of the universe.

Measuring distances in space

Astronomers investigate and describe both the smallest objects in the universe, such as atoms and subatomic particles, and some of the largest structures, such as galaxy clusters. To describe and understand the universe at these two extreme scales, astronomers use many different units of measurements.

We can use kilometres when talking about some distances within our own solar system. For example, we are about 150 million kilometres from the Sun. However, things quickly get complicated once you start dealing with distances bigger than this, such as the distances between stars or other galaxies. To overcome this, scientists use the following units of measurement:

- **astronomical unit (AU)** – based on the distance from Earth to the Sun
- **light-year** – the distance that light travels in one year, at a speed of 300 000 km/s, equivalent to 9.4607×10^{12} km
- **parsec** (or parallax second) – the equivalent to exactly 3.26 light-years.

astronomical unit (AU)
approximately 150 million kilometres, equivalent to the distance between the centre of Earth and the centre of the Sun

light-year
an astronomical unit of distance equivalent to the distance that light travels in 1 year; approximately 9.4607×10^{12} km

parsec
an astronomical unit of distance equivalent to 3.26 light-years

Astronomers use astronomical units to measure distances within the solar system. A light-year is used to measure distances between stars in the same galaxy, while a parsec is used to measure the truly massive distances in the universe, such as the distance between galaxies. For example, the nearest star to Earth after the Sun is Proxima Centauri. It is 40.17 trillion kilometres away. This is equivalent to 268 770 AU or 4.3 light-years, a much easier number to deal with. The closest known galaxy to our solar system at present is the Canis Major Dwarf Galaxy, which is 25 000 light-years away, which is equivalent to 7669 parsecs.

Figure 5.2.6 shows the extent of the known universe, which has a diameter of 93 billion light-years or 28 billion parsecs, a distance so big it's impossible for us to really comprehend it. Can you imagine how big this number would be if you converted this distance into kilometres? The view of the universe in Figure 5.2.6 is shown from our perspective, with our solar system at the centre.

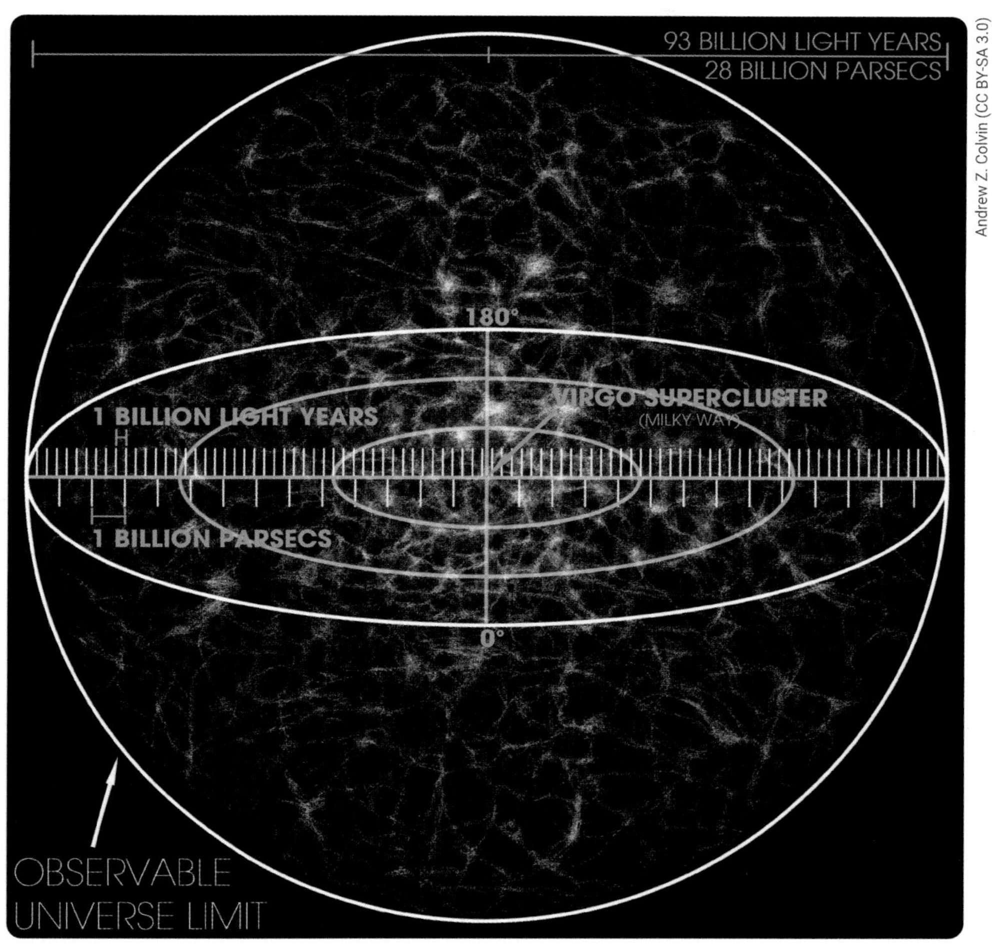

▲ **FIGURE 5.2.6** The size of the universe

5.2 LEARNING CHECK

1 **Identify** the different types of galaxies, according to their shape.
2 **Explain** why it is necessary to use special units of measurement when calculating distances in space.
3 **Calculate** the distance in kilometres to the Canis Major Dwarf Galaxy.

5.3 The life cycle of stars

BY THE END OF THIS MODULE, YOU WILL BE ABLE TO:

- ✓ describe how stars form
- ✓ explain the life cycle of stars.

Video activity
Life cycle of a star

Interactive resource
Drag and drop: Star life cycle

GET THINKING

After reading this module, you will be asked to compare the life cycle of a star to how a flowering plant develops. As you read through the material, think about the similarities that exist between the two.

How do stars form?

nebula
a cloud of cosmic gas and dust

constellation
a group of stars that looks like a picture or pattern

All stars are formed in much the same way. They start out in large clouds of interstellar dust and gas known as **nebulae**. Figure 5.3.1a shows a photo of one of the better-known nebulae, which can easily be seen from Earth in the belt of the **constellation** Orion.

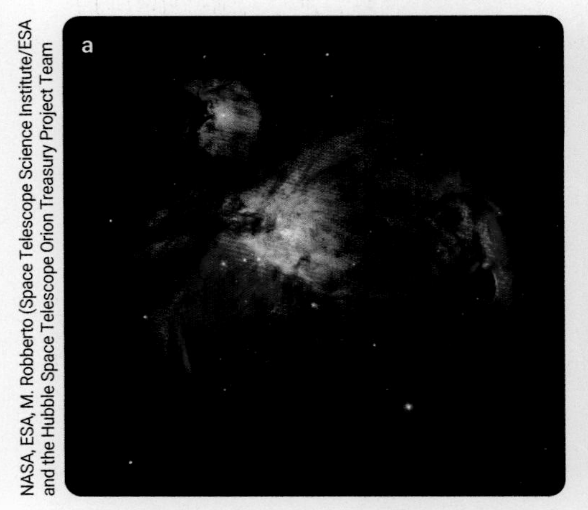

NASA, ESA, M. Robberto (Space Telescope Science Institute/ESA and the Hubble Space Telescope Orion Treasury Project Team

NASA, ESA, CSA, STScI

▲ **FIGURE 5.3.1 (a) A photo of the Orion Nebula taken with NASA's Hubble Space Telescope. (b) A young, star-forming region called NGC 3324 in the Carina Nebula, as seen through the James Webb Space Telescope**

In 2022, the James Webb Space Telescope produced spectacular images of the Carina Nebula, which is visible from the southern hemisphere (see Figure 5.3.1b).

protostar
an object in the process of forming a star, before nuclear fusion begins

nuclear fusion
the process of combining two nuclei to make a third nucleus

The gas in a nebula, which is mostly hydrogen, and the dust are attracted to each other by the force of gravity and collide. Once there is sufficient mass, the cloud of gas and dust starts to collapse. The centre of this mass gets hotter and begins to form a **protostar**. Eventually the protostar gets hot enough for the hydrogen nuclei to start fusing together. When two hydrogen nuclei fuse together, they form a helium nucleus in a process known as **nuclear fusion**. This is how a star is born! Energy from the fusion reaction keeps the core of the star hot. The time it takes for a star to form varies, depending on the size and the mass of material forming the star. A star the size and mass of our Sun can take about 50 million years to form.

The Sun's life cycle

The Sun, the star at the centre of our solar system, started out in much the same way as other stars.

1 A protostar formed just over 4.6 billion years ago from a swirling cloud of dust and gas known as a solar nebula.
2 It became hot enough for nuclear fusion reactions to begin. These continue to happen in the Sun today, keeping it hot.
3 When all the hydrogen has transformed into helium in about five billion years' time, the Sun will expand and become a **red giant**.
4 When these nuclear reactions eventually stop, gravity will take over and will force the Sun to contract, leaving behind a **planetary nebula** (see Figure 5.3.2), with a cooler, less bright **white dwarf** at its centre.
5 As it continues to cool, the Sun will eventually become a **black dwarf** star.

red giant
a dying star in the final stages of its life cycle

planetary nebula
a ring-shaped nebula formed by an expanding shell of gas around an ageing star

white dwarf
a dim star that has exhausted its fuel and is at the end of its burning stage

black dwarf
a star at the end of its life cycle that no longer emits light

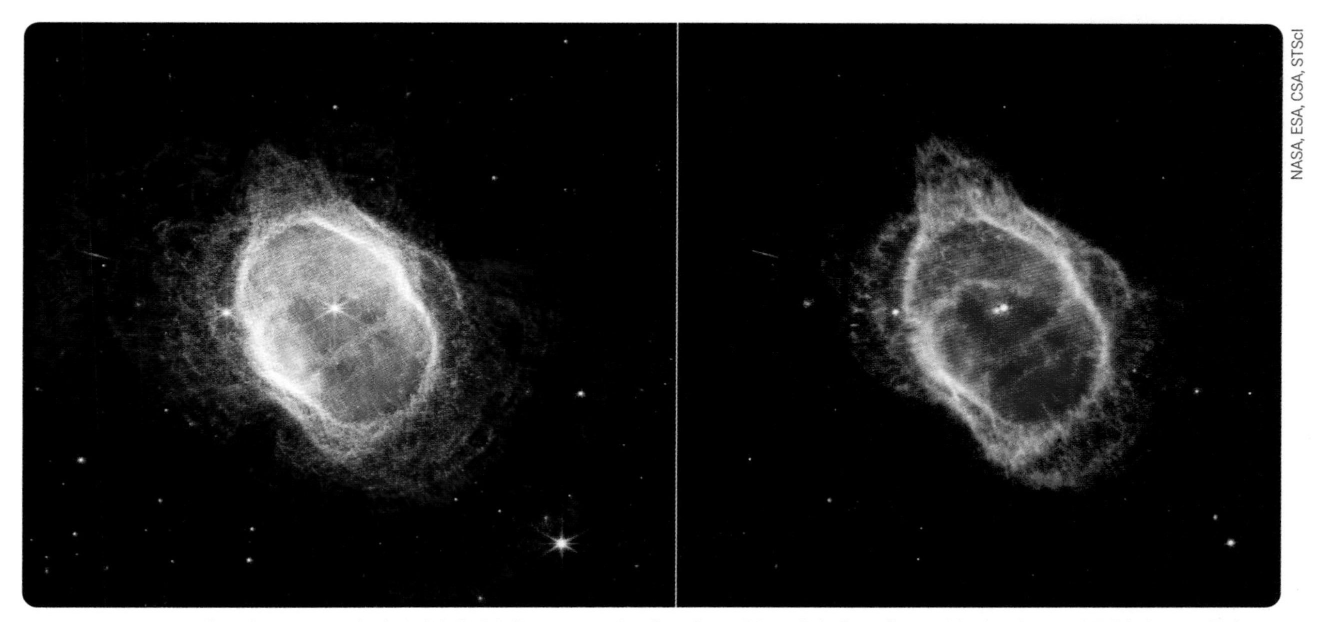

NASA, ESA, CSA, STScI

▲ **FIGURE 5.3.2** The planetary nebula NGC 3132, known as the Southern Ring Nebula, taken with the James Webb Space Telescope

The life cycle of larger stars

Stars that have a much greater mass than the Sun have a different life cycle.

1 Larger stars begin as protostars.
2 They continue to expand and become hotter until they form a **red supergiant**.
3 Eventually the red supergiant will explode in what is called a **supernova** event. These types of stars use up their fuel very quickly, so it can take only 10 million years for it to explode. Supernovas are the largest type of explosions in space.

red supergiant
similar to a red giant but much bigger; the biggest stars in the universe

supernova
a massive explosion caused by a massive star suddenly collapsing; new atomic nuclei are formed

black hole
a cosmic body of extremely intense gravity from which nothing, not even light, can escape

4 When this happens, depending on the original mass of the star, either a **black hole** or a **neutron star** will form.
5 Most neutron stars become **pulsars**, which rotate and emit pulses of radiation at regular intervals.

The life cycle of stars is shown in Figure 5.3.3.

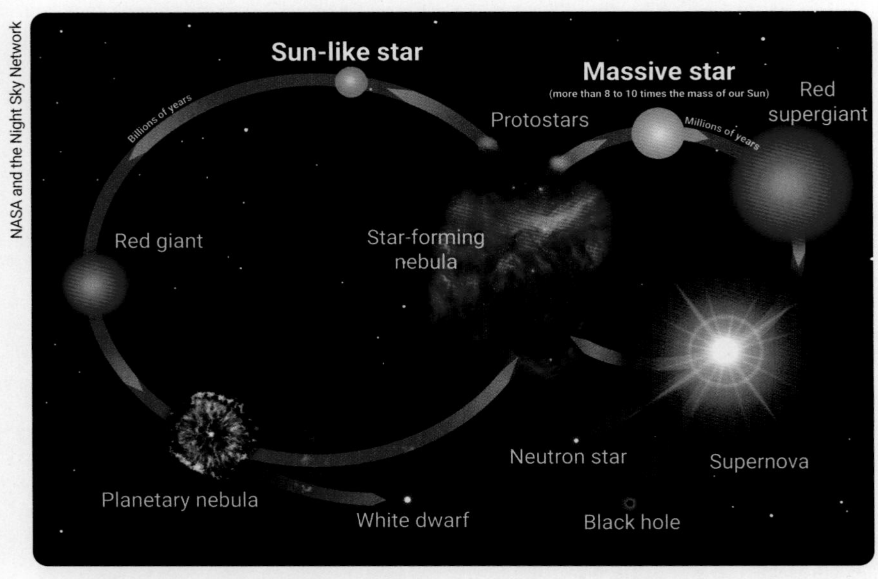

▲ **FIGURE 5.3.3** The life cycle of stars

Black holes

Black holes are areas in space where the pull of gravity is so strong that not even light can escape. Black holes can be as small as an atom or as wide as a solar system, but we cannot see them. Scientists know they are there because they can see the effects that the gravity of a black hole has on the stars, dust and gases around it. Black holes can arise when a massive star dies.

There are three types of black holes: primordial, stellar and supermassive black holes.

Primordial black holes are the smallest. They are thought to have been formed in the early days of the universe.

neutron star
a small, super-dense star that has spent all its fuel and is at the end of its burning life

pulsar
a highly magnetised rotating neutron star that emits pulses of radiation at regular intervals

Stellar black holes are the most common type of black hole. They can have a mass that is up to 20 times greater than the mass of our Sun, but they could fit into a space only 15 km across. Astronomers believe there may be dozens of these types of black holes in our own galaxy. Stellar black holes can form when a giant star dies.

As their name suggests, supermassive black holes are the biggest type of black hole. They are more than one million times the mass of our Sun. Astronomers believe that there is a supermassive black hole in the centre of every large galaxy. These black holes are believed to be the same age as the galaxy in which they exist.

5.3 LEARNING CHECK

1 **Describe** how stars form.
2 Different stars have different life cycles. **Explain** the major factor that determines how long a star will 'live'.
3 An analogy is when we make a comparison between two different things to explain an idea or concept. For example, we can use the analogy of the development stages of flowering plants to explain the life cycle of a star. **Explain** how you can use this analogy to model the life cycle of a star.

5.4 Starlight

BY THE END OF THIS MODULE, YOU WILL BE ABLE TO:

✓ explain how scientists use light spectra and the brightness of a star to identify its composition, movement and distance from Earth.

GET THINKING

When we look at a star at night, the first thing we notice is the light coming from it. As you work through this module, think about why the brightness of stars varies so much. What could make some stars really bright, and others appear quite dull? How can we use the light from stars to help us understand more about them?

Interactive resources
Simulation: Blackbody spectrum
Crossword: Starlight

Stars: an investigation of light

At night, most of the light we see in our sky comes from stars. We can't see this light during the day because the light from the Sun is so bright. We can analyse the light that reaches us from stars to help us to understand how stars move (movement), what they are made of (composition) and how far they are from Earth (distance).

spectrum
the different bands of colour that are visible when white light is refracted through a prism, as seen in rainbows

Analysing a star's light can involve examining the brightness of the light that comes from individual stars. It can also involve analysing the light's **spectrum**, which is all the different wavelengths that make up light. This is done using an advanced spectrometer, which is an instrument that measures the properties of light. The light is passed through a grating to break the light received into its component wavelengths, producing a spectrum on the detector (see Figure 5.4.1).

We can also learn a lot from examining the spectrum of the Sun for comparison.

▲ **FIGURE 5.4.1** A spectrometer

The composition of stars

Scientists can use a spectrometer here on Earth to analyse the composition of stars. This is because every element (such as hydrogen, helium and lithium) produces a unique light spectrum, like a fingerprint. The spectrum of hydrogen is shown in Figure 5.4.2 as both an absorption and emission spectrum to show the difference between these types of spectra.

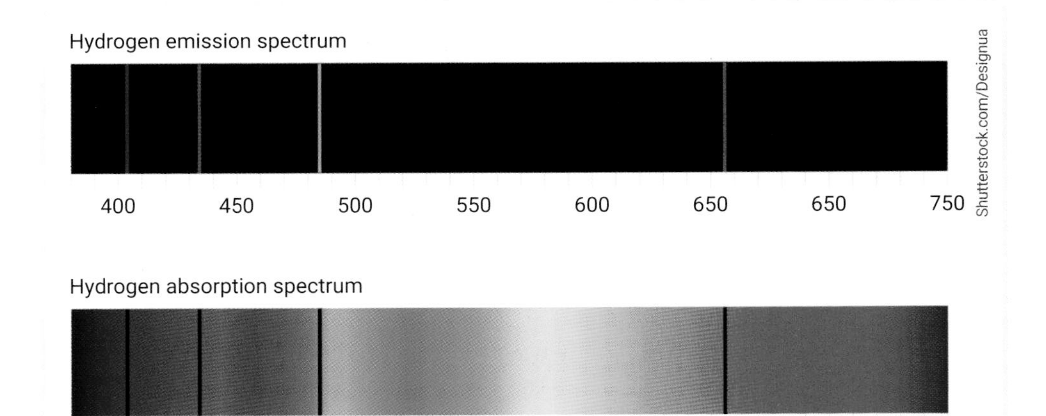

▲ **FIGURE 5.4.2** The absorption and emission spectra of hydrogen

continuous light spectrum
a spectrum in which all radiation 'colours' are emitted

absorption spectrum
the dark lines characteristic of an element or compound in a continuous spectrum

frequency
the number of repeated cycles per unit of time; expressed in hertz (Hz)

redshift
the shift of spectra from local stars moving away from Earth towards the red end of the spectrum

blueshift
the shift of spectra from local stars moving towards Earth towards the blue end of the spectrum

Using a very high-powered telescope, such as the Hubble Space Telescope, emitted light from a distant star is collected to enable scientists to compare the star's light emission spectrum to a known element's **continuous light spectrum** to determine which elements are present in the star.

Movement

Like the planets, stars move around (orbit) objects with very large masses. We know from our own solar system that planets orbit around a star. In contrast, a star usually orbits the centre of the galaxy it occupies. This is because the greatest concentration of mass in a galaxy is usually in the centre. As we read in Module 5.3, astronomers believe this may be due to the presence of a supermassive black hole at the centre of large galaxies.

Light from stars travels a long way before it reaches Earth. Since it takes so long to reach us, the light spectrum our telescopes receive is affected by the movement of stars. The light spectrum we observe depends on which direction a star is moving in relation to us here on Earth. The order of colours we observe in the light remains the same, which is the order of colours you see in a rainbow (red, orange, yellow, green, blue, indigo and violet). However, depending on whether a star is moving towards us or away from us, the wavelength of light we observe can shift slightly.

The wavelength of red light (700 nanometres, or nm) is longer than the wavelength of blue light (450 nm) as can be seen in the **absorption spectrum** of Figure 5.4.2. If we are observing light from a star that is moving away from Earth, the spectrum we see will have a 'shift' towards the red end (lower **frequency**). This is called a **redshift**. But if the star is moving towards Earth, the light we see will have a 'shift' towards the blue end (higher frequency). This is called a **blueshift**. Astronomers also use redshifts and blueshifts to study galaxies (see Figure 5.4.3).

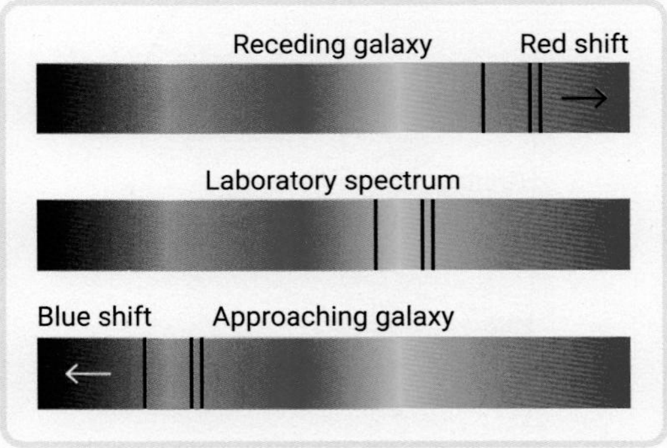

Receding galaxy
Red shift
Laboratory spectrum
Blue shift
Approaching galaxy

▲ **FIGURE 5.4.3** Comparing a redshift and blueshift. The arrows show the redshift of the absorption line spectrum from a light source moving away from Earth and the blueshift of a light source moving towards the Earth. The central spectrum is an unshifted absorption line spectrum from a light source that is not moving.

Measuring the temperature of stars

Max Planck (1858–1947) was a Nobel Prize–winning German theoretical physicist and a close colleague of Albert Einstein. He is often referred to as the 'father of quantum physics'. One of his discoveries determined that all hot objects, including stars, radiate energy. He found the intensity of the radiated energy was related to its wavelength, which we observe as its colour. Therefore, the colour we see when observing a star can be used to give us a direct measure of its surface temperature. The hottest stars are blue, and stars range in temperature downwards through white, yellow and orange, and finally to red stars, which have the lowest surface temperature.

Measuring the distance of stars

There are many ways to measure the distance between stars and Earth. The simplest way is to observe the star's brightness as seen from Earth and then compare this to a star's actual brightness as determined by the star's surface temperature.

Another method astronomers use to measure a star's distance from Earth is by determining the intensity of the light received on Earth, which is measured using a specially designed instrument called a photometer. A photometer measures the intensity of electromagnetic radiation in the range of ultraviolet to infrared, including visible light. This measured light intensity is used to measure distance in the following way.

1 A special scale, called the **magnitude scale**, is used to convert the measured light intensity of a star into an **apparent magnitude** of brightness as viewed from Earth. The apparent magnitude is given a number between −25 (brightest) and +25 (dimmest).
2 Because brightness changes with distance, the apparent magnitude of brightness must be converted mathematically into an **absolute magnitude**. This treats all objects as if they were viewed from the same distance, 10 parsecs from Earth, and is represented by a number between −15 (brightest) and +15 (dimmest).
3 The difference between the apparent magnitude and the absolute magnitude determines a number that can be converted to an actual distance between Earth and the star using a graph (see Figure 5.4.4).

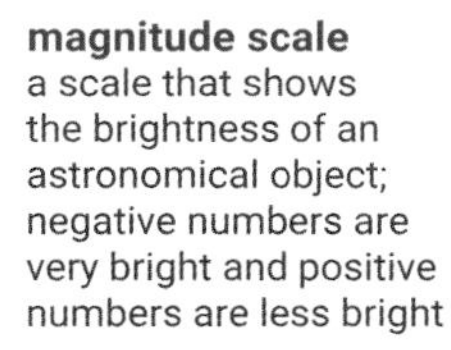

magnitude scale
a scale that shows the brightness of an astronomical object; negative numbers are very bright and positive numbers are less bright

apparent magnitude
the brightness of an astronomical object, as seen from Earth

absolute magnitude
the brightness of an astronomical object, as though all objects were 10 parsecs from Earth

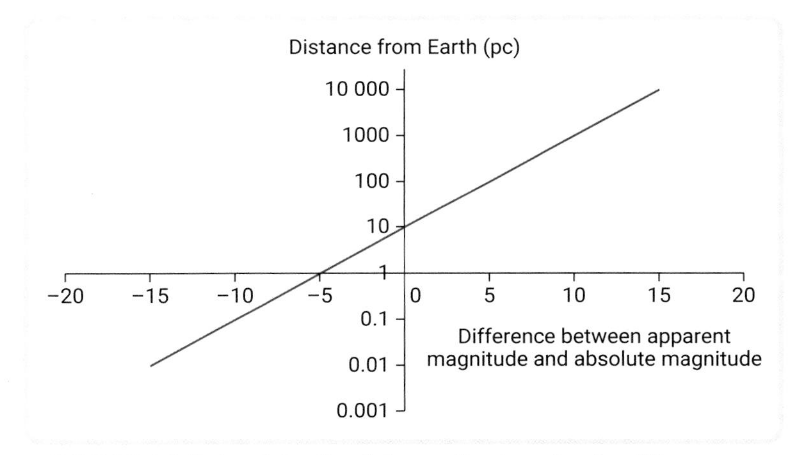

▲ **FIGURE 5.4.4 Determining the distance of a star using the difference between apparent and absolute magnitude**

The second brightest star in the night sky is Canopus. It has an apparent magnitude of −0.62 and an absolute magnitude of −5.5. Using the method described, the difference is $(-0.62 - -5.5) = 4.88$. The graph indicates this would put it about 95 parsecs from Earth, which is a close estimate to the actual value of 95.88 parsecs.

5.4 LEARNING CHECK

1 **State** what Max Planck discovered about hot objects and energy. Then **explain** how this relates to stars.
2 **Outline** why the brightness of stars varies so much when we view them from Earth.
3 **Describe** how we use light to determine the composition of stars.
4 **Explain** how we can use the light from a star to determine how it is moving relative to Earth.
5 Using Figure 5.4.4, **estimate** the distance to Rigel, which has an apparent magnitude of 0.18 and an absolute magnitude of –6.69.

5.5 The Big Bang

BY THE END OF THIS MODULE, YOU WILL BE ABLE TO:

- ✓ describe the Big Bang theory and explain how the Big Bang theory is used to describe the origin and evolution of the universe
- ✓ construct a timeline to demonstrate the changes in the universe from the Big Bang until the present day.

Video activity
Big Bang theory

Interactive resource
Drag and drop: Stages of the Big Bang

GET THINKING

The term 'Big Bang theory' gives the impression that the universe originated from a massive explosion. As you complete this module, think about whether this is the best way to describe how the universe formed. Why do you think it could be misleading? See if you can think of an alternative name for this theory that might help to correct the misconception that the name implies.

The origin of the universe

The best-supported theory by scientists of the origin of the universe centres on an event known as the Big Bang. It is generally accepted that the universe came into existence approximately 13.8 billion years ago. This theory explains that the universe originated from an extremely small point called a **singularity** that underwent a massive expansion in an infinitely small period called **inflation**. The singularity contained everything that now makes up the universe: energy, mass, space and time.

singularity
a point where density and matter are infinite

inflation
the initial rapid expansion of space-time just after the Big Bang

Big Bang theory
the generally accepted theory for the formation of the universe

The term '**Big Bang theory**' suggests that the universe began with an explosion. However, this idea is actually incorrect. The use of the term 'Big Bang' began in 1949 with eminent British astronomer Fred Hoyle, pictured in Figure 5.5.1. Hoyle was one of many astronomers of his time who made fun of the idea that a massive explosion was how time and space began when the universe was formed. Yet the Big Bang theory wasn't Hoyle's theory – he was pushing forward his own, alternative idea, known as the Steady State theory.

Alamy Stock Photo/Liam White

▲ **FIGURE 5.5.1** Astronomer Fred Hoyle

Many scientists since have contributed a lot of evidence to the Big Bang theory that proved Hoyle's Steady State theory was incorrect. You will look at this evidence in more detail in Module 5.6.

The stages of the Big Bang

The Big Bang theory suggests that there were two main stages in the universe's evolution, representing key events that shaped the universe. The first stage was dominated by radiation and lasted for about 300 000 years. The second stage was dominated by matter, which has lasted most of the 13.8 billion years. These main stages are further broken down into key events, which are outlined in Figure 5.5.2. These key events have shaped the universe from the beginning of time until the present.

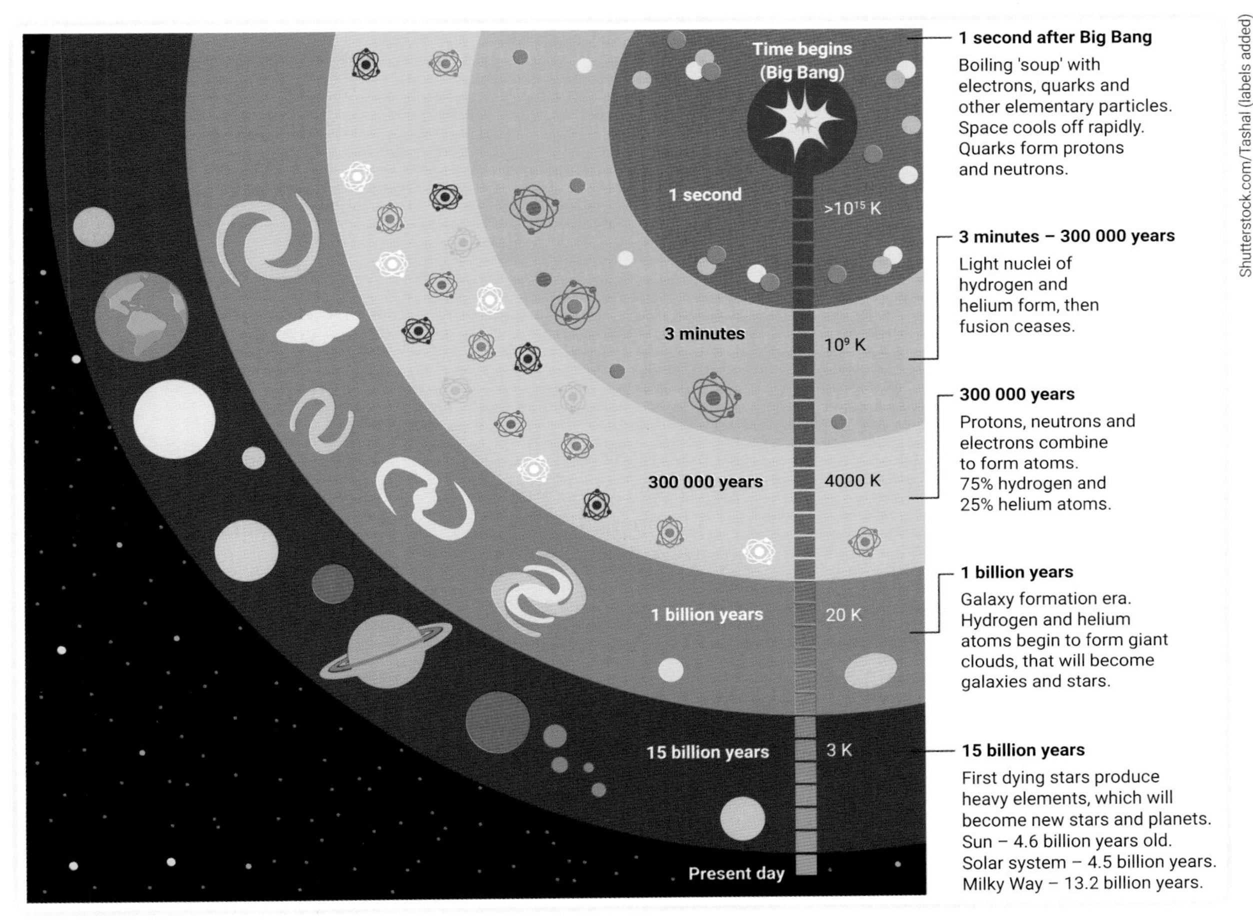

▲ **FIGURE 5.5.2** The timeline of the Big Bang

Within the first second of the Big Bang

The universe came into existence from a singularity that contained everything that was formed during the Big Bang. The singularity was immensely dense and hot, and in a 10 million-trillion-trillion-trillionth part of one second (10^{-43} s) it began to expand. At this point, the universe, time and space came into existence. As the universe continued to expand, the temperature decreased.

quark
a type of elementary particle that is the fundamental component of matter

It took less than 1 millisecond after the Big Bang for the temperature to cool sufficiently to form the fundamental particles, known as **quarks**. These combined to form the subatomic particles of protons and neutrons. The conditions present at this time resulted in fusion, which formed the nuclei of the elements hydrogen, deuterium (an isotope of hydrogen), helium and a small amount of the larger nuclei of lithium.

Three minutes to 300 000 years after the Big Bang

Three minutes later, fusion ceased, leaving a universe composed of a fog mostly made up of 75 per cent hydrogen nuclei and about 25 per cent helium nuclei, with a small number of lithium nuclei. For the next 300 000 years, the temperature of the universe gradually cooled, but conditions were still extremely hot compared with the universe today.

300 000 years after the Big Bang

photon
a form of elementary energy particle

Around 300 000 years after the Big Bang, temperatures dropped to 3000–4000 K (2726–3726°C). This allowed nuclei to capture electrons and form atoms, and **photons** were free to move. When this happened, the universe became transparent to light and entered the stage where matter took over from radiation to dominate the universe.

As the universe continued to expand, the force of gravity started to pull these newly formed atoms together to form clouds of hydrogen and helium gas, which became denser and collapsed under gravity, becoming hot enough to start nuclear fusion. We do not know exactly when and how the first stars and galaxies formed, but there is evidence of star and early galaxy formation around 500 million to a billion years after the Big Bang.

A billion years after the Big Bang

Gravity continued to attract stars to form the first primitive galaxies. Galaxies clustered, and swirling clouds of dust and gas formed more stars as the universe continued to cool and expand. Our Sun formed along with our solar system some 4.6 billion years ago.

absolute zero
the lowest possible temperature (0 K or −273.15°C); the point at which there is no particle movement and no energy

Today, the universe is still cooling and expanding. The temperature of the universe is currently 2.725 K, just above **absolute zero**. The universe is a sphere that has expanded from a point where the distance from Earth to the edge of the visible universe is calculated to be about 46.5 billion light-years, or about 14.26 billion parsecs in any direction.

Studying the history of the universe

Astronomers use different ways to try and understand exactly what happened in the early stages of the universe.

The European Council for Nuclear Research (known as CERN) constructed the Large Hadron Collider to replicate the conditions thought to be present shortly after the Big Bang (see Figure 5.5.3). The Large Hadron Collider is the world's longest particle accelerator. It consists of a 27 km ring of superconducting magnets that keeps charged particles, like protons, moving at near the speed of light, and uses controlled collisions between these particles to investigate matter. Scientists use the Large Hadron Collider to understand more about what makes up matter, and how matter formed in the moments after the Big Bang.

Another way that astronomers study the early moments of the universe is by using powerful telescopes. For example, astronomers use deep imaging observations from the James Webb Space Telescope and Hubble Space Telescope (see Figure 5.5.4) to help answer questions about when the earliest stars formed (see Figure 5.5.5).

▲ **FIGURE 5.5.3** The Large Hadron Collider

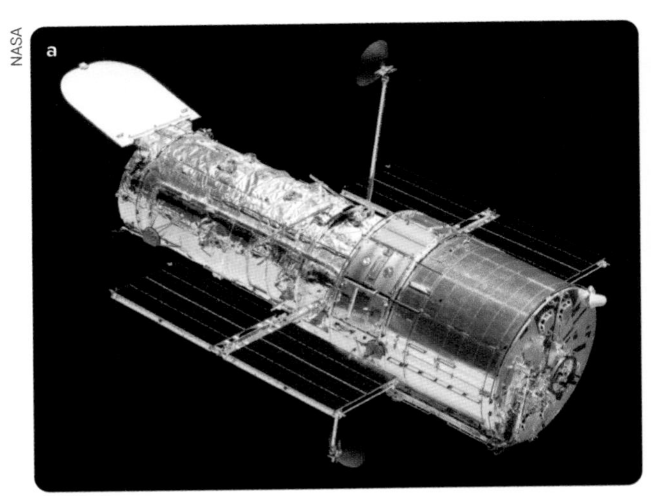

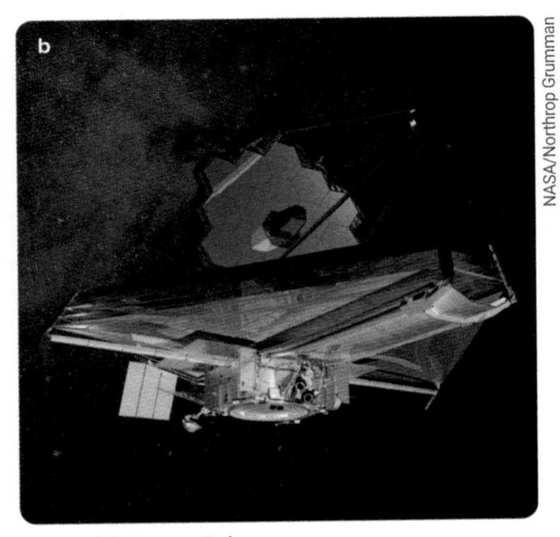

▲ **FIGURE 5.5.4** **(a)** The Hubble Space Telescope and **(b)** the James Webb Space Telescope

ESA/Hubble, M. Kornmesser, and NASA

▲ **FIGURE 5.5.5** An artist's impression of some of the earliest formed stars

5.5 LEARNING CHECK

1 **Describe** how the universe came into existence, according to the Big Bang theory.
2 Create a table to **summarise** the stages of the Big Bang listed in Figure 5.5.2. For each stage, provide information on:
   - when the stage occurred
   - what the temperature was
   - how different components of the universe were forming.
3 **Explain** why the fundamental force of gravity is so important in the formation of the universe.
4 **Describe** what the Large Hadron Collider is and **explain** how it relates to understanding the Big Bang theory.
5 Using what you know from reading the module, **explain** how you would rename the Big Bang theory to better reflect the formation of the universe.
6 **Describe** the differences between the two main stages of the Big Bang.
7 Hydrogen and helium formed early in the evolution of the universe.
   a **Describe** where all the other elements that make up the universe formed.
   b **Explain** how these elements were formed.
8 True or false? The Big Bang theory is an accurate description of how the universe formed. **Justify** your response using the information you have learned in this module.

5.6 Evidence for the Big Bang

BY THE END OF THIS MODULE, YOU WILL BE ABLE TO:

✓ describe the evidence used to support the Big Bang theory and how it is obtained.

GET THINKING

As you complete this module, think about how the work of scientists is validated by others. Think about the significant effects that discoveries have on refining previous ideas and theories.

Video activity
Big Bang evidence

Quiz
Big Bang evidence

Extra science investigation
Expanding universe

What evidence is used to support the Big Bang theory?

There are two key pieces of evidence that support the Big Bang theory:

1. Edwin Hubble's discovery in 1924 showing the relationship between the velocity at which a galaxy is receding from Earth and its distance from Earth
2. Arno Penzias and Robert Wilson's accidental discovery of **cosmic microwave background radiation** (CMBR) in 1964.

cosmic microwave background radiation
remnant radiation from the early stages of the universe in the microwave wavelength; emitted approximately 300 000 years after the Big Bang

An expanding universe

In 1924, American astronomer Edwin Hubble (see Figure 5.6.1a) first measured the distance to the Andromeda Galaxy. He discovered that it was another galaxy, not a part of the Milky Way, and that it was further away than previously thought. In later observations, using a method developed by Henrietta Leavitt (see Figure 5.6.1b), Hubble noticed that almost all the galaxies he observed were moving away from Earth. He concluded that the universe was much larger than described by other scientists, including Albert Einstein and Harlow Shapley. Hubble used redshift data to graph recessional velocity (the rate at which an object moves away from an observer) with distance to determine what was to become **Hubble's law**: the greater the distance to the galaxy, the faster it is moving away from Earth. Georges Lemaître (see Figure 5.6.1c) reviewed the work of Hubble, and in 1927 he concluded that the universe is expanding, a key piece of evidence used to support the Big Bang theory.

Hubble's law
the law that states that the observed velocity of a receding galaxy is proportional to its distance from the observer

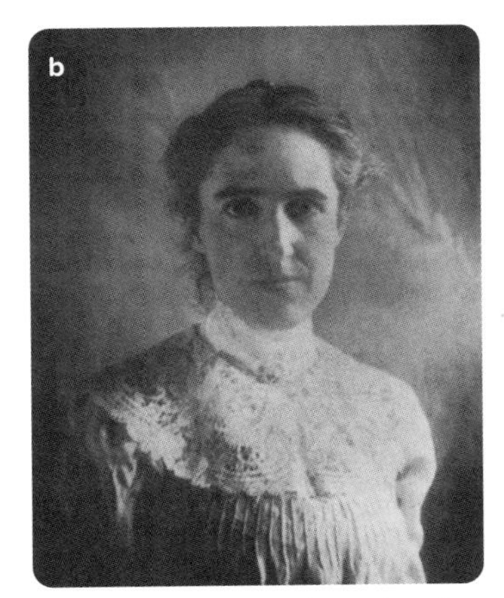

Left to right: Alamy Stock Photo/Science Photo Library; Getty Images/Boston Globe; Alamy Stock Photo/Granger/Historical Picture Archive

▲ **FIGURE 5.6.1** (a) Edwin Hubble, (b) Henrietta Leavitt and (c) Georges Lemaître

Cosmic microwave background radiation

Cosmic microwave background radiation is electromagnetic radiation left over from the formation of the universe. It was first discovered in 1964 when Arno Penzias and Robert Wilson were working with a very sensitive radio telescope for Bell Telephone Laboratories. They were looking for a signal that represented neutral hydrogen. As they were using the receiver of the radio telescope, they kept picking up what they thought was interference. A fortuitous meeting with another group of scientists looking for evidence of cosmic microwave background radiation brought to light that Penzias and Wilson had in fact discovered cosmic microwave background radiation.

Many organisations and agencies, including the European Space Agency and NASA, have since mapped cosmic microwave background radiation. Figure 5.6.2 is an image of cosmic microwave background radiation as seen from Planck, a European Space Agency space-based observatory.

NASA/WMAP Science Team

▲ **FIGURE 5.6.2** The temperature variations in cosmic microwave background radiation, as seen from the Planck satellite

Note: The greenish areas are average temperatures of 2.725 K (−270.4°C), the reddish areas are slightly above average temperatures, and the blue areas are slightly below average temperatures. The variation between the hotter areas and the colder areas is only 0.000 01 K.

anisotropies
small variations in temperature in the universe due to cosmic microwave background radiation that are not consistent in all directions

The data shown above was collected in 2013 and shows the temperature differences in cosmic microwave background radiation, which are used to identify **anisotropies** (small variations in temperature). These temperature fluctuations are thought to be due to fluctuations in the density of matter in the early universe and are used to determine detail about the physical processes that occurred at that time. These temperature fluctuations help scientists identify precursors to the large-scale structures that we know exist, such as filaments, voids and galaxy clusters.

Redshifts and blueshifts

You were introduced to redshifts and blueshifts in Module 5.4, which discussed light being used to determine the direction of movement of stars. A redshift indicates that the light source is moving away from Earth and a blueshift indicates it is moving towards Earth. When we look at distant galaxies, they all show redshifts. This indicates that distant galaxies are moving away from Earth, which is evidence that the universe is expanding, supporting the Big Bang theory.

The deep field image from the James Webb Space Telescope released in July 2022 (see Figure 5.6.3) shows the most distant objects ever identified. The galaxies seen in this image belong to the early universe. The redshifts indicate that the light from the farthest galaxy shown here travelled 13.1 billion years before it reached Webb's mirrors.

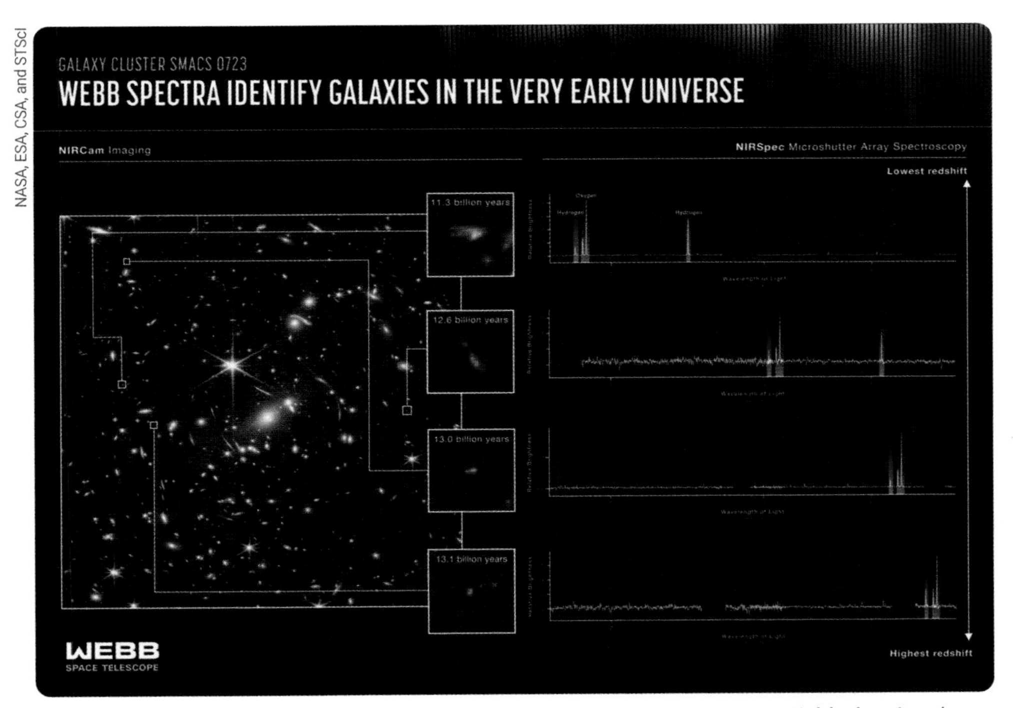

NASA, ESA, CSA, and STScI

▲ **FIGURE 5.6.3** A detail from the James Webb Space Telescope First Deep Field, showing the emission spectra for some very old and distant galaxies

The proportion of matter in the universe

As the universe evolved, there was an abundance of the light elements of hydrogen and helium, which formed in the early stages of the Big Bang. The heavier elements were formed via nuclear fusion within stars and then ejected into space when a star finally died in a supernova event or were cast off as a planetary nebula. When spectral analysis (the analysis of a star's spectrum) is used to determine the composition of older stars, they are found to contain mostly hydrogen and helium. Similarly, when the composition of **main sequence** stars is analysed, they are found to have an abundance of hydrogen (74%) and helium (25%), as well as heavier elements such as lithium and carbon (1%), which fits with the **proportion** predicted by the Big Bang theory.

main sequence
stars that fuse hydrogen to helium

proportion
the relative abundance of elements dispersed throughout the universe

Alternative theories to the Big Bang

There have been several alternative theories about the formation of the universe. The most prominent of these was the Steady State theory. This was a theory to rival the Big Bang, first put forward in 1948 by the British astronomers Sir Hermann Bondi, Thomas Gold and Sir Fred Hoyle. Hoyle was a vocal opponent of the Big Bang theory. In the Steady State theory, the expansion of the universe is explained by the continuous creation of matter and space. The discovery of cosmic microwave background radiation placed the Steady State theory out of favour, as it could not explain this remnant radiation, nor could the theory predict a finite age of the universe. Over time, this and other theories have been disproven by astronomers. This occurs as advances in technology allow astronomers to gather and analyse more evidence, which have so far supported the Big Bang theory.

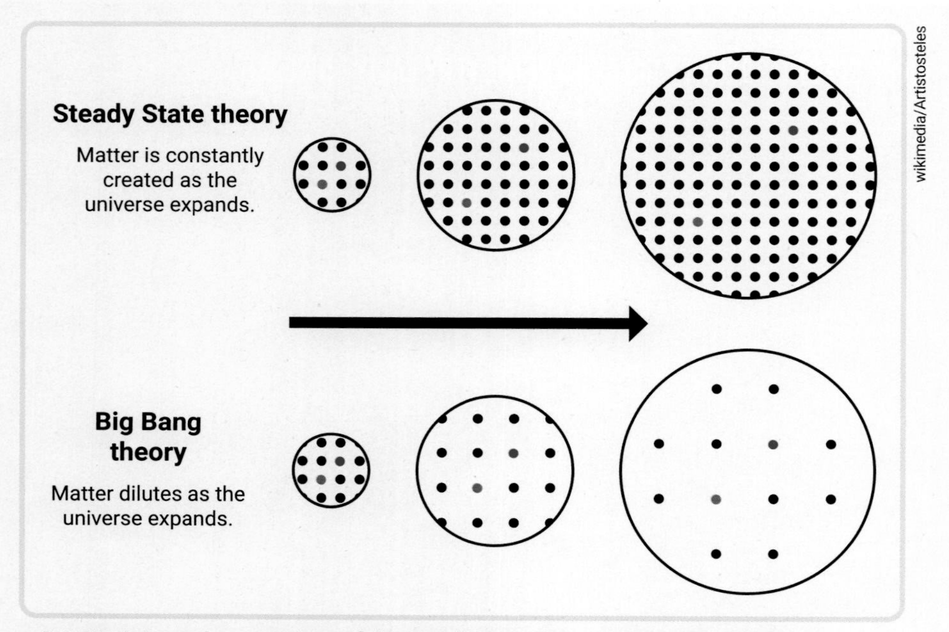

▲ **FIGURE 5.6.4** A comparison of the Steady State theory with the Big Bang theory.

5.6 LEARNING CHECK

1. **List** the evidence that supports the Big Bang theory.
2. **Explain** how cosmic redshifts are used as evidence for the expansion of the universe.
3. **State** the key findings that disproved the Steady State theory.
4. **Explain** why the discovery of cosmic microwave background radiation could be considered accidental.
5. **Justify** why the work of Edwin Hubble was so important to our understanding of the universe and its formation.

5.7 New discoveries

BY THE END OF THIS MODULE, YOU WILL BE ABLE TO:

✓ describe how new discoveries make it possible to refine previous theories

✓ explain the implications of recent discoveries concerning gravitational waves, dark matter and dark energy on our understanding of the universe.

GET THINKING

As you work through this module, think about how new discoveries refine our ideas about previous theories. Why did Einstein's prediction of the existence of gravitational waves take so long to prove? Why is technology so important to the future refinement of theories?

Video activity
Dark matter
Gravitational waves

Recent discoveries

Prior to the work of Hubble, scientists, including Einstein, supported the Steady State theory of the universe. However, our ideas and theories in cosmology have evolved over time due to the work of notable scientists. Australians have made significant contributions to some of the most recent discoveries in cosmology. Their contributions have included work on mapping the universe, **dark energy** and **gravitational waves**.

dark energy
a kind of negative pressure that is thought to be responsible for the accelerating expansion of the universe

gravitational waves
ripples in space-time produced when a massive body is accelerated or otherwise disturbed

Dark energy

In 1994, Australian physicist Professor Brian Schmidt (see Figure 5.7.1) and American astronomer Nicholas Suntzeff formed an international group of astronomers known as the High-Z SN search team. They were searching for Type 1a supernovas, which occur when a white dwarf is destroyed in a violent and bright explosion. It had been discovered earlier that these objects all have about the same brightness and could be used as a reference to accurately determine the distance to distant galaxies. In 1998, the team discovered that these supernovas occurred when the universe was much younger, and they were also found to be fainter than expected. This discovery led them to conclude that the universe is expanding at a faster rate than it did in the past. This was an important discovery, because according to Albert Einstein's famous theory of relatively (published in 1915), the universe's rate of expansion should be decreasing due to the effect of gravity. So, what is causing the expansion of the universe to speed up? One explanation is the existence of dark energy.

Although dark energy can't be seen, calculations imply that dark energy makes up 68 per cent of the universe. It is hypothesised that dark energy is a force that acts opposite to gravity, which is why it may be responsible for accelerating the expansion of the universe. Although dark energy has not been observed, observations of gravitational interactions between cosmic objects suggest that it exists.

Alamy Stock Photo/epa european pressphoto agency b.v.

▲ **FIGURE 5.7.1** Australian Nobel Prize winner Brian Schmidt, who jointly discovered that the expansion of the universe is accelerating

Dark matter

Dark matter does not absorb, reflect or emit light like normal matter, which is why it cannot be seen. Therefore, its existence can only be inferred from the gravitational effect it has on visible matter. Calculations indicate that dark matter makes up 27 per cent of the universe. The existence of dark matter and the descriptive term were first proposed by Dutch astronomer Jan Oort in 1932. An international team of researchers known as the Dark Energy Survey Collaboration have only recently started creating a detailed map of dark matter distribution throughout the universe, with surprising results (see Figure 5.7.2).

N. Jeffrey, Dark Energy Survey

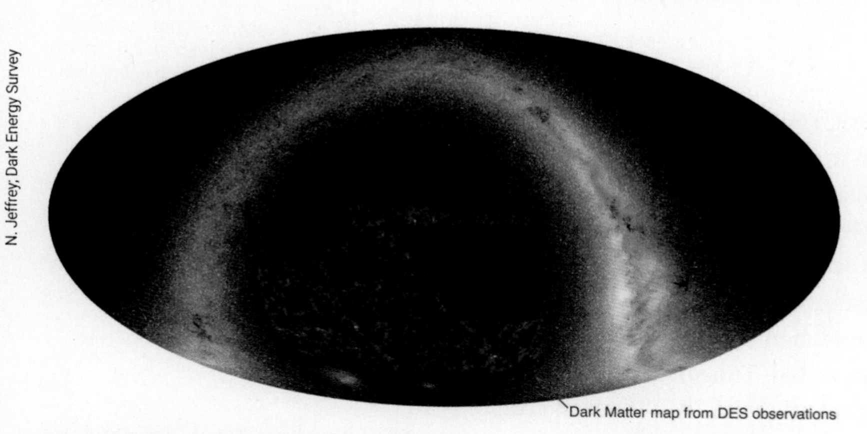

▲ **FIGURE 5.7.2** The largest ever map of dark matter

The research team analysed the distortion of light from 100 million distant galaxies to identify the location of dark matter. The greater concentrations of dark matter created the greatest distortion of light. The black areas of the map are voids: vast areas where nothing appears to exist. Concentrated dark matter is indicated by bright areas called 'haloes'. In among the haloes are brightly shining galaxies. From the map constructed, the team concluded that those galaxies are part of a larger invisible structure.

Before gathering the data to create the map, the Dark Energy Survey Collaboration created a model to predict where they would find dark matter in the universe. They used cosmic microwave background radiation data collected previously by the Planck satellite and assumptions based on Einstein's theory of relatively. Unexpectedly, their results appear not to match the model, throwing up more questions! Was Einstein wrong, or is there more about his theories that need to be investigated when applying them to the scale of the universe? What are these large-scale structures that appear to be related to the distribution of galaxies? Why are there vast areas of voids and what are they?

Gravitational waves

Gravitational waves are disturbances or ripples in **space-time**. Einstein predicted the existence of gravitational waves in his general theory of relativity. He described interactions between massive accelerating objects as the cause of such disturbances. For example, gravitational waves occur when objects such as neutron stars or black holes orbit one another. When more violent or cataclysmic events take place, such as colliding black holes or neutron stars, or supernovas, the most energetic gravitational waves are produced. These interactions disrupt space-time and emit large quantities of energy as a wave. Evidence that such gravitational waves existed was first recorded in 1974. Astronomers discovered a binary pulsar system, where a pulsar orbits around a neutron star almost 21 000 light-years from Earth, as shown in Figure 5.7.3.

space-time
the interconnected nature of the dimensions of space and time

Astronomers observed the binary pulsar system for 8 years. The results showed the system behaved exactly the way Einstein predicted it would. As the stars orbited, they started getting closer to each other at a rate that was only possible if they were emitting gravitational waves. Similar systems have been discovered and observed that behave in the same way.

However, it wasn't until 14 September 2015 that the purpose-built Laser Interferometer Gravitational-Wave Observatory recorded the first extremely faint gravitational waves, almost 100 years after Einstein predicted their existence.

With the recent launch of the James Webb Space Telescope, astronomers are discovering more insights and posing new questions about what happened in the first seconds after the Big Bang.

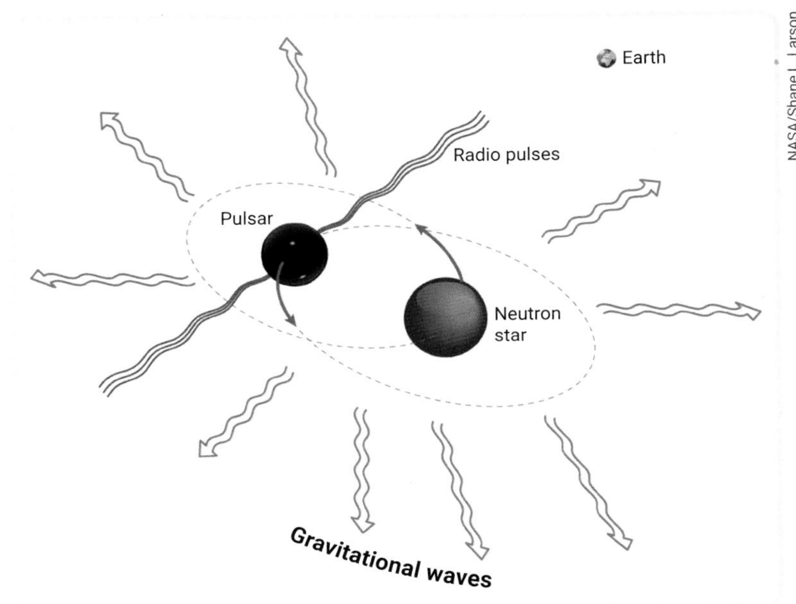

▲ **FIGURE 5.7.3** The Hulse–Taylor binary system

5.7 LEARNING CHECK

1. **Describe** what is significant about Type 1a supernovas.
2. **Outline** the contribution Professor Brian Schmidt and his research partners made to our knowledge of the universe.
3. **Explain** why the term 'dark' is used for naming dark energy and dark matter.
4. **Explain** why it took almost 100 years to prove the existence of the gravitational waves first predicted by Einstein.
5. **Evaluate** the impact of new discoveries such as dark matter, dark energy and gravitational waves. In what ways have they made cosmologists rethink existing theories and models of the evolution of the universe?

FIRST NATIONS SCIENCE CONTEXTS

5.8 First Nations Australians' perspectives on the origin of the universe

IN THIS MODULE YOU WILL:

✓ explore First Nations Australians' cultural narratives about the origin of the universe.

What is ontology?

Ontology is the branch of philosophy that explores complex concepts such as existence, reality, becoming and being. Many cultures have different beliefs and ontologies. Creation narratives from many cultures explain the origin of the universe, Earth and of life. Creation narratives hold important information about culture, beliefs, morals, values and living things.

Observations of celestial bodies and astronomical movements have provided the framework for many First Nations Australians' understanding of the origin of the universe. The creation narratives about the beginning of the universe held in First Nations Australians' cultural narratives contain important lessons, information and scientific understandings about the formation of the land and origin of the universe. The structure of these narratives as engaging oral traditions help to make the information memorable and reinforce the lessons and understanding on retelling.

There are more than 200 distinct First Nations Australian cultural groups. Narratives about the origin of the universe differ between groups; however, there are also similarities. For example, the Celestial Emu (or the Emu in the Sky) shown in Figure 5.8.1, has featured in the cultural narratives of many First Nations Peoples across Australia.

▲ **FIGURE 5.8.1** The Celestial Emu is a feature of cultural narratives of many First Nations Peoples across Australia.

9780170463041

First Nations Australians' creation narratives

The cultural narratives described below summarise some well-known oral traditions about the origin of Earth and the universe shared by First Nations Australians.

Wolfe Creek Crater

Wolfe Creek Crater is the site of a meteorite impact in the Great Sandy Desert in Western Australia (see Figure 5.8.2). It is the second biggest meteor impact site in the world. The site is known as Kandimalal in the Jaru language. Kandimalal means 'no potatoes', as the local people noticed that bush potatoes did not grow in the area around the crater. Jaru Elder, Jack Jugarie, shared the story of how the crater formed when the Moon and evening star passed very close to each other. He explained how the evening star became so hot that it fell. This caused an enormous explosion, which frightened people. But when people visited the crater site, they realised it was where the evening star had fallen to Earth. This knowledge was told in Jaru culture before the British colonisation of Australia.

iStock.com/nordseegold

▲ **FIGURE 5.8.2** Wolfe Creek Crater, Western Australia

The Rainbow Serpent

Many Australians have heard of the Rainbow Serpent. However, many people aren't aware that each First Nations Australian cultural group has their own specific Dreaming that explains the beginning of time. Many of these Dreamings feature giant serpents, but each group has a unique name and narrative.

Many creation narratives across much of Northern Australia share similarities. One aspect that they have in common is the tracing of the ancestral serpent's tracks back to the Wolfe Creek Crater. Many First Nations Australian cultural groups identify Wolfe Creek Crater as the sacred site where this great serpent entered Earth. Their narratives describe how it came from the sky and began criss-crossing the landscape, performing great feats such as creating landforms and enshrining laws for social and cultural order. The vast tracks left behind connect many First Nations Peoples of Australia, despite the immense geographical distances between them.

These cultural narratives are complex, and some aspects of them are known only to certain people within a respective culture. The Jingili cultural narrative depicted in Figure 5.8.3 contains information that can be shared with non-Jingili people.

In this painting, titled *Walamunda*, Jingili Elder and artist Uncle Bulugarri Sambono depicts the significant Dreaming for the ancestral serpent Walidjabudi. The larger circle in the centre represents Walamunda, a place of significance to the Jingili people, north-west of Elliott in the Northern Territory. The bright red colours signify the importance of this Country for ceremonies and lore. Uncle Bulugarri Sambono explains 'the Walidjabudi came from the sky in the west (shown in the left-hand side of the painting) and did sacred things on the way. He then came to Walamunda, where he performed sacred ceremonies and lores and continued on his way east, towards Queensland. No one knows where he went or where he settled'.

Source: Uncle Bulugarri Sambono

▲ **FIGURE 5.8.3** *Walamunda* by Jingili Elder and artist, Uncle Bulugarri Sambono

☆ ACTIVITY

Consider the two creation narratives that have been shared.

1. Why did potatoes not grow in the area around Kandimalal? List some potential reasons using your knowledge of the physical requirements for plant growth.
2. Draw a table with two columns. In the first column, list the main features of each narrative. In the second column, describe the scientific knowledge each feature communicates.
3. Connect with a local First Nations Australian community group to ask about cultural narratives from your area that can be shared. What do these narratives tell you about the local environment?

5.9 The Square Kilometre Array

SCIENCE AS A HUMAN ENDEAVOUR

BY THE END OF THIS MODULE, YOU WILL BE ABLE TO:

✓ explain how the development of technologies such as fast computers made the analysis of radio astronomy signals possible.

Radio telescopes

Optical telescopes use visible light to detect the objects we wish to investigate. Radio telescopes, as the name suggests, detect radio signals. Radio telescopes are more versatile than optical telescopes because most cosmic objects, including invisible gases, emit radio signals.

Video activity
SKA

Planned to be the world's largest radio telescope, the Square Kilometre Array (SKA), is an international effort that is still under construction. It will consist of two radio telescope arrays: one in Australia and the other in South Africa, and will eventually have a total collecting area of about 1 million square metres. There will be 130 000 radio antennae called 'dipoles' spread across outback Western Australia in Murchison, and 200 larger radio dishes in the Karoo region in South Africa (see Figure 5.9.1). The two telescopes will observe the sky by using different radio frequencies.

In Australia, the SKA's low-frequency antennae are designed to give astronomers greater insight into the formation and evolution of the first stars and galaxies after the Big Bang. The site at Murchison was chosen because of its isolation from sources of radio waves and vibrations that could interfere with the collection of data.

◀ **FIGURE 5.9.1** Artist impressions of the components of the SKA: (**a**) an array of antennae in Western Australia; (**b**) the dishes in South Africa, each 15 metres in diameter

Supercomputers

Advanced computer technology has enabled this project to go ahead. The SKA will use two special supercomputers, one of which will be housed in Perth, to analyse the huge amount of data that the array will collect every second. The ability of supercomputers to link the widely dispersed radio antenna and dishes will make the SKA the largest, most powerful and sensitive radio telescope ever built.

5.9 LEARNING CHECK

1. **Describe** why supercomputers are so important to the SKA.
2. **State** why the site for the Australian array is in outback Western Australia.
3. **Describe** why the array is referred to as the 'Square Kilometre Array'.
4. **Explain** why radio telescopes are more versatile than optical telescopes.

SCIENCE INVESTIGATIONS

5.10 Validating the work of scientists

SCIENCE SKILLS IN FOCUS

IN THIS MODULE, YOU WILL FOCUS ON LEARNING AND IMPROVING THESE SKILLS:

- recognising how the work of scientists is validated by other scientists
- evaluating conclusions based on the evidence collected
- evaluating scientific arguments based on ethical and global issues
- effectively communicating scientific concepts and ideas by applying evidence-based arguments.

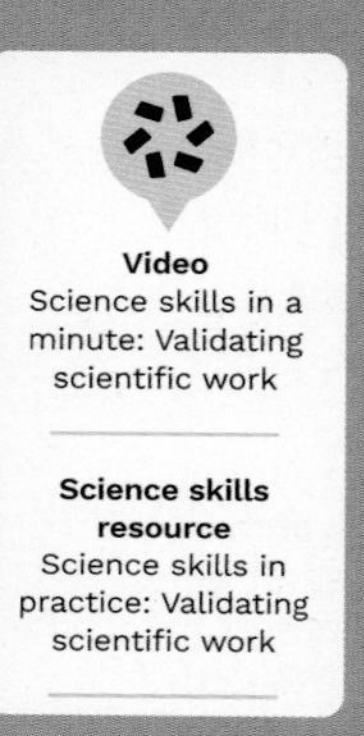

Video
Science skills in a minute: Validating scientific work
Science skills resource
Science skills in practice: Validating scientific work

▶ What is validity?

In the past, when scientists completed their research, they would usually communicate their findings to other scientists by publishing an article in a scientific journal, giving a presentation at a conference or writing a book. These days, scientists also use the Internet to spread information more quickly and easily to large audiences. Scientists publish their research so that others can critique what they have done and check whether their conclusions are accurate. Having their scientific methodology repeated by others increases the reliability and validity of the researcher's conclusions.

It is very important for science to be validated so that we can develop accurate theories, like the theory of gravity.

Determining validity is achieved by:

1. using the same type of apparatus
2. following the same procedure
3. utilising the same measuring devices
4. using the same variables (controlled, independent and dependent)
5. statistically testing the new results to compare them with the original results.

▶ Statistical testing

Statistical testing of the results considers the effect of errors that may happen outside the control of the researcher. Random errors (e.g. environmental conditions) can affect the precision of the results. The more the research is replicated, and the more similar statistical results that are produced, the more accurate the science. This is how scientists develop theories.

▶ The impact of technology on scientific theories

As technology improves, historic research should also be replicated. For example, there is now a different value for the Hubble constant, as recent investigations have found that the rate of the expansion of the universe is accelerating.

9780170463041

INVESTIGATION 1: VALIDATING THE HUBBLE CONSTANT

BACKGROUND

In determining the relationship between the recessional velocity of distant galaxies from Earth and the distance to these galaxies from Earth, Hubble constructed a graph. The value of the gradient of that graph is known as the Hubble constant. This is a measure of the rate at which the universe is expanding.

AIM

To validate the Hubble constant

MATERIALS

- ☑ graph paper
- ☑ calculator
- ☑ ruler

METHOD

1. Construct a graph with 'Recessional velocity' on the *y*-axis and 'Distance from Earth' on the *x*-axis.
2. Plot the values provided in Table 5.10.1 on your graph.
3. Draw a line of best fit that goes through the origin (0, 0).
4. Calculate the gradient of your line of best fit by dividing the rise by the run $(y_2 - y_1) / (x_2 - x_1)$.

▼ **TABLE 5.10.1** Hubble's original data concerning the distance and recessional velocity of galaxies

Galaxy	Distance from Earth in megaparsecs (Mpc)	Recessional velocity (km/s)
Small Magellanic Cloud	0.032	170
Large Magellanic Cloud	0.034	290
6822	0.214	−130
598	0.263	−70
221	0.275	−185
224	0.275	−220
5457	0.45	200
4736	0.5	290
5194	0.5	270
4449	0.63	200
4214	0.8	300
3031	0.9	−30
3627	0.9	650
4826	0.9	150
5236	0.9	500
1068	1.0	920
5055	1.1	450
7331	1.1	500
4258	1.4	500
4251	1.7	960
4382	2.0	500
4472	2.0	850
4486	2.0	800
4649	2.0	1090

EVALUATION

1 Compare your calculated value for the gradient of the Hubble constant with that of Hubble's original value (500 km/s per Mpc).

2 Do you think that the value you calculated is reasonable? Justify your answer.

3 Explain whether the value you calculated would add validity to Hubble's conclusion that there is a relationship between the recessional velocity of galaxies and the distance they are from Earth.

CONCLUSION

Hubble calculated the original value of the Hubble constant from limited observations in 1929. The most recent calculation of the Hubble constant is approximately 69.8 km/s per Mpc. Currently, there are three main ways to measure the Hubble constant:

- ☑ using astronomical measurements to look at objects nearby and see how fast they are moving
- ☑ using gravitational waves from collisions of black holes or neutron stars
- ☑ measuring the light left over from the Big Bang, known as cosmic microwave background radiation.

Explain why there is such variation between Hubble's original calculation and the more recent value.

INVESTIGATION 2: INVESTIGATING SPECTRA

Before starting this investigation, refresh your understanding of the differences and similarities between an absorption spectrum, an emission spectrum and a continuous spectrum, and how these are used to determine the composition of stars and galaxies. There are examples of different spectra in Module 5.4. Completing this investigation will give you a better understanding of the different types of spectra.

AIM

To investigate light spectra

MATERIALS

- ☑ safety glasses and a lab coat/apron
- ☑ spectroscope
- ☑ incandescent 55 W light globe
- ☑ 15 W energy efficient light globe
- ☑ overhead fluorescent light
- ☑ computer or tablet screen
- ☑ hydrogen discharge tube (if available)
- ☑ helium discharge tube (if available)
- ☑ high-voltage power source for discharge tubes
- ☑ Bunsen burner
- ☑ platinum wire loop
- ☑ test tubes
- ☑ test-tube rack
- ☑ 50 mL (0.5 M) solutions of copper(I) chloride and sodium chloride
- ☑ 50 mL (1.0 M) hydrochloric acid

METHOD

1. Construct a table to record the item being observed and the type of spectra, as well as any special observations about the spectra (e.g. main colours, black line).
2. Use the spectroscope to observe indirect sunlight, recording your observations. You may need to rotate the spectroscope or change the position of your eye to see the full spectrum.
3. Repeat step 2 with an overhead fluorescent light, an incandescent light globe, an energy-efficient globe, a computer screen and two discharge tubes.
4. Half-fill one test tube with hydrochloric acid, add 15 drops of copper(I) chloride to another and add 15 drops of sodium chloride to another and place them in the rack.
5. Clean the platinum wire by dipping it in the hydrochloric acid and placing it in the hottest part of the Bunsen burner flame.
6. Dip the wire in the sodium chloride solution, then conduct a flame test by placing it in the hottest part of the Bunsen flame and observing the colour of the flame. Repeat step 5 to clean the wire.

Ensure that you are wearing the safety glasses and lab coat/apron when using the chemical solutions. Be extremely careful when using the high-voltage power source and discharge tubes.

7. Dip the wire in the copper chloride(I) solution, place it in the hottest part of the Bunsen burner flame and observe the colour of the flame.
8. Have someone repeat steps 5–7 while you are looking at the wire with the spectroscope while it is in the flame. You may need to do this several times.

EVALUATION

1. Compare the spectra of the two discharge tubes with the spectra of light from the Sun. **Describe** any similarities or differences observed and **explain** why they occurred.
2. Explain how analysing the spectra from the Sun would be beneficial in analysing the spectra from other stars.
3. Compare the spectra observed from the fluorescent light source, the incandescent and energy-efficient globes and the screen. Describe any similarities or differences observed.
4. Compare the colour of the two chloride solutions used in the flame tests. Why do you think there was a difference in colour?
5. Compare the spectra of the two chloride solutions. If there was a difference, explain why you think this was the case.

CONCLUSION

State which of the observations resulted in an emission spectrum and which resulted in an absorption spectrum. What differences do you think are responsible for the production of an absorption or an emission spectrum?

REMEMBERING

1 **Define** 'nebula'.

2 **Define** 'nuclear fusion'.

3 **Describe** what is meant by the terms 'redshift' and 'blueshift'.

4 **Name** three units used to measure distances in space.

5 **Describe** the difference between the two spectra shown below.

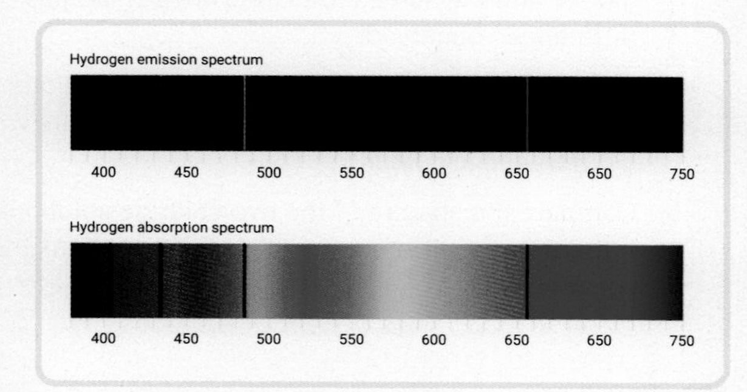

UNDERSTANDING

6 **Explain** the Big Bang theory.

7 **Describe** what is meant by the term 'protostar'.

8 **Describe** cosmic microwave background radiation.

9 **Explain** how cosmic microwave background radiation formed.

10 **List** three discoveries from the last 30 years that are currently challenging our ideas of the universe.

APPLYING

11 **Describe** what astronomers can learn about stars from analysing spectra.

12 **Compare** the spectrum of the Sun to a spectrum from a distant star in another galaxy. How are they the same? How are they different?

13 **Describe** the scientific phenomenon that features in Uncle Bulugarri Sambono's painting *Walamunda* (Figure 5.8.3) and what it has in common with other Rainbow Serpent cultural narratives.

14 **Calculate** the distance in kilometres to Wolf 359, a nearby star that is 7.9 light-years from Earth.

15 **Describe** the difference between absolute magnitude and apparent magnitude.

16 **Explain** how we know that dark matter exists in our universe.

17 **Explain** how spectral analysis is used to determine the distance to a star.

EVALUATING

18 **Explain** how Hubble determined that the universe was expanding.

19 **Describe** the main benefit of using radio astronomy rather than optical astronomy when investigating regions of the universe.

20 **Explain** why Local Group stars can produce both stellar redshifts and blueshifts, yet very distant stars will only produce cosmic redshifts.

21 a **Describe** the significance of Professor Brian Schmidt's contribution to our understanding of the universe.

b **Explain** the reasons why his findings were more accurate than those of Edwin Hubble.

22 **Explain** the main purpose of the SKA.

23 **Explain** why the temperature in the early stages of the Big Bang prevented the formation of atoms.

24 **Explain** the role of gravity in the formation of stars and galaxies.

25 **Predict** what would have happened if the Sun had been a much bigger star.

26 **Explain** why it took almost a century to prove that gravitational waves existed from when they were first proposed by Einstein.

CREATING

27 **Construct** a timeline that shows the key stages of the evolution of the universe.

BIG SCIENCE CHALLENGE PROJECT

1 Connect what you've learned

In Chapter 5 you learned a lot about cosmology and astronomy; one aspect of this was brightness, or more correctly, apparent magnitude. However, when you go outside to look at the night sky, it is becoming increasingly difficult to see the stars because of the amount of light pollution. To get a really good view of the night sky, you need to get well away from metropolitan areas. Globe at Night is an international citizen science campaign to raise public awareness of the impact of light pollution. You can join this campaign to measure your night sky brightness observations and submit your findings.

2 Check your thinking

Before you start the task, visit the Globe at Night website to get information about the campaign, such as campaign dates and which constellations you will focus on. Ensure that you understand what you are required to do before you start the project.

You will then need to install a night sky app on your mobile phone or tablet. There are many free versions available. Practise using the night sky app once you have installed it. It is best to use the app at least an hour after sunset (usually after 8.00 p.m.–10.00 p.m.). You can use the app to find out a lot about the night sky that you may find helps consolidate your understanding of the concepts in this chapter.

3 Get into action

You will need to spend some time engaging with the campaign. Plan how you will do this. During the project, keep a diary to make some notes about:

- the purpose of the campaign
- an explanation of the astronomy/science behind the campaign
- an explanation of what the campaign requires you to do
- what you are finding out as you participate in the campaign.

ESO/B. Tafreshi (twanight.org)

Looking through the centre of our galaxy, the Milky Way, as viewed from La Silla Observatory in Chile. This is just one of the estimated 200 billion galaxies in the observable universe.

4 Communicate

When you have finished your part of the campaign, construct a presentation that contains at least four slides covering each of the points outlined above. Use lots of images and ensure you acknowledge the source (e.g. website, photographer).

The purpose of your presentation is to share what you have done to encourage others to take part in this, or other, citizen science campaigns.

6 Climate change

BIG SCIENCE CHALLENGE

Dreamstime.com/Alexandra Barbu

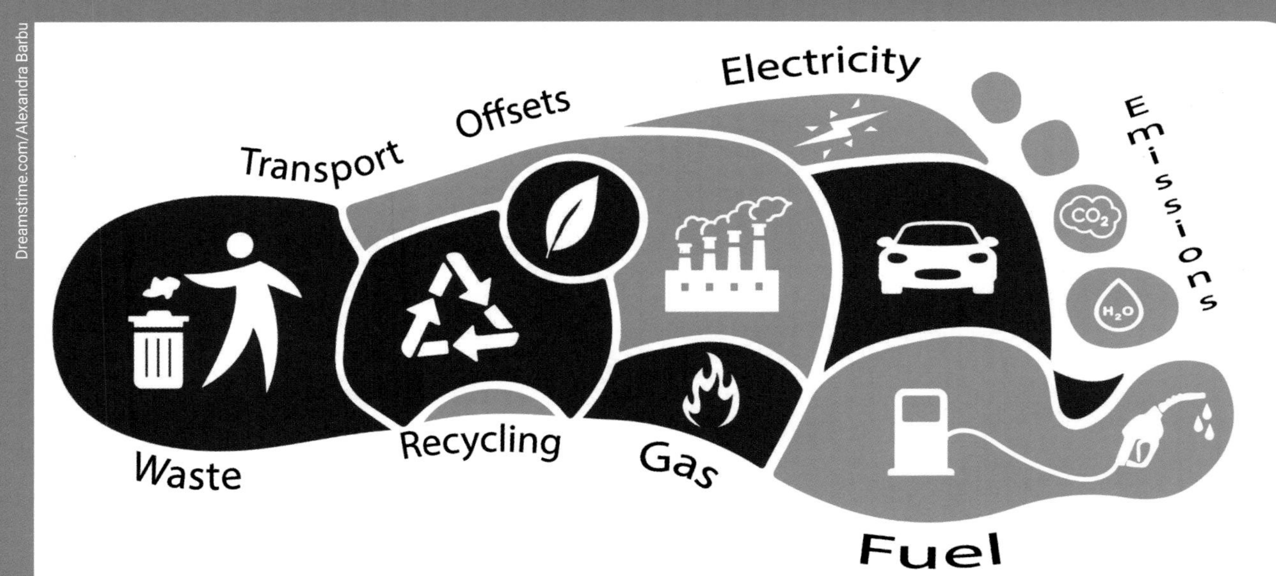

FIGURE 6.0.1 What is your carbon footprint?

We all contribute to climate change in many ways, whether through our lifestyle, behaviour or choices. We all need to do our bit to reduce the effects of climate change because the alternative is not good for our planet – or us. Even if everyone does just a little bit, the collective change can be significant.

▶ **What types of things could you and your family do every day to help reduce the effects of climate change?**

SCIENCE CHALLENGE ACCEPTED!

At the end of this chapter, you can complete the Big Science Challenge Project #6. You can apply the knowledge and skills you learn in this chapter to complete the project.

Assessments

- Prior knowledge quiz
- Chapter review questions
- End-of-chapter test
- Portfolio assessment task: Project

Videos

- Science skills in a minute: Trends and patterns in data **(6.10)**
- Video activities: Photosynthesis **(6.1)**; The carbon cycle **(6.2)**; What makes up the electromagnetic spectrum? **(6.3)**; The greenhouse effect **(6.3)**; Causes of climate change **(6.5)**; Climate cycles **(6.6)**; Natural climate change **(6.6)**; Ocean acidification **(6.7)**; Carbon farming **(6.8)**; Waste as a resource **(6.8)**; Climate modelling **(6.9)**

Science skills resources

- Science skills in practice: Trends and patterns in data **(6.10)**
- Extra science investigation: How the greenhouse effect impacts air and soil temperature **(6.3)**

Interactive resources

- Drag and drop: Evidence of climate change **(6.6)**
- Label: Interactions between Earth's spheres **(6.2)**; Climate change and human health **(6.7)**
- Crossword: Earth's spheres **(6.1)**
- Simulation: The greenhouse effect **(6.3)**

Nelson MindTap

To access these resources and many more, visit: **cengage.com.au/nelsonmindtap**

6.1 The four-sphere Earth system

BY THE END OF THIS MODULE, YOU WILL BE ABLE TO:

✓ describe Earth's atmosphere, biosphere, hydrosphere and geosphere.

GET THINKING

Earth is a complex planet. Scientists used observations to create a model that describes Earth as having four distinct spheres that together make the Earth system. As you read through this module, think of examples of different things around you that are part of each of the four spheres. Can you identify anything that is difficult to place into a single sphere?

climate
the average weather conditions in a particular region over an extended period

Earth system
Earth as a complex whole, made up of the four interacting spheres: atmosphere, biosphere, hydrosphere and geosphere

atmosphere
the gaseous layer surrounding Earth that is retained by Earth's gravity

biosphere
all aspects of Earth and the atmosphere that support life, including all living things

hydrosphere
the part of Earth containing all forms of water

geosphere
the rocks, minerals and landforms of the surface and interior of Earth

greenhouse gases
gases in the atmosphere that can trap heat and affect global surface temperatures and other aspects of the climate

photosynthetic
able to photosynthesise, converting carbon dioxide and water into sugar and oxygen

cyanobacteria
colonies of blue-green algae capable of photosynthesis

continental ice sheet
an extensive sheet of permanent ice covering a large area of the land surface

The Earth system

To understand our planet's **climate**, we first need to understand the nature of the **Earth system**. We describe Earth as being made up of four main 'spheres': **atmosphere**, **biosphere**, **hydrosphere** and **geosphere**. The four spheres interact continuously and no single sphere acts in isolation without having an effect on the others.

Atmosphere

The atmosphere is composed of the gases that make up the air surrounding the planet. The force of Earth's gravity keeps the atmosphere in place. Chemical analysis shows that the atmosphere is made up of 78 per cent nitrogen and 21 per cent oxygen, with the remaining 1 per cent made up of trace gases such as argon and a group collectively known as the **greenhouse gases**.

The atmosphere of 3 billion years ago was very different from the one we breathe today. Evidence from the fossil record and ancient rocks indicates that oxygen started to accumulate in the atmosphere about 2.4 billion years ago. **Photosynthetic** microbes called **cyanobacteria**, some types of which still exist today, were responsible for this increase in oxygen (see Figure 6.1.1). They used the Sun's energy to convert carbon dioxide and water into sugars in a process called photosynthesis.

We know that the composition of the atmosphere changed significantly in its early history, but for the past 2 million years it has been relatively stable. Evidence for the composition changing over time is seen in the chemistry of rock and mineral deposits, such as the iron ore deposits in northern Western Australia, and ancient bubbles of atmospheric gases that have been trapped in ice. Ice cores from Antarctica's **continental ice sheet** provide a record of the atmospheric composition going back as far as 800 000 years. Gases are frozen in time as bubbles of air trapped within the ice. The ice cores allow scientists to examine the composition of the atmosphere before human activities and compare this with the atmosphere today. Ice cores can also identify major events such as volcanic eruptions (see Figure 6.1.1).

▲ **FIGURE 6.1.1** (a) An Antarctic ice core containing trapped air and a dark line indicating a major volcanic eruption that deposited a layer of material on the ice. (b) A close-up of an ice core with trapped bubbles of atmospheric gas.

Heidi Roop/National Science Foundation (NSF)

NASA's Goddard Space Flight Center/Ludovic Brucker

Increases in carbon dioxide concentration coincides with the Industrial Revolution. Societies became more urbanised and industrial, moving away from agriculturally based economies. Inventions such as the steam engine, the development of rail networks and electricity generation led to an increase in the use of coal. Since the 1850s, human activities around the world have continued to increase the amount of carbon dioxide and other greenhouse gases in the atmosphere.

Video activity
Photosynthesis

Interactive resource
Crossword: Earth's spheres

Biosphere

climate change
a change in global or regional climate patterns

The biosphere contains all of Earth's living organisms – an amazing array of species. The different zones of the biosphere contain suitable habitats and ecosystems for the survival of different animals, plants and other species. Some species can live almost anywhere, but others are very restricted to specific locations. For example, Galápagos land iguanas are found on only a few small islands in the Pacific Ocean 1000 km from South America and lay their eggs in burrows in moist sand.

While new species are still being discovered regularly around the world, many others are facing extinction. We know from the fossil record that mass extinctions have occurred in the past. Several of these, including the extinction at the end of the Ordovician period that occurred 445 million years ago, are thought to be climate related. In the Ordovician extinction, 60–70 per cent of all species died out due to a short but intense ice age.

In 2019, scientists declared the Bramble Cay melomys to be the world's first mammal to become extinct due to **climate change** (see Figure 6.1.2). It lived on a single island in the Great Barrier Reef that was subjected to rising sea levels.

naturepl.com/Bruce Thomson

▲ **FIGURE 6.1.2** The Bramble Cay melomys is the world's first mammal to become extinct due to climate change.

Today, scientists have identified more than 700 species of animals and plants that are under threat from climate change. There are many endangered species in Australia, some of which visit or live in some of our most iconic regions like Ningaloo Reef in Western Australia and Queensland's tropical rainforests (see Figure 6.1.3).

▲ **FIGURE 6.1.3** Two different parts of the biosphere in Australia: (a) Ningaloo Reef in Western Australia; (b) a tropical rainforest in Queensland

Hydrosphere

permafrost permanently frozen soil, sediment or rock

The hydrosphere includes all forms of water on the planet, whether in solid, liquid or gaseous form. The frozen water in glaciers and ice caps is sometimes considered a separate sphere, known as the cryosphere. In this chapter, we have included the cryosphere as part of the hydrosphere. Water occurs as fresh water and saltwater. Only 2.5 per cent of the world's water is fresh, and most of this is inaccessible to people because it is locked up in the form of ice and **permafrost**. Humans use fresh water for drinking, bathing, recreation, agriculture and industry, but we can only access about 0.3 per cent of all fresh water on Earth, as shown in Figure 6.1.4.

Fresh drinking water is required for the survival of most living things. It is a very precious commodity in Australia and in many parts of the world. However, many organisms live in saltwater, such as marine organisms. Some species can tolerate moving between saltwater and freshwater environments.

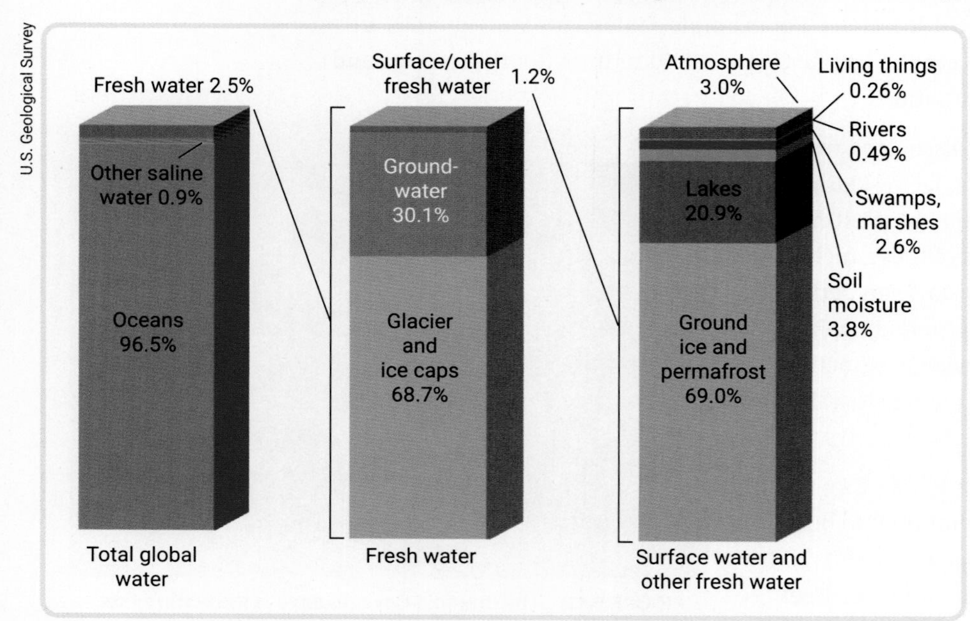

▲ **FIGURE 6.1.4** The distribution of Earth's water

Geosphere

The geosphere includes the rocks, minerals and landforms that make up Earth's surface; the sediments under the sea, lakes and rivers; and all the material that makes up the interior of Earth (see Figure 6.1.5). Land makes up about 29 per cent of the total surface area of Earth, with the rest (71 per cent) being water, mainly the oceans. The geosphere interacts with the other spheres; for example, it overlaps with the biosphere because many species live on or in this sphere. Although it appears static, the geosphere interacts with the flows of matter and energy from one sphere to another.

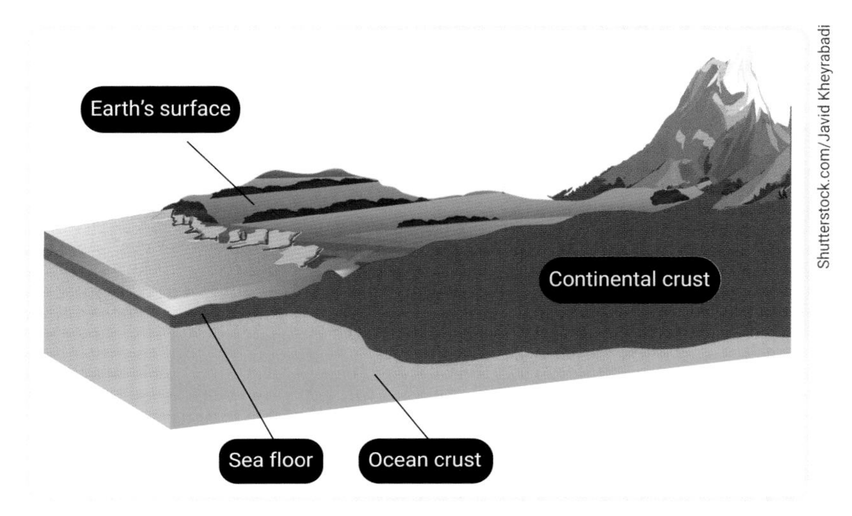

▲ **FIGURE 6.1.5** The geosphere

6.1 LEARNING CHECK

1 **List** the four spheres that make up the Earth system.
2 **Describe** the composition of gases within the atmosphere.
3 **Describe** how the concentration of oxygen in the atmosphere increased billions of years ago.
4 **Explain** the importance of ice cores to scientists' understanding of the atmosphere.
5 **Explain** why it is so important to take care of naturally occurring fresh water.

6.2 Interactions between spheres

BY THE END OF THIS MODULE, YOU WILL BE ABLE TO:

- ✓ describe how energy flows through the Earth system
- ✓ explain the cause of deep ocean currents and how they affect climate and marine life.

Video activity
The carbon cycle

Interactive resource
Label: Interactions between Earth's spheres

GET THINKING

We experience weather every day and can recognise patterns in the seasons. But have you ever looked at a satellite photo or zoomed around in Google Earth? The swirling clouds over the oceans and continents give you a clue about Earth's complexity, with many things interacting with one another. As you read this module, focus on the interactions taking place between the four Earth spheres and how these link to our changing climate.

Energy and matter flow through Earth's four spheres in complex ways through natural processes and chemical cycles. This means that the four spheres are connected and interact with each other. The interactions between spheres are also affected by the direct or indirect activities of people.

Energy from the Sun

radiation
a type of energy generally related to the electromagnetic spectrum

open system
a system in which energy and matter can be exchanged with their surroundings

Most of the energy on Earth comes from the Sun in the form of electromagnetic **radiation**. Earth's energy flow is an **open system** because energy can be exchanged with its surroundings. More of the Sun's energy reaches the equator and the area between the Tropics of Cancer and Capricorn than the polar regions. This is because less energy per unit area is received at the surface at higher latitudes due to Earth's spherical shape.

Of the energy coming from the Sun, 30 per cent is reflected back out to space without being absorbed by Earth. The remaining 70 per cent of the energy flows through Earth's four spheres, as shown in Figure 6.2.1.

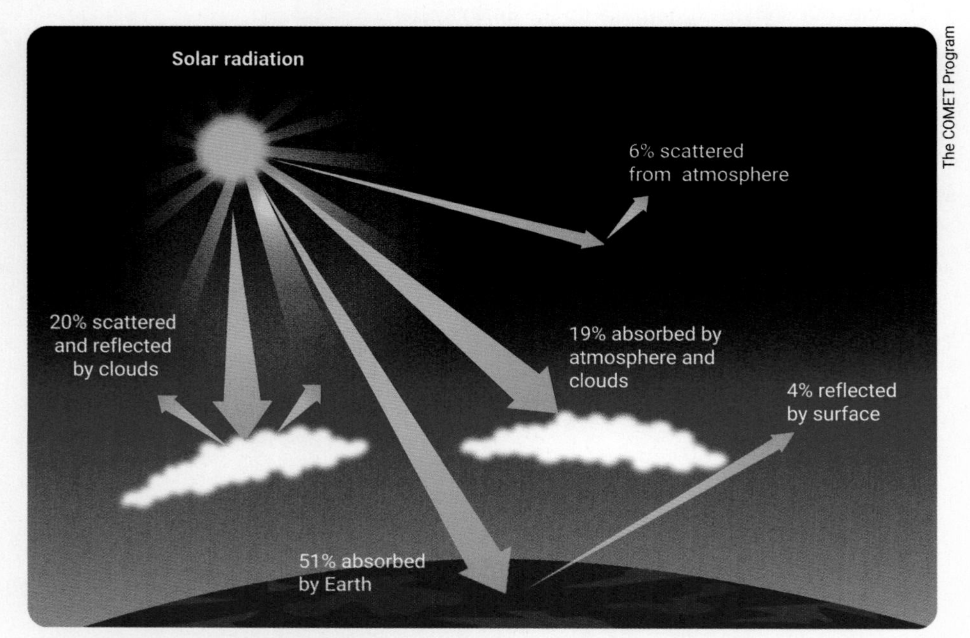

▲ **FIGURE 6.2.1** Earth's energy budget

Energy and matter flow between the spheres

A large amount of energy is absorbed by Earth's surface and atmosphere. This absorbed energy is captured within each of the spheres, some of which ends up back in the atmosphere, re-radiated as heat.

Processes in the spheres redistribute energy through naturally occurring cycles, such as the water, nitrogen and carbon cycles; **deep ocean currents** (discussed later in this module); and wind and **weather** patterns. These interact across multiple spheres and are all driven by the energy of the Sun (see Figure 6.2.2). The cycling of energy is complex and often includes the cycling of matter.

deep ocean currents
the water movement occurring below a depth of 400 m caused by changes in water temperature and salinity

weather
the conditions in the lower part of the atmosphere at a given time or place

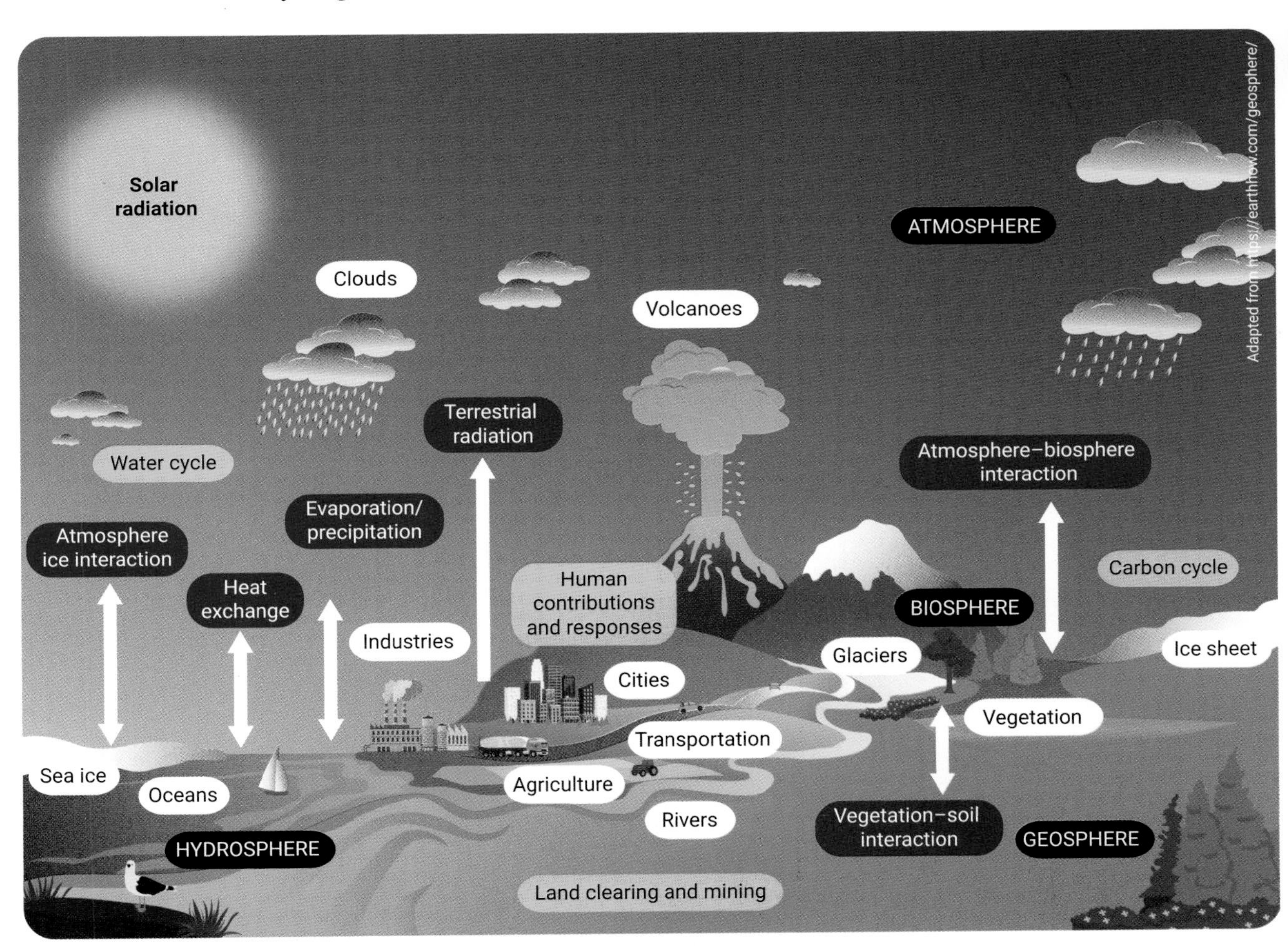

▲ **FIGURE 6.2.2** The interaction between Earth's four spheres

One example of a cycle that interacts with Earth's four spheres is the water cycle (see Figure 6.2.3). Energy from the Sun causes water to circulate through the atmosphere, biosphere, hydrosphere and geosphere.

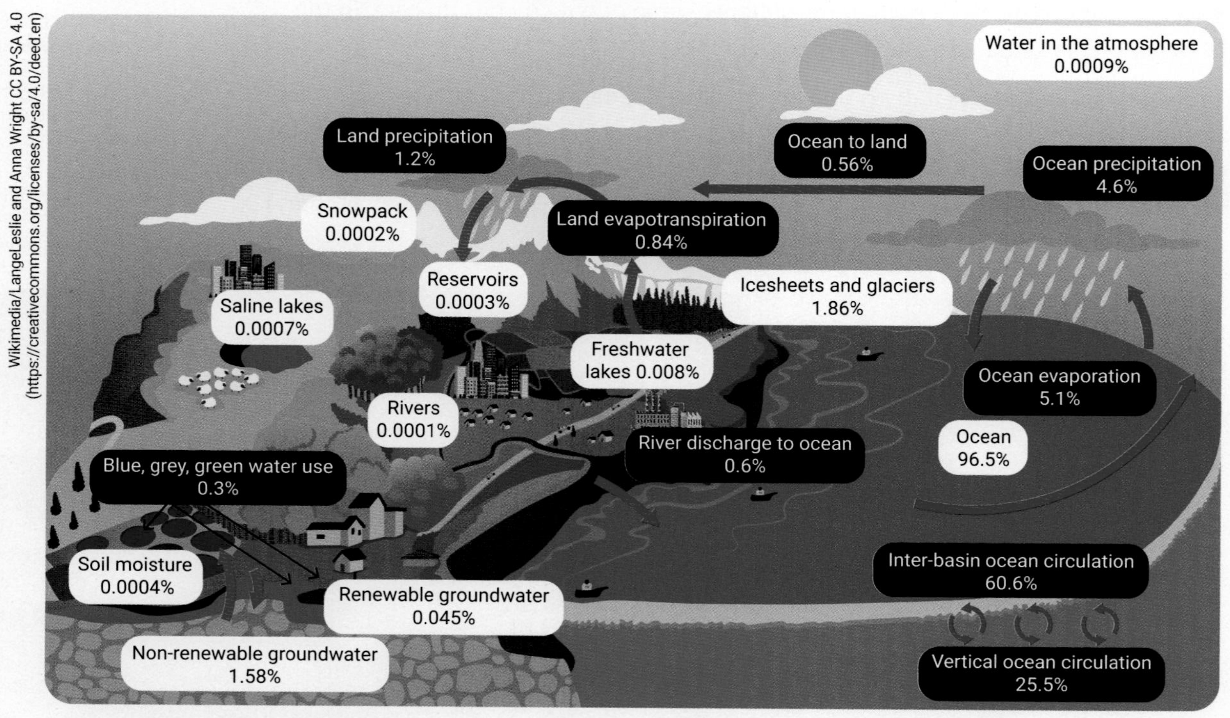

▲ **FIGURE 6.2.3** The water cycle

The carbon cycle is another example of a cycle that distributes matter and energy through the Earth's four spheres (see Figure 6.2.4). You learned about the carbon cycle in Year 9.

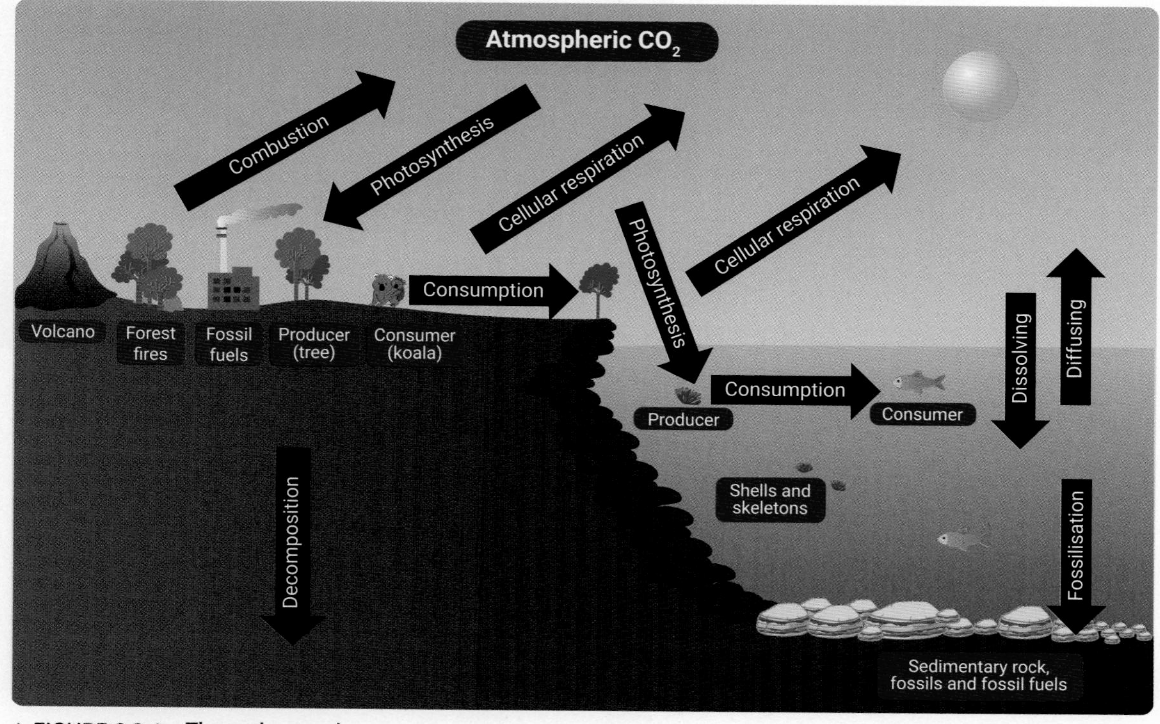

▲ **FIGURE 6.2.4** The carbon cycle

9780170463041

The impact of human activities on the Earth system

Human activities affect the natural resources and cycles within each sphere and, in some cases, disrupt the normal flow of energy and matter. Agriculture, aquaculture, mining and industrialisation have all had significant effects. For example, Australians have cleared almost 50 per cent of native forests in 200 years. **Land clearing** in Queensland and New South Wales is a serious threat to koalas and many other species that rely on forests for habitat. Land clearing increases the surface runoff of sediment into rivers and coastal regions, damaging coral reefs and marine ecosystems such as seagrass beds.

land clearing
the removal of natural vegetation and habitats, such as forests

▲ **FIGURE 6.2.5** (a) Land clearing, and (b) mining, in Australia destroys the habitat of different species and can disrupt Earth's natural cycles. (c) Land clearing has greatly reduced the habitat of koalas in New South Wales and Queensland; (d) Sediment runoff damages ecosystems such as seagrass beds.

Deep ocean currents – an energy conveyor belt

salinity
the concentration of dissolved salt

upwelling
the movement of water from the depths of the ocean to the surface, bringing nutrients and carbon dioxide with it

Deep ocean currents, many hundreds or thousands of metres below the ocean's surface, distribute energy around Earth. These are known as the thermohaline circulation, often referred to as the global conveyor belt. All deep ocean currents are controlled by **salinity**, temperature and density. The currents bring warm water to the poles from the equator and transfer colder water to the equator from the poles.

The deep currents begin in polar regions of the North Atlantic. As sea ice forms in the cold water, it leaves behind salt. This makes the water more saline and, thus, denser than the warmer, less saline water, causing it to sink.

The direction of the deep ocean currents is affected by the continents. At the equator, the water becomes warmer and, therefore, less dense, causing the currents to rise to the surface as an **upwelling**. The currents then loop back and eventually head north, returning to the North Atlantic to continue the cycle (see Figure 6.2.6). These currents move large volumes of water and energy very slowly – a few centimetres per second, compared with surface currents driven by the wind, which can move up to one metre per second. It can take up to 1000 years for the deep ocean current to complete its journey.

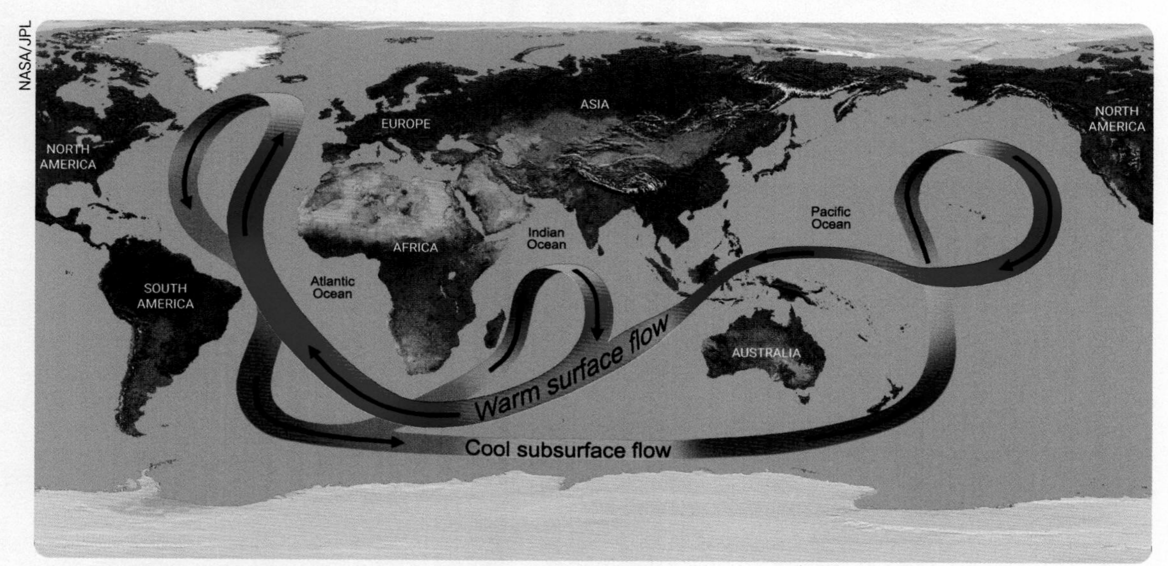

▲ **FIGURE 6.2.6** Global ocean currents

The effect of deep ocean currents on climate and marine life

surface temperature
the measure of the relative hotness or coldness of Earth's surface

Deep ocean currents have major effects on the climate. As the currents move, they gradually transfer heat to the surrounding water and the atmosphere, helping to regulate water and **surface temperatures**. As ocean water evaporates, it moderates the temperature and increases the humidity of the surrounding air. This in turn directly affects the climate near these currents, making Earth more liveable by moderating the climate and making it less extreme.

The deep ocean currents are also important for marine life. The water in these deep currents is rich in nutrients and carbon dioxide. As the currents move, they bring nutrients and carbon dioxide to the warmer surface layers. This helps the growth of algae and seaweed, supporting food chains in the upper layers of the oceans where most marine organisms live (see Figure 6.2.7).

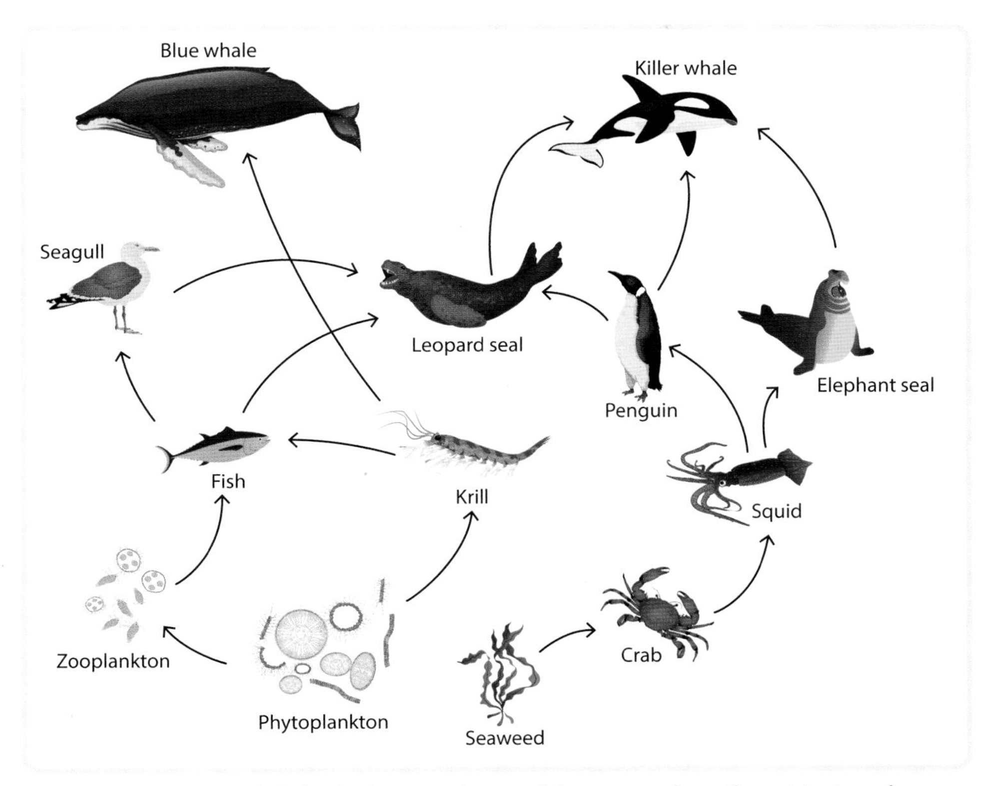

▲ **FIGURE 6.2.7** Food chains in the upper layers of the ocean rely on the nutrients and carbon dioxide moved by deep ocean currents.

6.2 LEARNING CHECK

1 **Describe** how the energy from the Sun drives the water cycle.
2 **List** the types of human activity that interrupt the natural flow of matter and energy in Earth's cycles.
3 **Describe** the path that deep ocean currents take in circling the world.
4 **Describe** an example of how Earth's spheres interact with each other.
5 **Explain** the cause of deep ocean currents.
6 **Explain** how deep ocean currents influence climate.
7 **Compare and contrast** the water and carbon cycles. Consider how they cycle through different spheres.
8 **Create** a mind map that links all the ways Earth's spheres interact with each other, and human activities that disrupt or change these interactions and cycles.

6.3 The greenhouse effect

BY THE END OF THIS MODULE, YOU WILL BE ABLE TO:

- ✓ use a labelled diagram to show the greenhouse effect
- ✓ explain how the greenhouse effect warms Earth.

greenhouse effect
a natural process that traps energy within the atmosphere, raising Earth's surface temperature

electromagnetic spectrum
the range of electromagnetic waves sequenced by energy, frequency and wavelength from gamma rays to radio waves

Video activities
What makes up the electromagnetic spectrum?
The greenhouse effect

Interactive resource
Simulation: The greenhouse effect

Extra science investigation
How the greenhouse effect impacts air and soil temperature

GET THINKING

As you complete this module, think about the presence of greenhouse gases in the atmosphere and their effect on Earth's climate. What would be the situation if these gases were not present at all, or present in much lower concentrations? Why do you think carbon dioxide and methane are so important to Earth's climate?

The greenhouse effect

The **greenhouse effect** is a natural process caused by a group of gases in the atmosphere commonly referred to as 'greenhouse gases'. These gases include water vapour, carbon dioxide, methane, nitrous oxide and ozone. Humans have caused an increase in all of these (other than water vapour) and added new, synthetic ones such as chlorofluorocarbons.

In Year 9 you were introduced to the **electromagnetic spectrum**. The spectrum is made up of electromagnetic waves that have different wavelengths but all travel at the speed of light (see Figure 6.3.1). All types of radiation can enter the atmosphere. Some of this radiation is reflected back into space. Shortwave radiation includes wavelengths within the infrared, visible and ultraviolet parts of the electromagnetic spectrum. Longwave radiation falls within the infrared region of the electromagnetic spectrum.

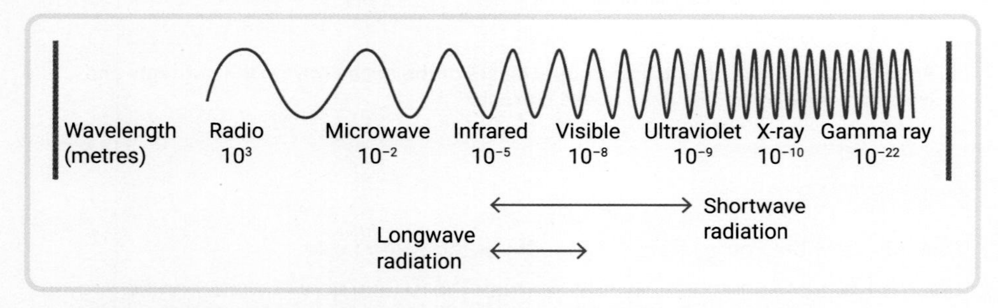

▲ **FIGURE 6.3.1 Shortwave and longwave radiation of the electromagnetic spectrum**

Most of the radiant energy from the Sun, in the form of high-energy shortwave radiation, passes through the atmosphere to Earth's surface. As it passes through the atmosphere, some of this radiation is absorbed by clouds and gases. Earth's oceans, rivers, land and plants also absorb energy. Earth's atmosphere and surface re-radiate this back into the atmosphere as longwave radiation or heat. The greenhouse gases absorb and trap the re-radiated energy, warming the atmosphere (see Figure 6.3.2).

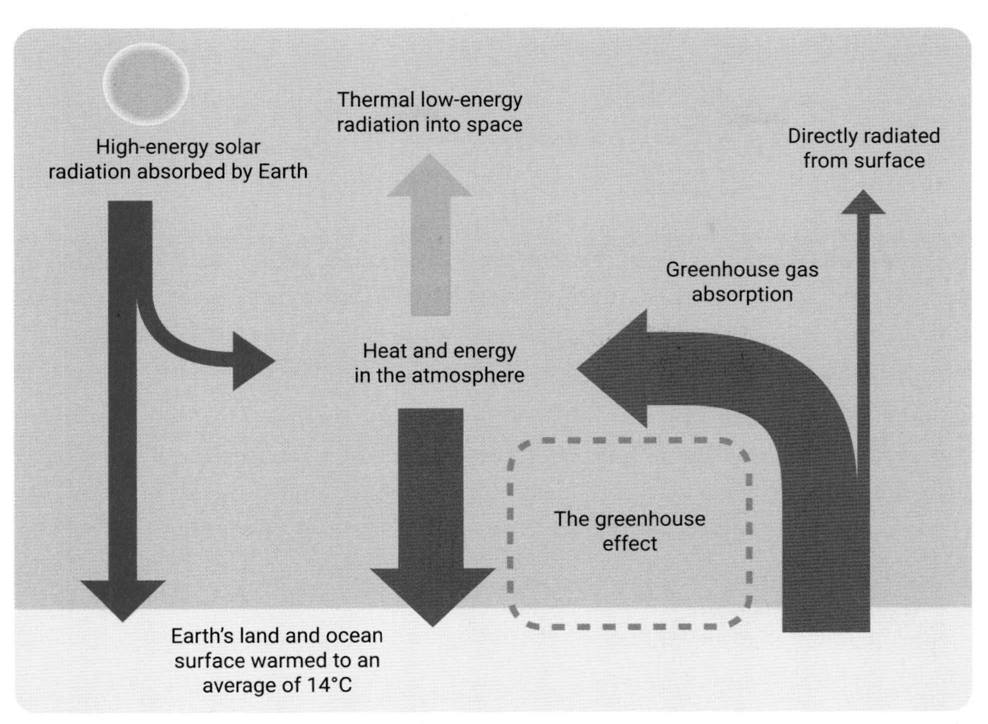

▲ **FIGURE 6.3.2** How the greenhouse effect is generated

Modelling the greenhouse effect

☆ ACTIVITY

Materials

- 2 × 2 L soft drink bottles (or similar)
- 2 thermometers (or temperature probes, if available)
- plastic wrap
- elastic band
- masking tape
- 100 W light globe and lamp
- scissors or craft knife

Warning

To complete this activity, you will need to cut the top off two soft drink bottles using scissors or a craft knife. Take care, as the cutting tool and the top of the cut soft drink bottles will be very sharp and could easily cut you if handled incorrectly.

Method

1. Cut the top of the soft drink bottles off at the 'shoulder', ensuring that they are the same height.
2. Stick the thermometers to the inside of the soft drink bottles, with the thermometer bulb or probe about 5 cm from the base. Ensure that it is easy to read the temperature from the outside of the bottles.
3. Place the two soft drink bottles each 15 cm from either side of the light globe.
4. Place the plastic wrap on top of one container and use the elastic band to hold it in place. Immediately take and record a temperature reading from both bottles.
5. Turn on the light, and then take and record measurements every 2 minutes for 30 minutes.
6. Compare the temperatures in the two soft drink bottles.

Evaluation

Was there a difference between the two bottles? If there was a difference, explain why this happened.

The natural greenhouse effect

The greenhouse effect can be thought of in two ways: the naturally occurring greenhouse effect and the human-induced **enhanced greenhouse effect**.

enhanced greenhouse effect
strengthening of the natural greenhouse effect due to an increase in greenhouse gases caused by human activities

The naturally occurring greenhouse effect is responsible for keeping conditions on Earth suitable for habitation. Without it, the temperature of Earth's surface would be, on average, about 33°C cooler. The greenhouse effect requires the presence of carbon dioxide and other greenhouse gases in the atmosphere. Without these gases, Earth would be unable to hold onto any heat and it would escape into space. This would leave the planet as an icy wasteland and life as we know it would not exist.

The extent of the natural greenhouse effect depends on the concentration of gases that make up the atmosphere. Too low a concentration of greenhouse gases and Earth loses too much energy. Too high a concentration and the greenhouse gases result in more energy and heat being trapped, raising Earth's average surface temperature.

The enhanced greenhouse effect

The enhanced greenhouse effect is caused by the increased concentration of greenhouse gases in the atmosphere as a direct result of human activities, such as the burning of fossil fuels. The gases carbon dioxide and methane are of particular concern.

The measured concentration of carbon dioxide in the atmosphere has increased by approximately 50 per cent since the beginning of the Industrial Revolution, from around 280 parts per million (ppm) to more than 420 ppm. What is alarming to scientists is the rate of change in concentration. Before the Industrial Revolution, changes in carbon dioxide concentrations occurred over thousands of years. Figure 6.3.3 shows the concentration levels of atmospheric carbon dioxide over the last 800 000 years.

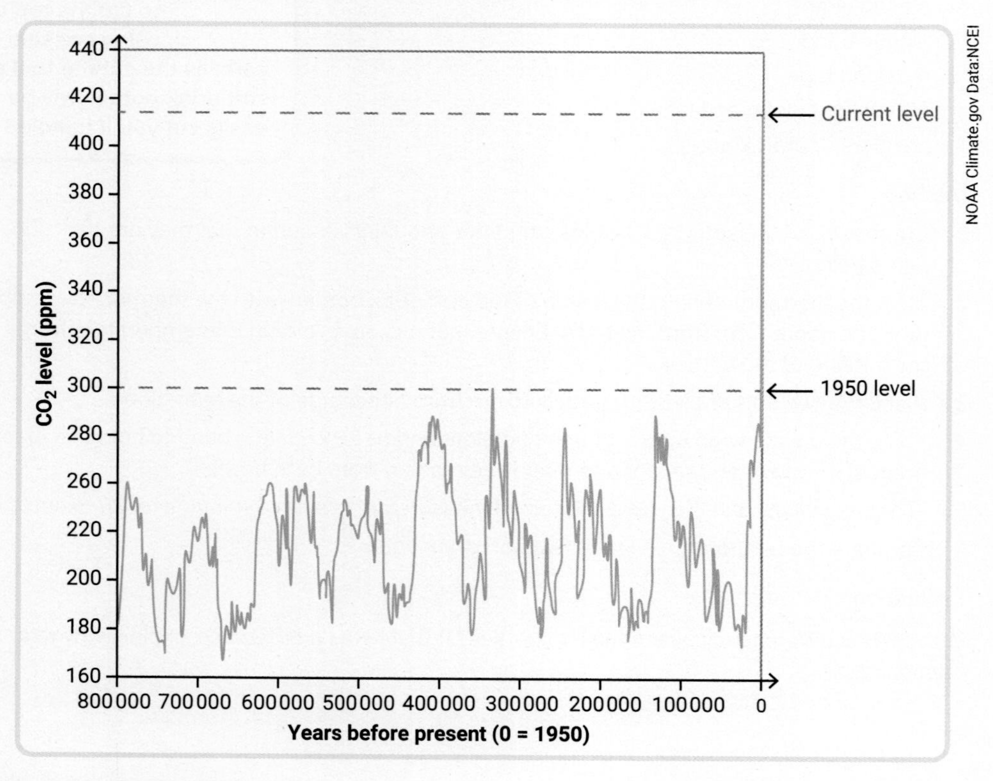

▶ **FIGURE 6.3.3** Atmospheric carbon dioxide concentration (in ppm) over the last 800 000 years

Measurements reported by the Intergovernmental Panel on Climate Change (IPCC) in 1986 showed it had taken 200 years for carbon dioxide levels to increase by 25 per cent. More worrying are the measurements reported for 2021, which show that carbon dioxide levels have increased 21 per cent in the last 35 years.

Methane concentrations have increased by nearly 160 per cent over the same 235-year period. This is mostly due to the increased farming of livestock, coal and gas production, and waste disposal in landfills.

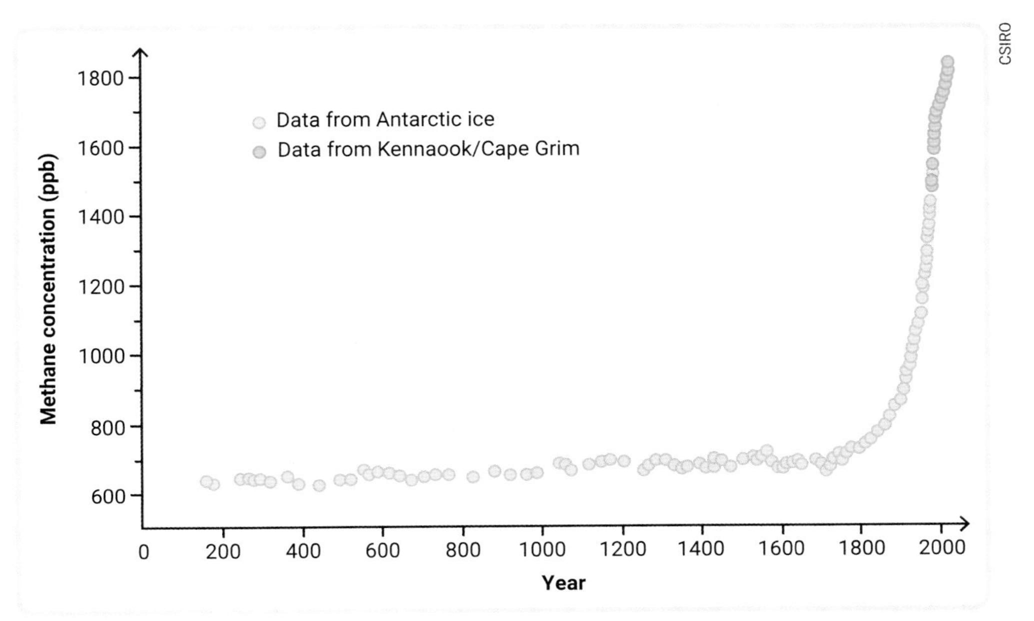

▲ **FIGURE 6.3.4** Atmospheric methane concentrations (in parts per billion) over the last 2000 years

As we emit more greenhouse gases into the atmosphere, the enhanced greenhouse effect is becoming stronger, and the average global temperature is increasing.

6.3 LEARNING CHECK

1. **List** five greenhouse gases.
2. **Describe** characteristics and features of shortwave and longwave radiation.
3. **Explain** why the natural greenhouse effect is so important to life on Earth.
4. **Discuss** how the enhanced greenhouse effect contributes to warming Earth.
5. **Compare** and **contrast** natural greenhouse effect with the enhanced greenhouse effect.
6. Imagine you have to explain the greenhouse effect to Year 7 students. **Create** an infographic or animation to show how the greenhouse effect is generated. Keep your explanations simple and concise, using language your intended audience will understand.

6.4 Global climates

BY THE END OF THIS MODULE, YOU WILL BE ABLE TO:

✓ label a map with the locations of Earth's different climate zones
✓ explain why different locations experience different climates.

precipitation
liquid or solid water in the form of rain, snow, sleet or hail

Köppen climate classification
a system used since 1900 that classifies climate zones based on temperature, amount and type of precipitation, and vegetation

dry climate
a climate zone where seasonal evaporation exceeds seasonal precipitation

temperate climate
a climate zone with moderate temperature and precipitation that exists between the extremes of tropical climates and polar climates

polar climate
a climate zone with ice, snow and temperatures too low to support most vegetation

Coriolis effect
the apparent deflection of large masses of air and water due to Earth's rotation on its axis

GET THINKING

As you work your way through this module, consider the climate where you live. What are the characteristics of your climate? How does the weather vary over the course of the year? Is the weather similar from year to year, or have there been significant changes over the past decade? How would your parents or grandparents respond to the same question?

Climate zones

Climate refers to long-term weather conditions in a particular region. It includes the average temperature, humidity, type and amount of **precipitation**, and seasons that occur over extended periods of time (usually 30 years or more). Weather is a description of the day-to-day variation that may occur in a location.

There are many ways to categorise different climate zones. In this chapter, we use the **Köppen climate classification**, which has five main climate zones: tropical climate, **dry climate**, **temperate climate**, continental climate and **polar climate**.

Factors that influence and create climate zones

Three main factors influence the different types of climate zones: Earth's spherical shape, the moisture in the air and the **Coriolis effect**.

Differences in solar radiation

The general distribution of climate zones (see Figure 6.4.1) roughly follows a pattern that seems related to latitude. This is due primarily to these locations receiving different amounts of the Sun's energy per unit of area because of Earth's spherical shape, with more energy received at the equator and less energy received at the poles.

Differences in Earth's surface temperature affect the air temperature and how the air moves around the planet. Hot air rises and cold air sinks, which produces areas of low and high air pressure, respectively. Air moves from areas of high pressure to low pressure, creating wind.

NOAA

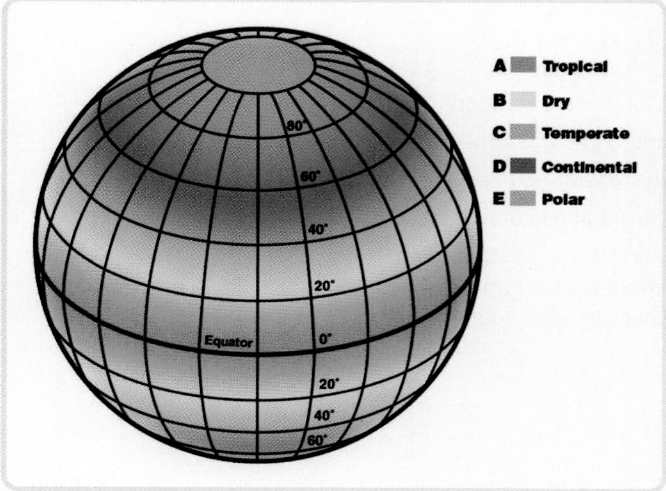

▲ **FIGURE 6.4.1** The distribution of climate zones

Moisture and precipitation

The second factor that influences climate is the moisture in the air and the resulting precipitation. The amount of water vapour the air can contain depends on the temperature, which is why latitude also indirectly affects precipitation in climate zones. The amount of water vapour in the air in comparison to the maximum amount the air could hold at that temperature is described as the **relative humidity**. The rate and type of precipitation in a climate zone depend on both temperature and relative humidity. The rate and type of precipitation are key distinguishing factors between different climate zones.

Quiz
Main climate zones

relative humidity
the amount of moisture that air holds compared with the amount it could hold if saturated at a given temperature

The Coriolis effect

The third influence is the Coriolis effect. This effect appears as the deflection of moving objects, such as large air masses, due to the rotation of Earth on its axis. The deflection appears to move towards the left (clockwise) in the southern hemisphere and to the right in the northern hemisphere. The movement of air is affected by the Coriolis effect, resulting in air flowing in different directions at different points on Earth.

Global convective cells

The combination of these three factors creates three types of **global convective cells** in each of the hemispheres: the Hadley cell, the Ferrel cell and the polar cell. These cells make the air circulate around Earth in directions that depend on the location (see Figure 6.4.2). The movement of air within these cells is due to **convection currents**.

global convective cells
three large atmospheric pressure cells that occur in both the northern and southern hemispheres

convection current
a flow of materials (such as air, water or molten rock) and energy caused by differences in densities due to temperature differences

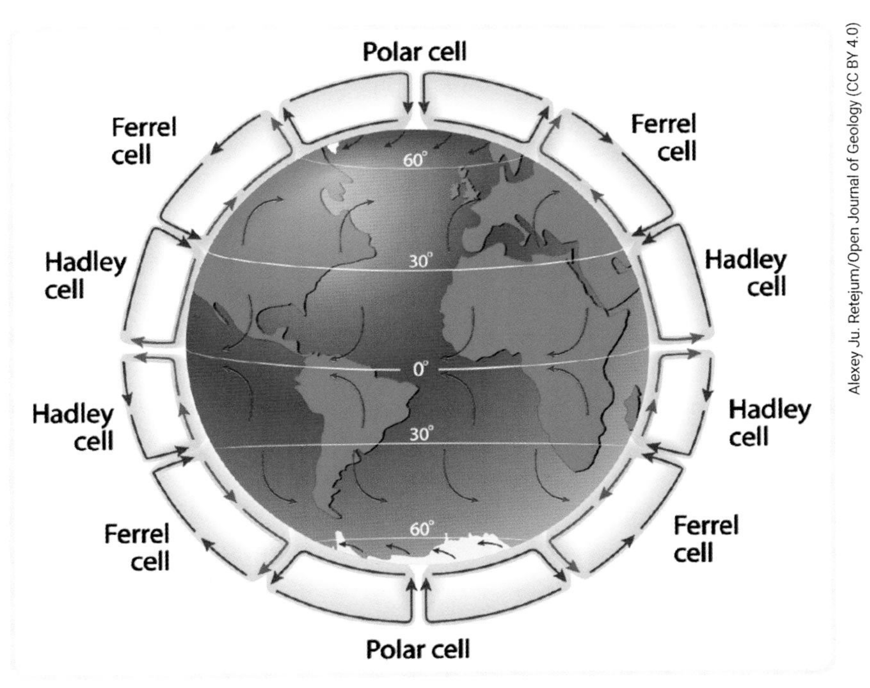

▲ **FIGURE 6.4.2** Global atmospheric circulation

The global convective cells influence average rainfall and average temperatures. They also circulate energy through the atmosphere in a similar way to the deep ocean currents, bringing warm or cool air to different parts of Earth. The global convective cells also interact with the ocean currents, move moisture and influence the precipitation in different climate zones.

The characteristics of the climate zones

The five climate zones have been categorised by the features of temperature, frequency, amount and type of precipitation, and the variety of vegetation (see Table 6.4.1 and Figure 6.4.3).

▼ **TABLE 6.4.1** The five main climate zones

Climate zone	Temperature	Precipitation	Vegetation type
Tropical	• Average above 18°C all year round	• More than 1500 mm of rain per year	• Tropical rainforest
Dry	• Large daily temperature range due to reduced cloud cover • Maximum temperatures can be above 40°C • Low night temperatures can get below 0°C	• Unpredictable • Less than 350 mm of rain per year • High rates of evaporation	• Mostly grasslands • Xerophytic (adapted to survive on limited water)
Temperate	• Variable; experience four seasons • Mean temperature above −3°C and below 18°C in its coldest month	• Moderate: 750–1500 mm of rain per year • Averages 800 mm of rain per year	• Mixed forests • Low shrub • Eucalypt forest, heath and mallee in Australia
Continental	• Large temperature variation • Summer average above 10°C • Coldest month average below −3°C	• Irregular • Averages 600–1200 mm of precipitation per year, mostly as snow in winter	• Coniferous and deciduous forests
Polar	• Summer temperature rarely above 10°C • Winter temperatures often reach −30°C	• Less than 250 mm of precipitation per year, mostly as snow	• Very low-lying types • Lichens dominant • Some moss and liverworts

The distribution of climate types in Australia

The Australian Bureau of Meteorology uses a variation of the Köppen climate classification. It divides some Köppen climates and combines others to better reflect the climate conditions and the ecological zones humans experience in these diverse parts of Australia (see Figure 6.4.3 and Figure 6.4.4).

You probably know that Australia has many diverse zones of climate. As shown in Figure 6.4.4, Australia has a tropical climate in the north, with temperate climate zones in the south. Within these broad climate zones there is significant ecological variation due to the natural landscape that the Australian Bureau of Meteorology modified climate classification takes into account.

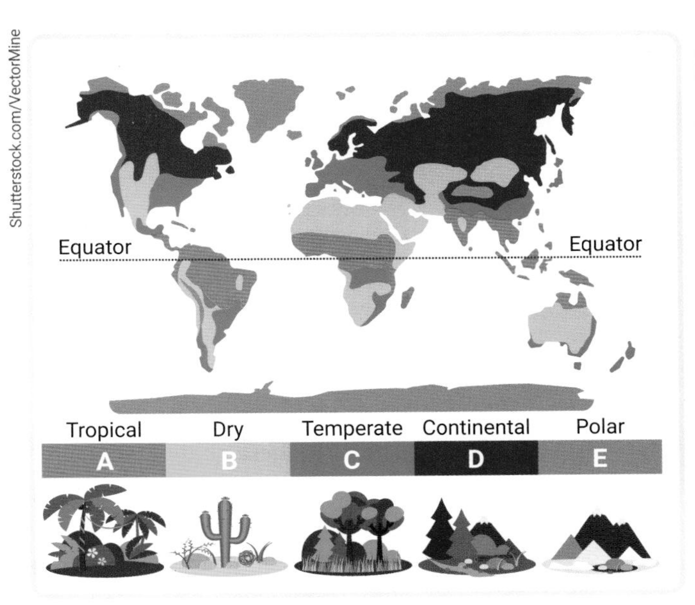

▲ **FIGURE 6.4.3** The distribution of Köppen climate types

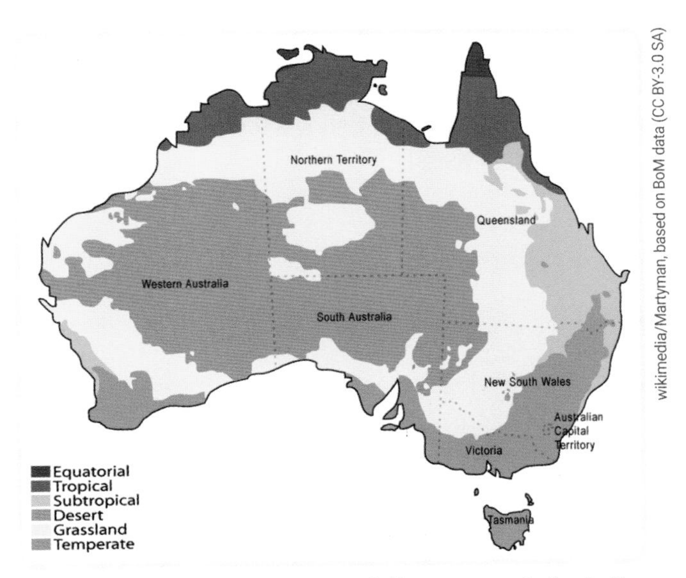

▲ **FIGURE 6.4.4** The distribution of climate zones in Australia

6.4 LEARNING CHECK

1. **List** the five main climate zones according to the Köppen classification system.
2. **Explain** the difference between weather and climate.
3. **Describe** where the different climate zones are located around Earth.
4. **Explain** the role that atmospheric convection currents and ocean currents have on the distribution of climate zones.
5. **Explain** how the energy from the Sun has an effect on climate.

6.5 Changes in global climate

BY THE END OF THIS MODULE, YOU WILL BE ABLE TO:

✓ explain the causes of climate change.

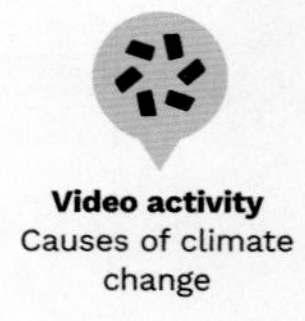

Video activity
Causes of climate change

GET THINKING

As you complete this module, think about what is causing changes to the climate. Think about the way we live and how that is causing the rate of climate change to accelerate. Start thinking about the simple things we can do to slow the progress of climate change.

Climate change is a term used to describe alterations to regional or global climate patterns. These changes can either occur naturally or relate directly to human-induced changes that have become apparent in the past 40 years. Almost all climate scientists agree that human activities are the cause of the climate change we are currently experiencing. As a result, we now generally use the term 'climate change' to describe climate change caused by human activities.

The causes of climate change

Climate data collected from observations and records such as tree rings and ice cores clearly shows that the atmospheric concentration of carbon dioxide and other greenhouse gases has increased significantly since around 1850. These are causing an enhanced greenhouse effect. Figure 6.5.1 shows the combined greenhouse gas concentrations over the last 2000 years.

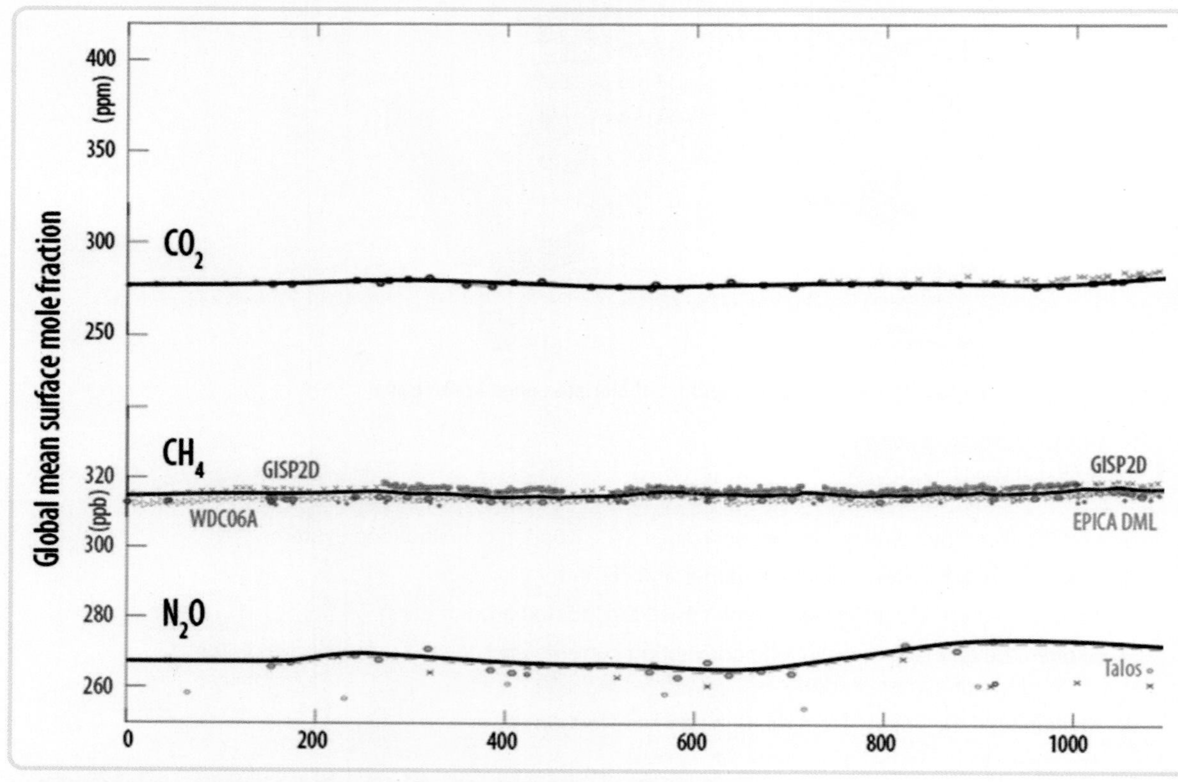

▲ **FIGURE 6.5.1** Greenhouse gas concentrations over the last 2000 years

9780170463041

Human activities are increasing the level of greenhouse gases in two main ways.

1 Activities such as the burning of fossil fuels and livestock farming are adding extra greenhouse gases to the atmosphere.
2 The removal of trees and vegetation is reducing the natural **carbon sinks** or reservoirs that take carbon dioxide out of the atmosphere.

carbon sink
something that naturally stores large quantities of carbon dioxide

Sources of greenhouse gas emissions

A major cause of climate change is the increase of carbon dioxide from burning fossil fuels for energy and transport.

Approximately 90 per cent of the world's carbon emissions are due to the burning of fossil fuels: coal, oil and natural gas. Australia is no different, with the majority of our emissions coming from the production of electricity and the use of fossil fuels in transport, agriculture and industrial processes. Australia ranks 14th in the world in carbon emissions, contributing 1 per cent of global emissions.

Methane concentrations in the atmosphere have also increased, mainly due to the farming of livestock, waste disposal in landfills and gas and oil production. For example, the methane produced by food waste in landfills across the world contributes 11 per cent of all greenhouse gases. In Australia, methane from food waste in landfills contributes the equivalent of 6.8 million tonnes of carbon dioxide.

Another greenhouse gas, nitrous oxide, is increasing because of agriculture, fuel combustion, wastewater treatment and industrial processes.

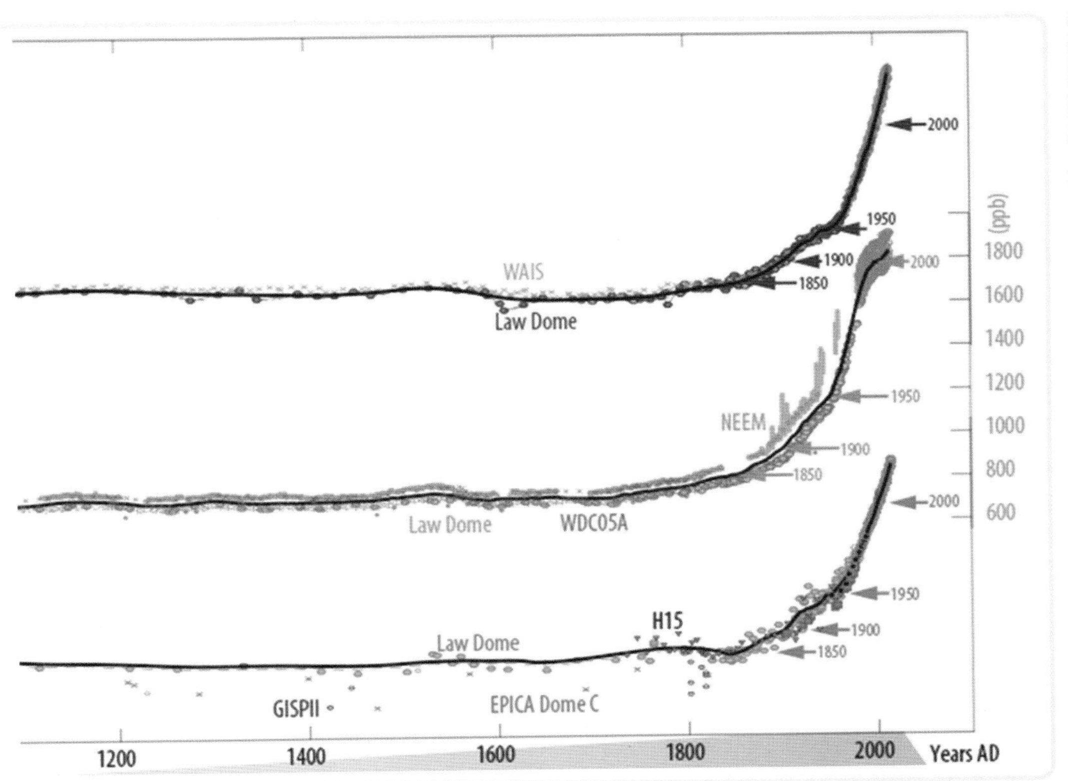

Source: M. Meinshausen et al.: Historical GHG concentrations for climate modelling (CMIP6) (CC BY 3.0)

Deforestation

Large areas of forest continue to be removed to meet the worldwide demand for timber and forest products and to increase the amount of land available for agriculture (see Figure 6.5.2).

Some of the main causes of deforestation are:

- agriculture – clearing land for agriculture, including stock grazing and the planting of crops
- urbanisation – clearing land for housing developments and infrastructure such as roads and dams
- logging – the harvesting of trees for wood and paper products
- forest fires.

deforestation the removal of large areas of forest to enable the land to be used for other purposes

Trees and vegetation store and use carbon dioxide, taking it out of the atmosphere. Fewer trees and plants mean Earth has fewer carbon sinks to help regulate the levels of carbon dioxide in the atmosphere. A further problem with **deforestation** is that when trees are removed or destroyed (e.g. by burning), the carbon stored in them can be released, adding even more carbon dioxide to the atmosphere. Deforestation also leads to further land degradation and other environmental problems when deforested areas are subjected to heavy rainfall and flooding. The Amazon rainforest and forests in Asia have the greatest amounts of land clearing.

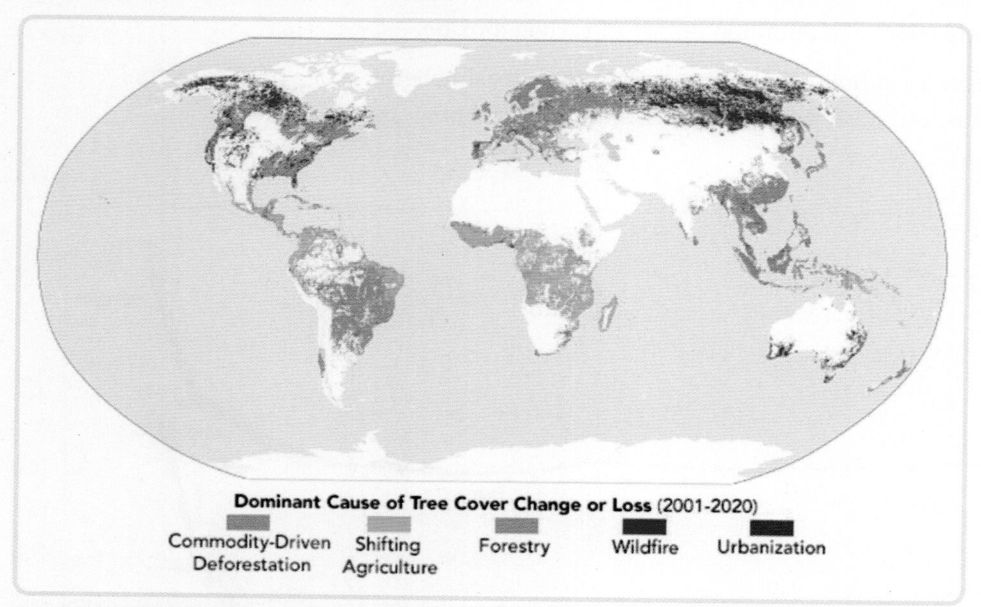

Adapted from NASA Earth Observatory images by Lauren Dauphin, using data from Curtis, P.G., et al. (2018), data from Goldman, Elizabeth, et al. (2020), and Landsat data from the U.S. Geological Survey.

▲ **FIGURE 6.5.2** The reasons for deforestation

6.5 LEARNING CHECK

1. **Describe** the difference between the natural greenhouse effect and the enhanced greenhouse effect.
2. **Describe** how food waste disposal contributes to greenhouse gases.
3. **Describe** the main causes of deforestation.
4. **Explain** why deforestation is significant for the concentration of atmospheric carbon dioxide.

9780170463041

6.6 Evidence for climate change

BY THE END OF THIS MODULE, YOU WILL BE ABLE TO:

✓ explain how changes in atmospheric and oceanic temperature, sea levels, biodiversity, permafrost and sea ice are indicators of climate change.

GET THINKING

In recent years, there have been regular reports in the media on sea ice melting in the Arctic Ocean, droughts and floods in South Africa, and coral bleaching on the Great Barrier Reef. Are these natural events or indicators that the climate is changing? Think about other things that you may have seen in the media that might be either natural variability or indicators of climate change.

Video activities
Climate cycles
Natural climate change

Interactive resource
Drag and drop: Evidence of climate change

Globally, there is an accumulation of different types of scientific evidence that all show the same clear picture: our climate is changing, and Earth is getting warmer.

Who finds the evidence for climate change?

In Australia, the Bureau of Meteorology and the Commonwealth Scientific and Industrial Research Organisation play a key role in monitoring, analysing and communicating observed and future changes in Australia's climate. Around the world, there are many research scientists working in meteorological and government organisations to monitor changes in climate. The IPCC is responsible for assessing the latest knowledge on human-induced climate change, its impacts and how to respond.

Scientists use many types of evidence to measure and understand climate change, including:

- increases in the concentrations of greenhouse gases
- increases in atmospheric and ocean temperature
- increases in ocean acidity
- increases in global sea levels
- reductions in sea ice and permafrost
- reductions in **biodiversity.**

biodiversity
the variety of all living organisms in an ecosystem

Atmospheric and ocean temperature

Over the past million years, global average surface temperatures have varied by up to 5°C, as Earth experienced cycles of ice ages and warmer periods approximately every 100 000 years. What is concerning to climate scientists is the current increased rate of change. The enhanced greenhouse effect has caused Earth's average global temperature to increase by around 1.1°C compared with levels in 1850–1900. The year 2021 was the seventh consecutive year in which the average global temperature was more than 1°C higher than the post-industrial levels. The period 2014–2021 has been the hottest since records began.

In Australia, climate change means that extreme heat is becoming a more regular occurrence, with temperature records being broken in many places. On 13 January 2022, Onslow, Western Australia, equalled Australia's highest recorded temperature, reaching a record 50.7°C previously recorded in Oodnadatta in South Australia on 2 January 1960. It also means parts of Australia are subjected to unseasonal increases in rainfall and record flooding, as was the situation in south-east Queensland and coastal New South Wales in February and March 2022, in Victoria and New South Wales in October and November 2022 and in South Australia and Western Australia in January 2023.

Ocean temperatures are also important. NASA has determined that the oceans hold 90 per cent of the accumulated heat in the Earth system. Recent measurements of heat in the ocean showed that 2020 was the ocean's warmest year on record and that rates of ocean warming have shown strong increases in the past 20 years. Figure 6.6.1 shows how the amount of heat held in the ocean has increased dramatically. Increasing ocean temperatures cause effects such as coral bleaching and changing distributions of fish.

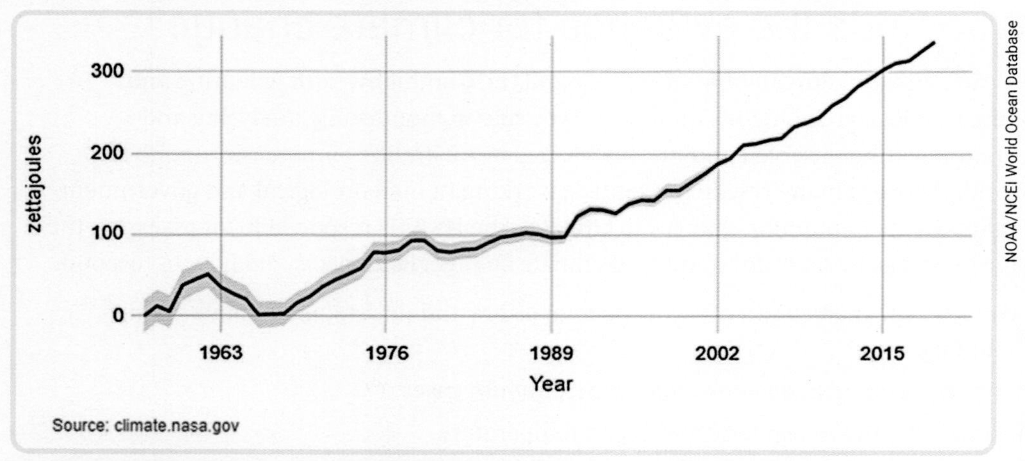

▲ **FIGURE 6.6.1** Estimated change in the global ocean heat content since 1955. The blue shading provides an indication of the confidence range of the estimate.

Ocean acidity

Like trees, the ocean is an important carbon reservoir that takes carbon dioxide out of Earth's atmosphere. The ocean absorbs about 23 per cent of annual human-produced carbon dioxide emissions. Increasing ocean temperatures are causing oceans to become more acidic. Measurements of pH show the ocean is the most acidic it has been for at least the past 26 000 years. As the ocean becomes more acidic, it absorbs less carbon dioxide from the atmosphere. Increased ocean acidity makes it difficult for marine calcifying organisms, such as coral and some plankton, to form shells and skeletons. Acidity also can lead to existing shells dissolving and becoming weaker.

Sea levels and sea ice

Higher atmosphere and ocean temperatures are causing the continental ice sheets in Antarctica and Greenland to melt. Higher temperatures are also causing glaciers to melt, adding more water to the oceans. The melting of land-based ice, combined with the **thermal expansion** of water in the ocean, is making sea levels rise. Since 1880, sea levels have risen 21–24 cm, with almost a third of that rise occurring in the past 25 years.

thermal expansion
an increase in the volume of materials as they get hotter

Satellites have been used since the early 1990s to determine changes in sea level remotely. Satellite data shows that the rate of sea-level rise is accelerating. Between 1993 and 2002, there was an average global increase in sea level of 2.1 mm per year. Between 2013 and 2021, sea levels increased at more than twice that rate, by an average of 4.4 mm per year (see Figure 6.6.2).

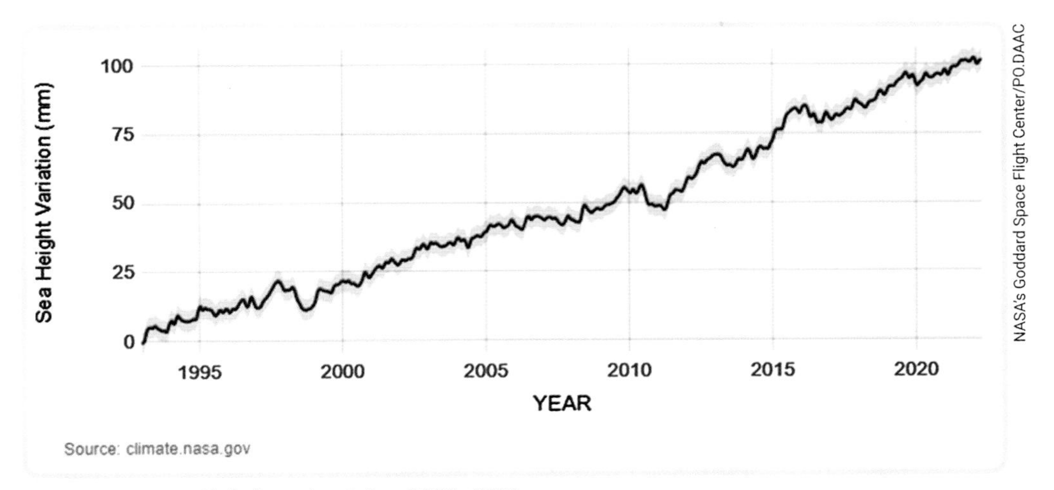

▲ **FIGURE 6.6.2** Global sea-level rise, 1993–2021

A further source of evidence of climate change is the reduction in sea ice in the Arctic region. Satellite monitoring shows that the amount of sea ice is now declining at a rate of about 13 per cent every decade.

Permafrost

Permafrost is ground that has been at a temperature of 0°C or below for at least two consecutive years. Permafrost accounts for about 17 per cent of Earth's surface and is very sensitive to rising temperatures. Data from boreholes drilled between 2007 and 2016 showed that in permafrost regions, the temperatures 10 m below the ground have increased by about 0.3°C.

It is predicted that the continued warming of the planet will ultimately melt permafrost areas. When this happens, the carbon released from these soils will add large amounts of carbon dioxide and methane to the atmosphere.

▲ **FIGURE 6.6.3** Permafrost and ice on Herschel Island, Canada, 2012

Biodiversity

It is estimated that one-eighth of the world's species are threatened with extinction. Climate change poses a significant risk to many of these species. The IPCC has confirmed that climate change and biodiversity are linked. Climate change and the loss of biodiversity are both driven by human activity, and each negatively affects the other.

Changes in climate have altered environmental conditions and habitats. If animals and plants are unable to adapt to these changes, their survival is at risk. Species that live within a restricted range are particularly vulnerable, such as the endangered Carnaby's black cockatoo that is endemic to south-west Australia. It is very susceptible to heat stress, so more frequent heatwaves are making its future even more precarious. Green turtles on the northern Great Barrier Reef are in danger because rising temperatures are causing 99 per cent of hatchlings to be female.

iStock.com/animalinfo

▲ **FIGURE 6.6.4** Carnaby's black cockatoo is endangered and susceptible to climate change.

Research has found that the top predators in ecosystems that are usually affected the most by climate change. For example, polar bears in the Arctic are declining in large numbers as both their food sources and available habitat are restricted by the reduction in sea ice.

6.6 LEARNING CHECK

1 **List** five sources of evidence that confirm climate change is occurring.
2 **Describe** the purpose of the IPCC.
3 **Describe** how the Great Barrier Reef is being affected by climate change.
4 **List** three examples that offer evidence of climate change in Australia.
5 **Explain** what causes the acidification of the oceans and suggest the likely outcome if it were to continue.

6.7 The effects of climate change

BY THE END OF THIS MODULE, YOU WILL BE ABLE TO:

✓ use a flow chart to represent the impacts of climate change

✓ explain that extreme weather events are occurring with greater frequency as a consequence of climate change.

GET THINKING

As you complete this module, think about recent examples of extreme weather events you have experienced or heard about; for example, increases in temperature, severe droughts or heavy rainfall, which lead to disasters such as fires and floods. Is there a link between these events and human activities? What is the role of climate change in these events?

Video activity
Ocean acidification

Interactive resource
Label: Climate change and human health

As we saw in Module 6.6, there is a lot of scientific evidence that clearly shows our climate is changing. But what does this mean in terms of impacts? How could climate change affect us in Australia?

Extreme weather

In Module 6.2 and Module 6.4, we looked at how the world's climate and weather patterns are closely linked to the circulation of energy via deep ocean and atmospheric currents. Climate change is affecting and modifying these currents, interrupting the flow of energy around the planet and causing further changes to the climate.

Unfortunately, we are already seeing the impacts of these climate changes in the severity and frequency of **extreme weather events** around the world. Extreme weather events include cyclones, flooding, heatwaves and drought. In Australia, bushfires are becoming more intense and are causing more damage. They are also burning in places that do not normally experience bushfires, such as in the ancient rainforests of south-east Queensland and north-east New South Wales (see Figure 6.7.1). Climate change is drying out these forests, making them more vulnerable to fires. This in turn destroys the habitat of many animals and plants, further threatening Australia's biodiversity.

extreme weather event
unexpected, unusual, severe or unseasonal weather

Shutterstock.com/Nick Fox

Alamy Stock Photo/Cavan Images

▲ **FIGURE 6.7.1** The Gondwana Rainforests in Queensland (a) before and (b) during the severe bushfires of early 2020

Human health

Extreme weather can have immediate impacts on human health, such as heat stroke leading to death. Each year, extreme heat kills more people in Australia than all other natural disasters combined.

Extreme weather can also have far-reaching negative effects on humans, by affecting our food supplies through crop failures or damage from weather events such as storms, floods and droughts. Other examples of the indirect effects of climate change on human health include an increase in air pollution from larger, more intense bushfires and droughts severely reducing our reserves of fresh water. Further links between climate change and human health are shown in Figure 6.7.2.

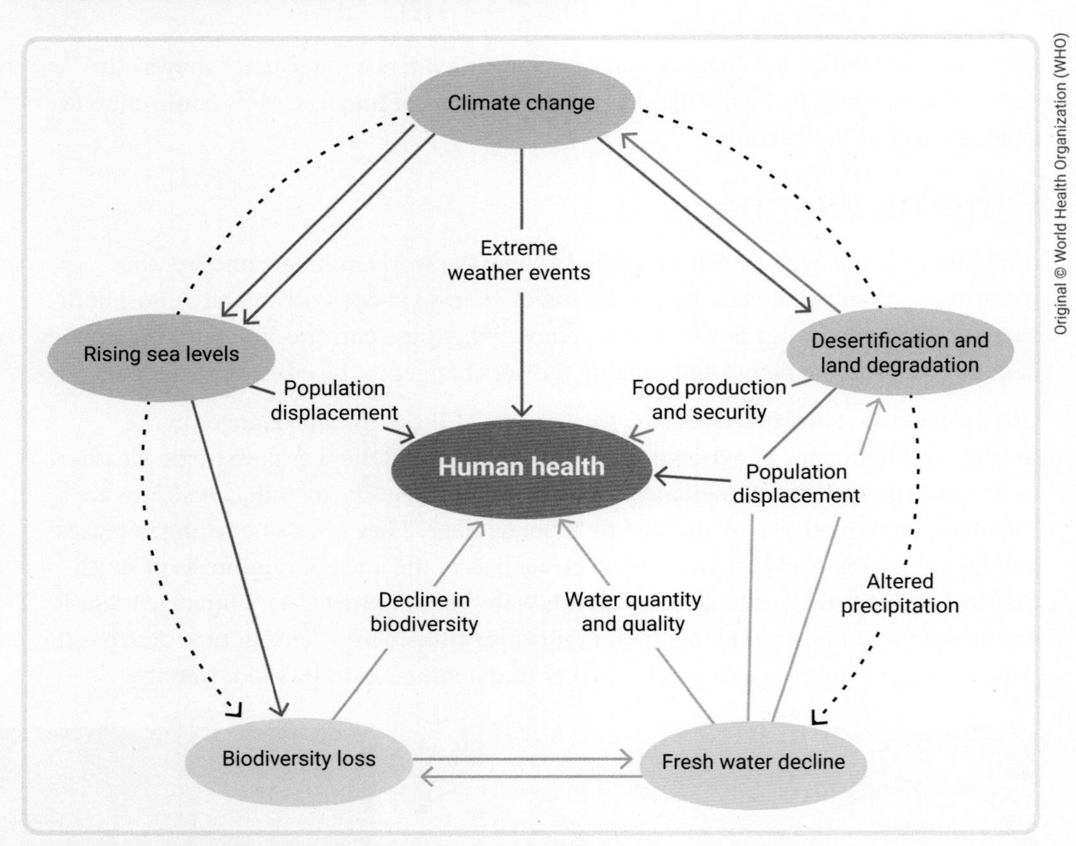

▲ **FIGURE 6.7.2** The links between climate change and health

Disappearing islands and coastlines

Rising sea levels due to climate change are posing a serious threat to millions of people living in low-lying islands and coastlines around the world. Islands in countries such as the Maldives, Solomon Islands, Tuvalu and Papua New Guinea are already dealing with the consequences of rising sea levels, with many people forced to leave their homes and relocate to higher ground or different islands.

On coastlines, very high tides during extreme weather events such as cyclones can result in **storm surges**. These cause destructive flooding, which can damage buildings, roads, beaches and vegetation (see Figure 6.7.3).

storm surge
a brief increase in sea levels due to the combined effect of storms, low air pressure, strong winds and tides

Getty Images/Brook Mitchell

▲ **FIGURE 6.7.3** Storm surges can cause huge amounts of damage during extreme weather events.

6.7 LEARNING CHECK

1 **List** five extreme weather events affecting Australia.
2 **Describe** how extreme weather events have an impact on biodiversity.
3 **Explain** how storm surges occur.
4 **Describe** how human health can be affected by climate change.
5 **Explain** why the effects of rising sea levels are a major concern.
6 **Create** a flow chart to show the links between human activities, climate change and the impacts of climate change.

6.8 Solutions to climate change

BY THE END OF THIS MODULE, YOU WILL BE ABLE TO:

- ✓ explain how different strategies can limit climate change or reduce its extent
- ✓ discuss the barriers that may stop climate change strategies from being implemented
- ✓ describe the key outcomes from the 26th Conference of the Parties to the United Nations Framework Convention on Climate Change (COP26).

Video activities
Carbon farming
Waste as a resource

GET THINKING

As you read through this module and hear about the outcomes of COP26, think about whether we can slow down or stop climate change. Think about strategies you and your family can use to help.

COP26
the 26th Conference of the Parties to the United Nations Framework Convention on Climate Change

net zero greenhouse emissions
when emissions of greenhouse gas do not exceed the amount of gases absorbed or stored

COP26 set the aim of achieving **net zero greenhouse emissions** by 2050. This was the level that modelling by scientists around the world has shown would avoid the most dangerous changes to climate and keep global temperature rises at no more than 1.5°C. In addition to aiming to achieve net zero emissions by 2050, 153 of the 196 countries represented at COP26 agreed to new 2030 emissions targets. To achieve these targets, it is up to all of us to do our part.

Strategies to reduce carbon dioxide emissions

The most important thing we all need to do is greatly reduce the amount of carbon dioxide and other greenhouse gases emitted into the atmosphere. There are many different ways we can do this.

Stop using fossil fuels to generate electricity

To reduce the amount of greenhouse gases we produce, we need to find and use alternative ways to generate electricity. This was identified as a priority at COP26.

Many countries have started to adopt these strategies by phasing out and replacing coal-fired power stations with renewable alternatives such as wind power, geothermal energy, hydroelectricity, hydrogen power and solar energy.

Shutterstock.com/FiledIMAGE; Shutterstock.com/myphotobank.com.au

▲ **FIGURE 6.8.1** Due to our sunny, windy climate, solar and wind energy are the most viable and common forms of renewable energy in Australia.

Find alternative ways to power vehicles

Vehicle emissions are a major source of greenhouse gases. Replacing traditional internal combustion engines with electric engines or alternatives that use hydrogen gas is an important step to reducing climate change. Governments in many countries are providing incentives to encourage people to upgrade to electric vehicles to reduce greenhouse gases and pollution in large cities.

Global shipping and aviation are responsible for 8 per cent of emissions of carbon dioxide. The solution for these modes of transport will be in new technologies, such as the current pilot program that is testing whether solar energy can be used to convert carbon dioxide and water into fuel.

Reducing methane emissions

Methane is a potent greenhouse gas, capable of holding up to 25 times more heat than carbon dioxide. Methane is released as a by-product of producing oil and gas and from the decay of material in landfill. However, most comes from farming livestock such as cows and sheep. Livestock is responsible for about 15 per cent of all greenhouse emissions, half of which come from cattle. Research has discovered that adding just 0.5 per cent of a particular seaweed, *Asparagopsis taxiformis*, to the feed for cattle reduces their emission of methane by up to 80 per cent (see Figure 6.8.2). The Commonwealth Scientific and Industrial Research Organisation has worked with organisations in Queensland to develop a feed for cattle using this seaweed.

Shutterstock.com/Matyas Rehak

▲ **FIGURE 6.8.2** Feeding livestock seaweed is one way to reduce methane emissions.

Another way to reduce emissions is to reduce our consumption of meat and junk food. Research has found that if people who frequently eat meat made small changes to their diet, such as substituting one serve of red meat with chicken or lentils per day, they could reduce their **carbon footprint** by almost half. In the next 15–20 years, agriculture could be responsible for more than half the world's emissions, with 70 per cent coming from meat and dairy alone.

carbon footprint
the amount of carbon dioxide emitted by a person, an organisation or an event

Carbon sequestration

As well as reducing emissions, we can increase the extraction of carbon from the air and store it away from the atmosphere. Australian researchers have come up with a commercially viable method of turning carbon dioxide back into solid carbon that can be used for building materials or other useful products. This is a form of **carbon sequestration**.

carbon sequestration
the process of capturing and storing atmospheric carbon dioxide

A different method of carbon sequestration is to remove carbon dioxide from the air at a large source of emissions, such as a power station, and inject it underground into geological formations as either a liquid or gas. Another important sequestration method is the reforesting of areas using quick-growing forests that trap carbon and can also produce wood as a resource. This helps save our old-growth forests, preventing further land degradation and maintaining biodiversity. Another way to remove carbon dioxide from the atmosphere is to increase the farming of kelp, which grows fast and is very effective at absorbing and storing large quantities of carbon. Some examples of carbon sequestration methods are shown in Figure 6.8.3.

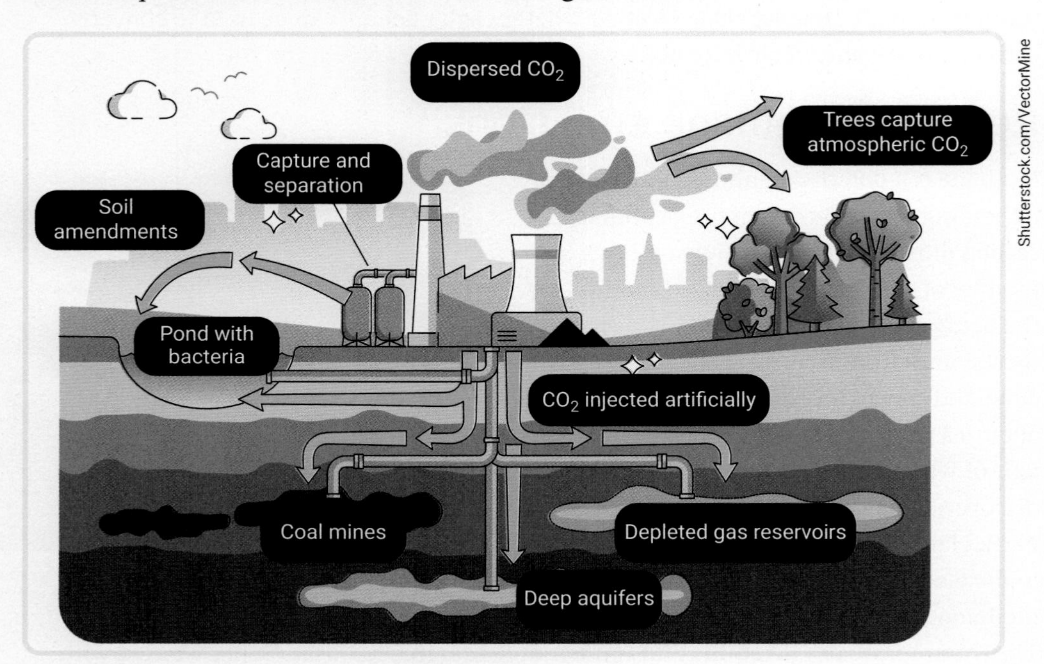

▲ **FIGURE 6.8.3** Carbon sequestration methods

Recycling

finite resource
a limited resource

The effective recycling of **finite resources** instead of manufacturing them from scratch helps reduce both the amount of energy used and the greenhouse emissions produced. Professor Veena Sahajwalla (see Figure 6.8.4), a waste and recycling technology researcher from the University of New South Wales and the 2022 New South Wales Australian of the Year, has created 'green steel' through a process called polymer injection technology. It produces steel from shredded recycled car tyres and other recycled polymers instead of coking coal. This green steel technology also contributes hydrogen to the process and produces more environmentally friendly and cost-effective high-quality steel, resulting in less energy needed and lower carbon emissions.

▲ **FIGURE 6.8.4** Waste and recycling technology researcher Professor Veena Sahajwalla

Barriers to achieving change

There is serious concern among the United Nations and governments around the world that we won't be able to reach our goal of having net zero emissions by 2050. Unless we can make large reductions to our greenhouse gas emissions in the coming decades, global temperature rises of up to 4.5°C could occur by 2100.

Two main barriers could stop us from reaching the goal of net zero emissions by 2050:

1 the economic cost of investing in new technologies and implementing strategies
2 the politics of **mitigation** policies.

mitigation
reducing the severity of an impact or event

There are substantial costs to finding and implementing solutions to address climate change. An additional complication is that because climate change is a problem affecting the whole planet, many nations and governments need to work together to find solutions. Some industrialised countries are able to deal with the costs and pressures caused by climate change. However, many countries will be slower to implement change because their economies are still developing and they rely on cheap energy, or they don't have the resources to manage the impacts of climate change.

Shutterstock.com/Greyboots40

▲ **FIGURE 6.8.5** Adults and school students protesting for climate action

In Australia, we need to do our bit to slow down or stop climate change. What do you think governments in Australia could do to address and find solutions to climate change?

6.8 LEARNING CHECK

1 **List** three strategies that individuals can use to reduce their carbon footprint and help reduce greenhouse gas emissions.
2 **Describe** the aim of COP26.
3 **Describe** three examples of carbon sequestration.
4 **Explain**, using an example, how research is being used to devise means of reducing greenhouse emissions.
5 **Explain** why strategies to reduce climate change may not be implemented.

6.9 Climate change models

SCIENCE AS A HUMAN ENDEAVOUR

BY THE END OF THIS MODULE, YOU WILL BE ABLE TO:

✓ explain how scientists model climate change.

Video activity
Climate change modelling

To understand and model climate change, climate scientists use data sets that contain measurements collected over long periods of time. Scientists use the long-term measurements to construct computer-generated climate models. These data sets provide baseline information that scientists use to compare with the results from climate modelling.

Climate models are based on similar physics, mathematics and processes to weather forecasting models, but make calculations over a much longer time period. Climate models produce projections of how average conditions will change decades into the future.

Climate scientists use technology to monitor climate change and to help develop models to simulate what will happen to Earth's climate in the future. Scientists monitor the climate using observations from satellites hundreds of kilometres above Earth's surface (see Figures 6.9.1 and 6.9.2), as well as measurements on the ground and in the ocean.

NASA/Goddard Space Flight Center

▲ **FIGURE 6.9.1** The Suomi National Polar-orbiting Partnership satellite is an example of a climate satellite.

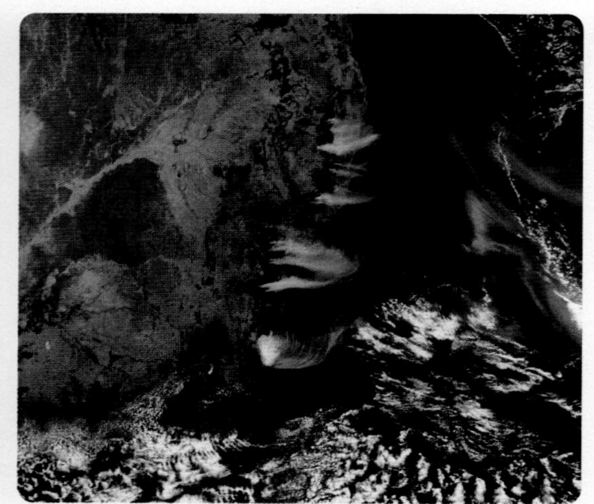

NASA

▲ **FIGURE 6.9.2** An image taken from the climate satellite Suomi National Polar-orbiting Partnership showing smoke and storms due to massive bushfires across eastern Australia in December 2019.

Other measurement activities include ice core drilling in Antarctica and Greenland, monitoring permafrost ground temperatures in deep holes in North America and measuring the extent of coral bleaching on the Great Barrier Reef.

9780170463041

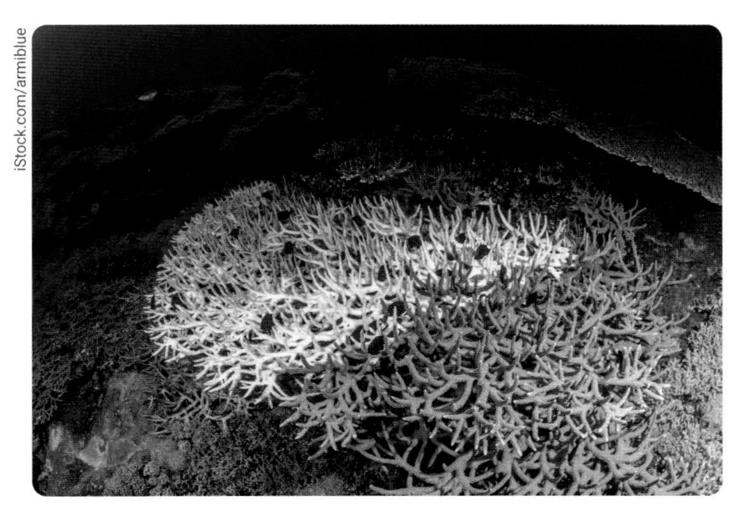

▲ **FIGURE 6.9.3** Data about coral bleaching extent and frequency is used in climate change models.

Climate models can simulate the transfer of water and energy in a climate system. Measurement data is entered into computer models to represent current climate conditions. Starting with this initial data, the model then performs thousands of mathematical calculations based on the physics of the Earth system to produce climate projections. Additional information can be added to the model to prepare different scenarios. For example, reducing methane emissions by 5 per cent in the model will show how that will affect things such as atmospheric and ocean temperatures and sea level.

The three most common types of climate model are:

- energy balance models – monitor changes in Earth's energy to determine the effects of heat accumulation in the oceans and atmosphere in a region or across a continent
- intermediate complexity models – similar to energy balance models but also include impacts from land, oceans and ice sheets. They detect changes in ocean currents, glaciation and atmospheric composition over longer periods
- general circulation models – the most complex and accurate models, used to model climate change by simulating geochemical cycles, atmospheric chemistry, glaciers, ocean circulation and other aspects of the Earth system.

There are many possible outcomes based on each of these models, depending on the information entered (such as future changes to greenhouse gas emissions) and how the models simulate aspects of the climate. Scientists use the range of model results accompanied by a measure of statistical probability.

6.9 LEARNING CHECK

1 **Describe** why climate scientists work with computer simulations to make predictions about the effects of climate change.
2 **Describ**e why climate scientists use remote sensing techniques and satellites to collect data.
3 **Explain** why the data sets used in simulations are from as long a timeframe as possible.
4 **Analyse** why the scenarios each model describes would be accompanied with a statistical probability of it occurring.

SCIENCE INVESTIGATIONS 

6.10 Analysing data to identify patterns and trends

SCIENCE SKILLS IN FOCUS

IN THIS MODULE, YOU WILL FOCUS ON LEARNING AND IMPROVING THESE SKILLS:

- posing questions and making predictions
- analysing representations of data to identify patterns and trends
- analysing models of climate change.

How to analyse data to identify patterns or trends

The purpose of any investigation is to test the validity of a hypothesis. Once the investigation is conducted and data is collected, you need to analyse the data.

1. Check the data. Is there sufficient data to come to a conclusion? How many trials were used? Was that sufficient? If not, collect more data.
2. If repeated trials were conducted, the data would need to be averaged to reduce the effect of variations between trials and make it easier to analyse the data.
3. The averaged data should be presented in a suitable table with column titles and units. This data is what must be analysed to determine whether there is a pattern or trend that will confirm or reject the hypothesis.
4. Plot the data on a graph. A line graph is often used in science, where the independent variable is plotted on the x-axis and the dependent variable data on the y-axis.
5. To determine if there is any causal relationship between the dependent and the independent variables, the data may need to be manipulated to create a straight line with the formula $y = mx + c$, where m is the gradient and c is the y-intercept (if there is one). For example, the graph that is plotted may appear similar to Figure 6.10.1.

▲ FIGURE 6.10.1 A curved line graph

To make it a straight line, it is essential to plot A v. $\frac{1}{t}$, as shown in Figure 6.10.2.

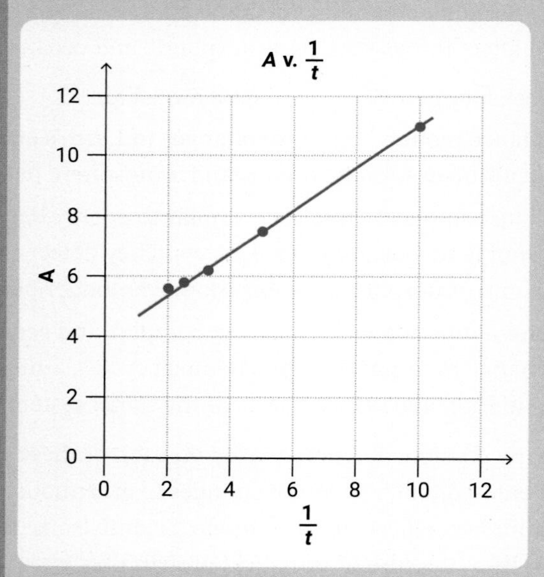

▲ FIGURE 6.10.2 A straight line graph

However, the data may not always appear linear. For example, when the data exhibits an exponential relationship, a smooth curve may be a better way of showing the trend.

Software such as Excel will help. You can use the 'chart menu' to display your data in several ways to help analyse for trends.

9780170463041

INVESTIGATION 1: ANALYSING ICE CORE DATA TO DETERMINE THE RATE OF CARBON DIOXIDE INCREASE

AIM

To analyse data to identify whether the rate of carbon dioxide concentration has increased

MATERIALS

- ☑ graph paper
- ☑ ruler
- ☑ pencil
- ☑ scientific calculator

METHOD

In this investigation, you will use the ice core data in Table 6.10.1. These have been collected from Antarctica by the Scripps Carbon Dioxide Program. The data set is an averaged subset of data that begins in 1959. It has been modified to 5-year intervals for purposes of graphing the data.

1 Look at the data in Table 6.10.1 and determine the scale you need to use on your graph.

Video
Science skills in a minute: Trends and patterns in data

Science skills resource
Science skills in practice: Trends and patterns in data

▼ **TABLE 6.10.1** The average atmospheric concentrations of carbon dioxide in ppm

Year	Carbon dioxide concentration (ppm)
2020	411.99
2015	399.19
2010	388.00
2005	378.16
2000	368.25
1995	359.52
1990	352.84
1985	344.77
1980	337.85
1975	330.34
1970	325.00
1965	319.72
1960	316.68

2 To make the graph easier to analyse, start your concentration axis at 300 ppm.

3 Plot the data and draw a line of best fit.

4 Determine the gradient of the line of best fit. This should be the average rate of increase in carbon dioxide concentration.

EVALUATION

1 Does a line of best fit suit the data or should a smooth curve be drawn instead?

2 If the data fitted a smooth curve rather than a line, what would be the best way of determining the rate of increase?

3 Does using 5-year data points affect the rate of increase of carbon dioxide concentration?

CONCLUSION

Write a statement based on what you were able to determine from this data.

INVESTIGATION 2: MODELLING SEA-LEVEL RISE

AIM

To model sea-level rise

MATERIALS

- 2 × 300 mL cups or beakers
- 80 mL beaker
- Blu-Tack
- marker pen
- 2 ice cubes

METHOD

1. Invert the 80 mL beaker and place it in the centre of the larger beaker (or cup). Use the Blu-Tack to fix it firmly in place. Label the large beaker as '1: continental ice'.
2. Fill the 'continental ice' beaker with enough tap water to just reach the height of the smaller inverted beaker.
3. Label the second large beaker as '2: sea ice'. Place the same amount of water in this beaker.
4. Place an ice cube on top of the 80 mL beaker in Beaker 1 so it is above the water, and the same sized ice cube into the water in Beaker 2 so that it floats in the water.
5. Immediately mark the level of the water in both beakers.
6. Allow the ice in each beaker to completely melt and observe the water level.

EVALUATION

What did you observe? Try to explain why this occurred.

INVESTIGATION 3: COMPARING THE FREEZING OF FRESH WATER AND SALTWATER

AIM

To compare the freezing of fresh water and salt water

MATERIALS

- 2 × 350 mL plastic bottles with screw lids (clear, cleaned fruit juice or milk bottles with screw lids would be suitable)
- measuring cylinder
- salt
- marker pen
- funnel
- tablespoon
- access to a freezer

METHOD

1. Place 250 mL of water into each bottle. Tightly screw the lid on one, marking it with the number '1'. Mark the level of the water in the bottle.
2. Use a funnel to add 2 tablespoons of salt to the other bottle. Screw the lid on tightly and mark this with the number '2'. Shake the bottle until all the salt has dissolved. Mark the level of the saltwater in the bottle.
3. Place both bottles in a freezer next to each other.
4. Observe each bottle every hour for 3–4 hours, if possible, then remove them from the freezer the following day and check the levels.

EVALUATION

What did you see? Try to explain why you think this has happened.

CONCLUSION

Write a statement based on what you observed in completing the modelling of sea-level rise.

9780170463041

6 REVIEW

REMEMBERING

1 **Define** 'greenhouse effect'.

2 **List** the four spheres that make up the Earth system.

3 **Describe** the difference between the greenhouse effect and the enhanced greenhouse effect.

4 **Describe** the difference between climate and weather.

5 **Describe** what is significant about the Bramble Cay melomys.

UNDERSTANDING

6 **Explain** why the greenhouse effect is important to the existence of life on Earth.

7 **Describe** the characteristics used by Köppen to classify climate types.

8 **Describe** the parts of the electromagnetic spectrum that are:

a shortwave radiation.

b longwave radiation.

9 **Construct** a table as shown below. List the sources responsible for the increased concentration of each gas in the atmosphere.

	Carbon dioxide	Methane	Nitrous oxide
Sources of increased atmospheric concentration			

10 **Explain** why the deep ocean current might be referred to as Earth's energy conveyor belt.

11 **Describe** what the water and carbon cycles are responsible for in the diagram below.

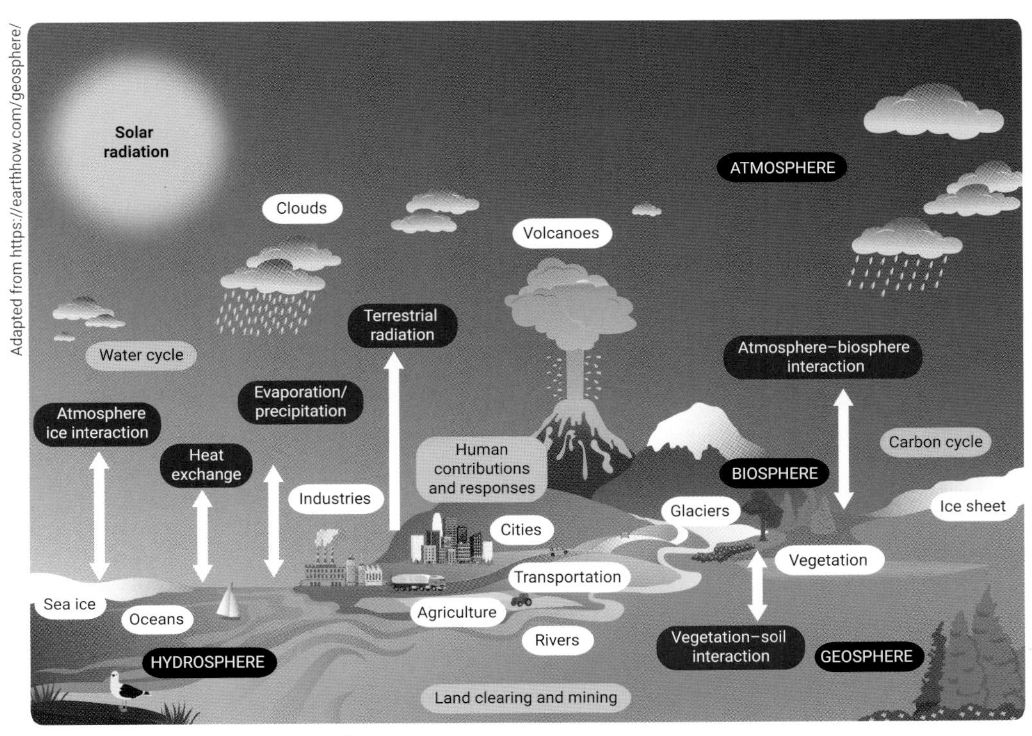

The interaction between Earth's four spheres

APPLYING

12 **Explain** why the acidification of the oceans is occurring.

13 **Compare** the effects of reducing atmospheric carbon dioxide and methane concentrations.

14 **Explain** why the following diagram shows no continental climate zone in the southern hemisphere.

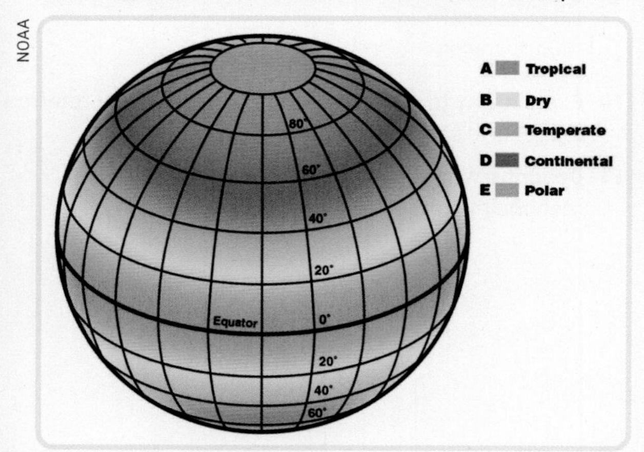

The distribution of climate zones

15 **Compare** deep ocean current circulation to atmospheric convection currents.

16 **Explain** how deep ocean currents affect marine life.

17 **Explain** why biodiversity and climate change should be considered a single problem rather than separate problems.

ANALYSING

18 **Describe** the likely impact on polar bears of continually reduced amounts of sea ice.

19 **Describe** what will happen beyond 2050 if net zero emissions is not achieved.

EVALUATING

20 **Explain** how individuals changing their behaviour can have a significant effect on climate change.

21 **Compare** the effects of melting permafrost compared to the effects of melting of continental ice sheets such as Greenland and Antarctica.

22 **Explain** why recycling would be beneficial to reducing global warming.

23 The following diagram shows the concentration of atmospheric carbon dioxide over 800 000 years.

a **Explain** why there are peaks and troughs in the earlier part of the graph.

b **Explain** why the orange line has risen so sharply.

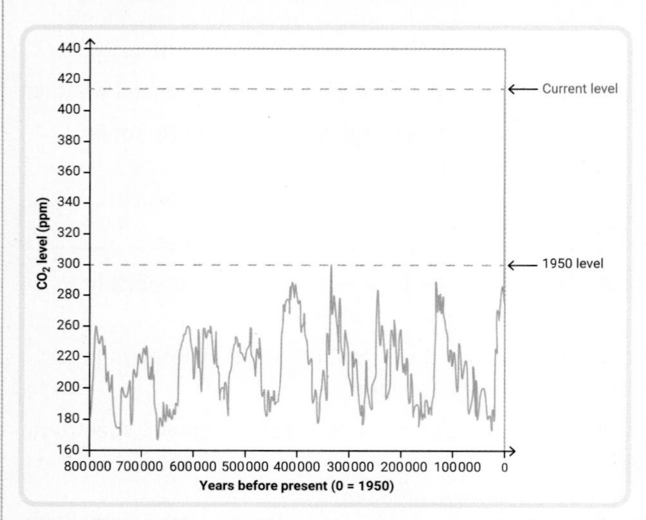

Atmospheric carbon dioxide concentration over the last 800 000 years

24 Consider the following photo. **Explain** how what you are seeing is related to climate change.

Storm surges can cause huge amounts of damage during extreme weather events.

25 **Explain** what some of the barriers are to achieving progress on climate mitigation strategies.

26 **Explain** why temperate climate zones are located close to coastal regions.

27 **Explain** why a dry climate zone occurs in central Asia.

9780170463041

BIG SCIENCE CHALLENGE PROJECT #6

1 Connect what you've learned

By completing the modules in this chapter, you have learned about the causes and effects of climate change and what can be done to slow down or reduce its severity. The global plan is to achieve net zero emissions by 2050, but will this be enough? Reducing climate change is everyone's responsibility. This challenge gets you to reflect on how your actions and behaviours contribute to climate change by determining your individual carbon footprint. You can find out more by asking your family and friends to also undertake a carbon footprint analysis. But let's start with you.

2 Check your thinking

The first thing you need to do is reflect on your daily activities that contribute to greenhouse gas emissions. Start by listing as many activities as possible that you are engaged in that you know are linked to greenhouse gas emissions.

- Is it possible to modify any of these activities to reduce your impact?
- What changes to your lifestyle are you willing and realistically able to make to reduce your individual carbon footprint?
- To help, there are many free online carbon footprint calculators that you can use to calculate your carbon footprint. Do a search online to locate a suitable Australian carbon footprint calculator.

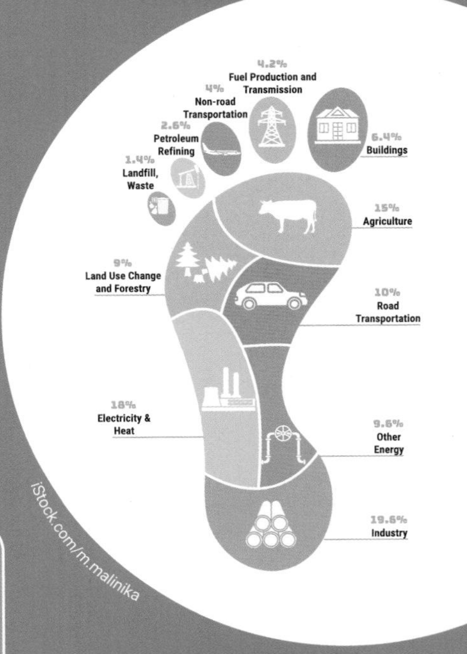

What is your carbon footprint?

3 Make an action plan

Use an online carbon footprint calculator to determine your current carbon footprint. Make a note of the result.

Now identify the changes you are willing to make. Redo your calculation and see the difference. You may wish to do this several times with a different mix of changes to see the difference it makes to your footprint.

Encourage your whole family to be involved and determine your family's carbon footprint.

4 Communicate

At the completion of the project, design a simple pamphlet to encourage other members of your class to think about reducing their individual carbon footprint. You may like to set up a process where other members of your class can indicate whether they wish to take part in an ongoing project to measure the collective change to your class's carbon footprint. There is nothing stopping you encouraging others beyond your class to get involved. The more people who participate, the better off the planet will be!

7 Motion

9780170463041

BIG SCIENCE CHALLENGE 

NASA/Joel Kowsky

With a mission to search for extraterrestrial life and another habitable planet, humans have been launching rockets and rovers in an attempt to explore Mars since the 1960s. The first exploration device to reach Mars landed in 1971. As of 2021, there are two United States rovers and one Chinese rover operational on Mars, surveying and studying its surface and atmosphere. NASA's most recent landing was the rover Perseverance, which has one key objective (of many) to search for signs of ancient microbial life. But how did they all get there? How far is it between Earth and Mars? How long did it take to fly there? How fast did they travel? How were these speeds achieved? Motion in outer space can be explained by the same principles of motion that apply here on Earth.

- **What do you already know about rockets and how they travel?**
- **How does a rocket's motion change when it is close to Earth compared to when it is far away?**
- **How many similarities and differences can you think of between a real rocket and a toy rocket launched from the ground with a firework?**

 **SCIENCE CHALLENGE ACCEPTED!**

At the end of this chapter, you can complete the Big Science Challenge Project #7. You can use the information you learn in this chapter to complete the project.

Assessments
- Prior knowledge quiz
- Chapter review questions
- End-of-chapter test
- Portfolio assessment task: Data test

Videos
- Science skills in a minute: Linear data **(7.11)**
- Video activities: Speed, velocity and acceleration **(7.4)**; Newton's laws of motion **(7.5)**; Objects falling in a vacuum **(7.7)**; Science of golf **(7.8)**; Female crash test dummies **(7.10)**

Science skills resources
- Science skills in practice: Analysing linear data **(7.11)**
- Extra science investigations: Using ticker timers **(7.4)**; Investigating Newton's second and third laws of motion **(7.6)**; Stopping distance **(7.8)**; Motion in traffic **(7.8)**; Crumple zones **(7.10)**

Interactive resources
- Label: Speed-time graphs **(7.3)**
- Simulation: Forces and motion **(7.5)**; Gravity force lab **(7.7)**
- Quizzes: Distance **(7.1)**; Calculating speed **(7.2)**

Nelson MindTap

To access these resources and many more, visit: **cengage.com.au/nelsonmindtap**

7.1 Distance travelled

By the end of this module, you will be able to:

- ✓ understand the measurement and calculation of distance travelled by moving bodies
- ✓ construct and interpret distance–time graphs to represent the motion of moving bodies.

GET THINKING

Flip through this module and make a list of the headings. Add these headings to a mind map and draw lines that link terms where you think there is a connection. Annotate each of the links with a description of why you think these terms are connected.

Quiz
Distance

SI units

In science, especially physics, there are many different types of units in which different quantities can be measured. This can be confusing at times, especially when you have to change units to make them compatible with one another in formulas. The International System of Units, also known as the SI Unit System, is a globally accepted list of measurement units that set a 'base' unit for all types of measurements. SI units are all compatible with one another and can be used together in mathematical formulas and physics equations. When all inputs to a formula are in SI units, the output for the equation will also be in SI units. Table 7.1.1 shows the SI units for some commonly used measurements, many of which we will use in this chapter about motion.

▼ **TABLE 7.1.1** The SI units for some commonly used measurements

Measurement	SI unit name	SI unit symbol
Length or distance	metre	m
Time	second	s
Temperature	Kelvin	K
Mass	kilogram	kg
Speed	metres per second	$m\ s^{-1}$ or m/s
Acceleration	metres per second squared	m/s^2
Force	Newton	N
Energy	joule	J

Distance as a quantity

body
an object, person or thing that has mass

position
the location of a body

In physics, we often refer to an object or thing that is moving as a **body**. A body could be anything made of matter and includes living things like humans and large objects like cars. A human, a ball and a car are all types of bodies.

When a body is moving or 'in motion', the **position** or location of the body changes. When a ball is thrown, the position changes from someone's hand to different points in the air to the landing site.

When a car travels down the street, its position changes from one end of the street to the other end. The body will travel a certain path to change its position and the length of this path is known as the **distance** travelled by the body. Distance also features prominently in many fields of athletics (see Figure 7.1.1).

distance
the length of the pathway taken by a moving body

7.1

▲ **FIGURE 7.1.1** In most races, the person who completes the same distance in the shortest amount of time is the winner.

The symbol used for distance in physics is usually a lowercase *d*. The SI unit for distance is metres (m) but can often be given in millimetres (mm) or centimetres (cm) for small distances, or kilometres (km) for bigger distances. Some conversions for distance measurements are shown in Figure 7.1.2.

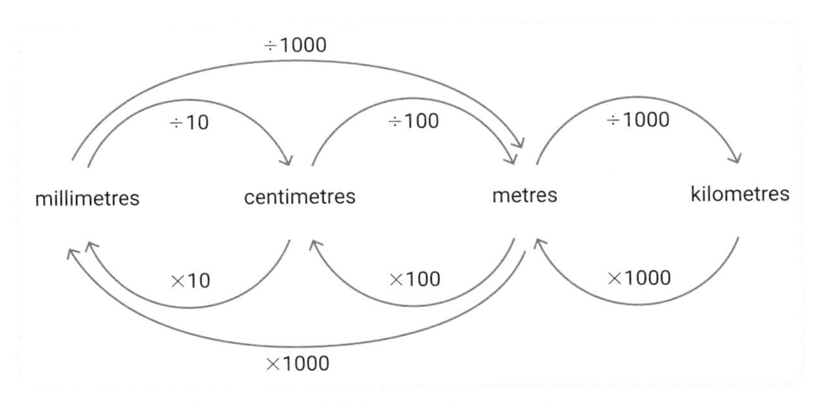

▲ **FIGURE 7.1.2** A summary of distance unit conversions

Measuring distance

Distances can be measured using a variety of tools. Shorter distances might be measured with a ruler or a tape measure, while longer distances could be measured with a measuring wheel (see Figure 7.1.3) or using GPS software. Laser tape measures allow you to point a laser from a start position to an end object and the straight-line distance between them can be determined by the device (see Figure 7.1.4).

Shutterstock.com/I AM NIKOM

▲ **FIGURE 7.1.3** Measuring wheels can measure longer distances by multiplying how many circumferences of the circle have been traced along the ground.

iStock.com/Dmytro Varavin

▲ **FIGURE 7.1.4** Laser measuring devices measure straight-line distances by means of reflected light.

When a body's movement has been in a straight line, the distance travelled is simply the length of the straight path. When the movement has been on a path that changes direction, the distance travelled by a body is always the total length of the path the body has taken during that time. For example, if you were to walk from one corner of a rugby field to the opposite corner along the field's boundary lines, the distance travelled is half the perimeter of the whole field, not the shorter distance across the field between the end position and the start position (see Figure 7.1.5).

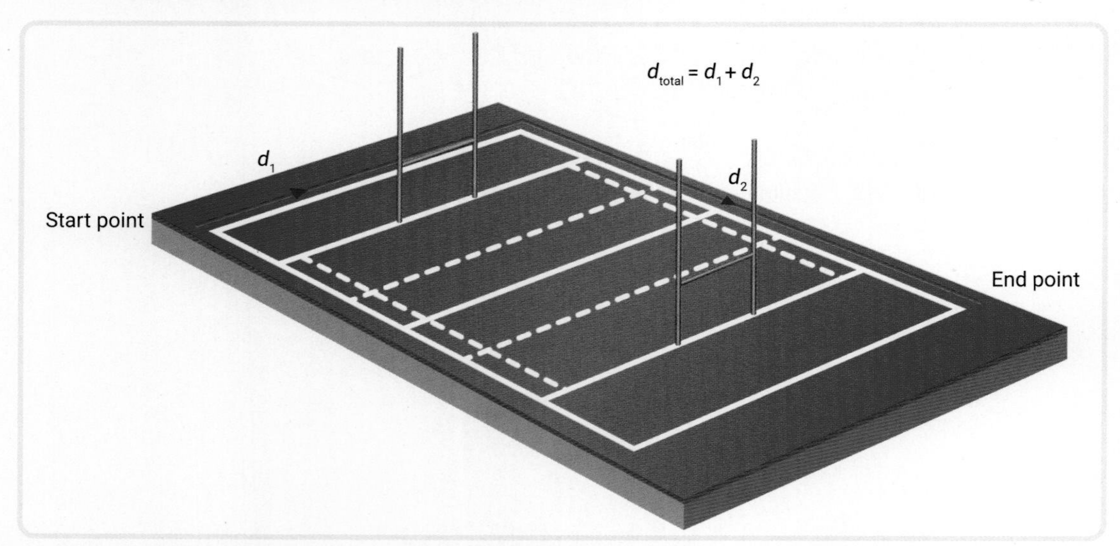

▲ **FIGURE 7.1.5** The distance travelled is the total length of the path taken.

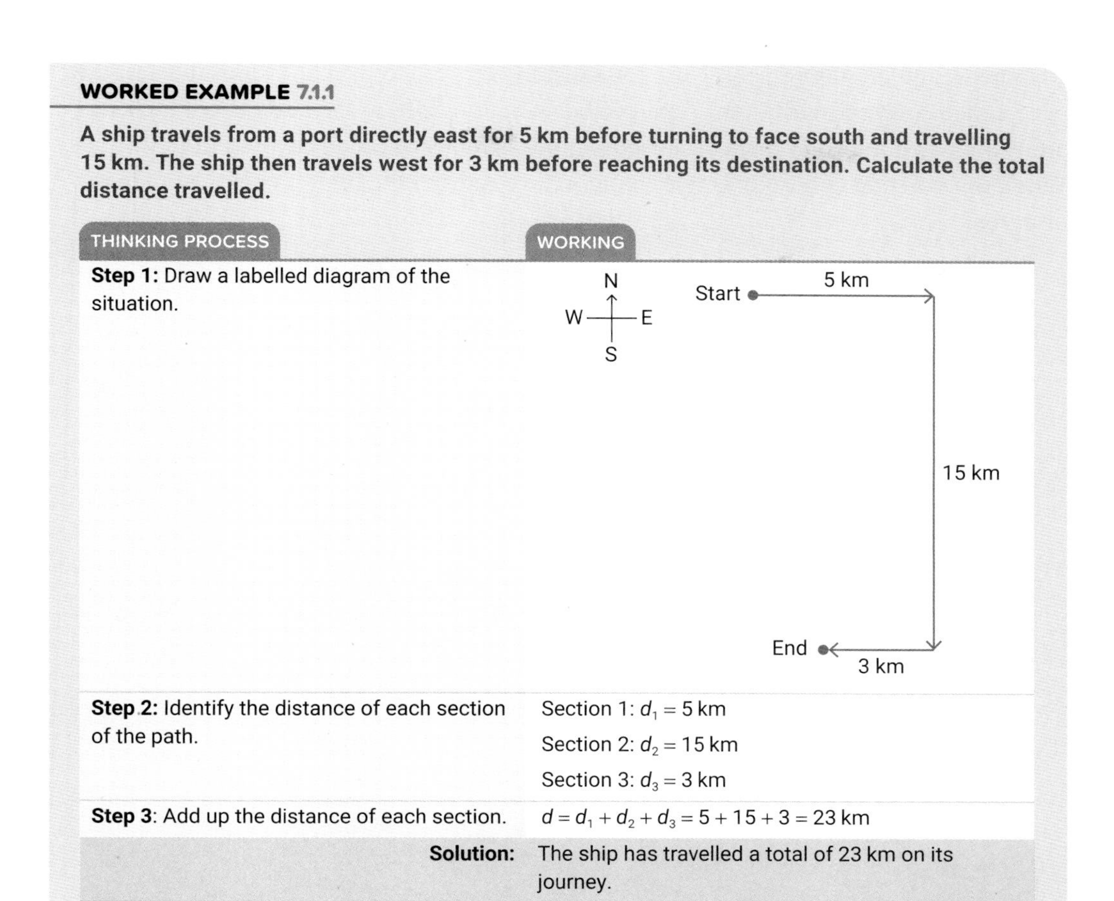

WORKED EXAMPLE 7.1.1

A ship travels from a port directly east for 5 km before turning to face south and travelling 15 km. The ship then travels west for 3 km before reaching its destination. Calculate the total distance travelled.

THINKING PROCESS	WORKING
Step 1: Draw a labelled diagram of the situation.	N, W, E, S; Start; 5 km; 15 km; End; 3 km
Step 2: Identify the distance of each section of the path.	Section 1: $d_1 = 5$ km Section 2: $d_2 = 15$ km Section 3: $d_3 = 3$ km
Step 3: Add up the distance of each section.	$d = d_1 + d_2 + d_3 = 5 + 15 + 3 = 23$ km
Solution:	The ship has travelled a total of 23 km on its journey.

Distance–time graphs

Visualising the journey of a moving body can be made easier by using different types of graphs. One graph that is particularly helpful is a distance–time graph (see Figure 7.1.6), which plots the distance travelled by a moving body against time. As time increases on the *x*-axis, the *y*-axis plots the total distance the body has travelled at that time.

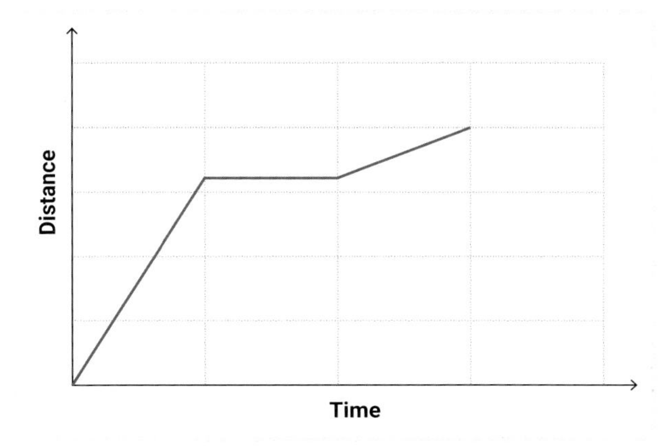

▲ **FIGURE 7.1.6** A distance–time graph for a moving body

stationary
keeping a constant position or not moving in any direction; at rest

rest
the state of being stationary, having a speed of 0 m/s

Distance–time graphs can show the total distance travelled by a body and the amount of distance covered in different time periods. Sections where the body is not moving are shown as horizontal lines because no more distance is being covered in that time. When the position of a body is not changing, we refer to the body as being **stationary**, or at **rest**.

WORKED EXAMPLE 7.1.2

Consider the distance–time graph in Figure 7.1.7, and then complete the questions below.

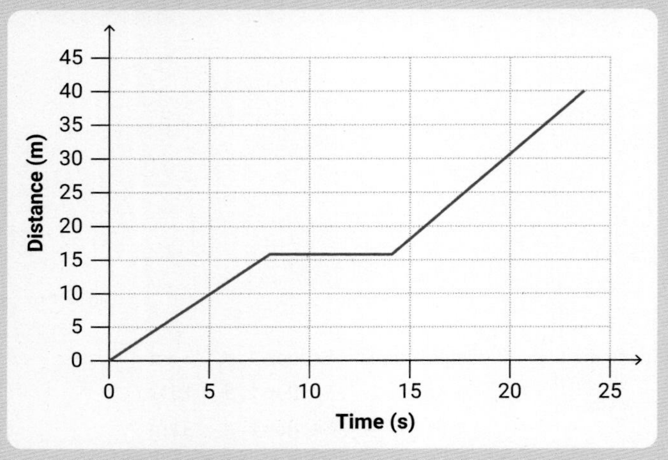

▲ **FIGURE 7.1.7** A distance–time graph for a moving body

QUESTIONS	WORKING
1 Identify the maximum distance travelled by the body in the time shown.	The total distance travelled in the period is 40 m
2 Determine the distance travelled by the body between 15 s and 20 s.	Distance travelled after 15 s = 18 m Distance travelled after 20 s = 31 m Distance travelled between 15 s and 20 s = 31 − 18 = 13 m
3 Identify the period when the body was stationary.	The body is not moving from 8 s to 14 s, shown by no increase in distance between these times.

7.1 LEARNING CHECK

1 **Convert** the following measurements into the units stated.
  a 0.987 km to m
  b 46 700 mm to m
  c 33.4 m to cm
  d 0.556 m to mm
  e 4.0 km to cm
  f 60 000 mm to km
2 **Determine** the total distance (in metres) travelled by each moving body.
  a Sal ran 100 m down to the local running track and did one lap of the 500 m track before running home again.
  b A plane flew 125 km east before turning and travelling 42 000 m north.
  c A dog ran the perimeter of a square yard three times. The yard is 6 m wide.
  d Charlie walked 400 m to the corner shop to buy an ice-cream and then another 320 m to the post office before journeying home along the same route.
3 **Construct** a distance–time graph for Charlie's journey in Question 2d.
4 For the distance–time graph shown in Figure 7.1.8:
  a **identify** the maximum distance travelled by the body in the time shown.
  b **determine** the distance travelled by the body in the first 25 minutes.
  c **identify** the time period when the body was stationary.

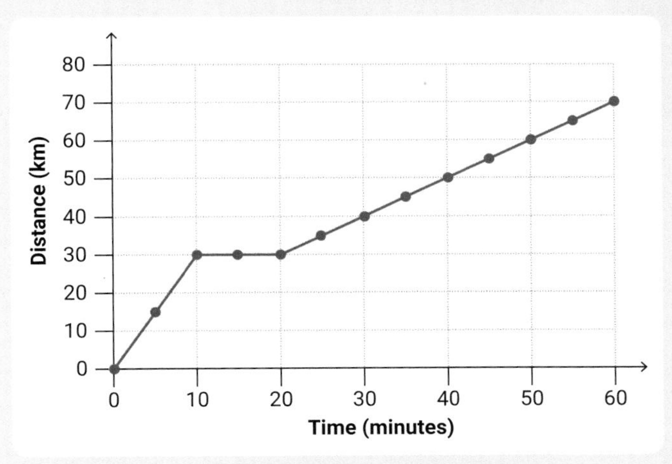

▲ **FIGURE 7.1.8** A distance–time graph for a moving body

7.2 Speed

BY THE END OF THIS MODULE, YOU WILL BE ABLE TO:

- ✓ understand the measurement and calculation of speed
- ✓ construct and interpret speed–time graphs to represent the motion of moving bodies
- ✓ interpret distance–time graphs and use them to calculate speed.

speed
a measure of how much distance a body covers per unit of time; typically metres per second (m/s)

GET THINKING

Make a list of words that you associate with the term 'speed'. As you work through this module, tick them off if they come up.

Shutterstock.com/Harry Collins Photography

▲ **FIGURE 7.2.1** Peregrine falcons dive vertically at a speed of up to 400 km/h, which is about 111 m/s.

What is speed?

Speed is a measurement that describes how fast a body is travelling. Speed is a rate measurement that indicates how much distance is covered per unit of time. It is usually represented by a lowercase *v*. Speed can commonly be measured in kilometres per hour (km/h or km h^{-1}) but the SI unit for speed is metres per second (m/s or m s^{-1}). If a body is travelling at a speed of 2 m/s, it covers 2 m of distance every second. Bodies that are at rest have no speed or a speed of 0 m/s and are not covering any distance. They are stationary. The speed of peregrine falcons is shown in Figure 7.2.1.

To convert between speed units, the following rules can be used:

$$\text{speed in}\frac{\text{km}}{\text{h}} = \text{speed in}\frac{\text{m}}{\text{s}} \times 3.6$$

$$\text{speed in}\frac{\text{m}}{\text{s}} = \text{speed in}\frac{\text{km}}{\text{h}} \div 3.6$$

The box below shows how these rules can be derived from smaller steps converting the distance and time measurements.

Quiz
Calculating speed

Converting units of speed

Convert metres to kilometres:

$$2\frac{\text{m}}{\text{s}} = 2 \div 1000\frac{\text{km}}{\text{s}} = 0.002\frac{\text{km}}{\text{s}}$$

Convert seconds to minutes:

$$0.002\frac{\text{km}}{\text{s}} = 0.002 \times 60\frac{\text{km}}{\text{min}} = 0.12\frac{\text{km}}{\text{min}}$$

Convert minutes to hours:

$$0.12\frac{\text{km}}{\text{h}} = 0.12 \times 60\frac{\text{km}}{\text{h}} = 7.2\frac{\text{km}}{\text{h}}$$

Combining steps into one step:

$$2\frac{\text{m}}{\text{s}} \div 1000 \times 60 \times 60 = 7.2\frac{\text{km}}{\text{h}}$$

$$2\frac{\text{m}}{\text{s}} \times 3.6 = 7.2\frac{\text{km}}{\text{h}}$$

Rules

To convert from m/s to km/h, multiply by 3.6.

To convert from km/h to m/s, divide by 3.6.

Speed–time graphs

Some bodies travel at a constant speed, such as a car travelling down a highway at the speed limit of 100 km/h (27.8 m/s). Or bodies can travel with a varying speed. Speed–time graphs are helpful tools to analyse the speed of a body during a journey. Figure 7.2.2 shows a speed–time graph for a 45-second journey where a body starts with a speed of 0 m/s, increases speed to a speed of 60 m/s in the first 10 seconds before remaining at a constant speed for 20 seconds and then slowing down to stop again.

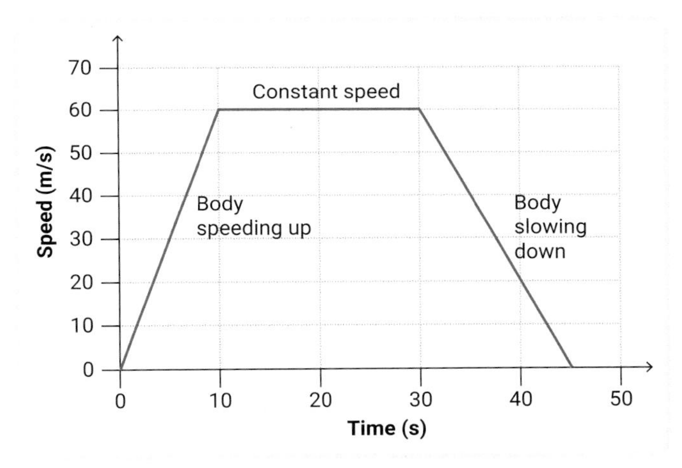

▲ **FIGURE 7.2.2** A speed–time graph for a moving body

Measuring and calculating speed

Speed is generally not measured directly as a single measurement. Speed can be calculated using the following formula:

$$\text{Speed} = \frac{\text{total distance travelled}}{\text{total time taken}}$$

$$v = \frac{d}{t}$$

where v is speed measured in metres per second (m/s), d is the total distance travelled measured in metres (m) and t is the total time taken measured in seconds (s).

WORKED EXAMPLE 7.2.1

Calculate the speed of a runner who runs 100 m in 25 s.

THINKING PROCESS	WORKING
Step 1: Identify known and unknown variables. Ensure known values are in SI units.	$v = ?$ $d = 100$ m $t = 25$ s
Step 2: Choose an appropriate formula.	$v = \frac{d}{t}$
Step 3: Substitute known variables into the formula.	$v = \frac{100}{25}$
Step 4: Rearrange using algebra (if necessary) and solve.	$v = 4\frac{\text{m}}{\text{s}}$
Solution:	The runner is running at a speed of 4 m/s.

This formula can also be used to calculate distance or time depending on the information given.

WORKED EXAMPLE 7.2.2

Calculate how long it would take a bee to fly 20 m if it were to fly at a speed of 0.25 m/s.

THINKING PROCESS	WORKING
Step 1: Identify known and unknown variables. Ensure known values are in SI units.	$t = ?$ $d = 20$ m $v = 0.25$ m/s
Step 2: Choose an appropriate formula.	$v = \frac{d}{t}$
Step 3: Substitute known values into the formula.	$0.25 = \frac{20}{t}$
Step 4: Rearrange using algebra (if necessary) and solve.	$0.25 \times t = \frac{20}{t} \times t$ $0.25t = 20$ $t = \frac{20}{0.25}$ $t = 80$ s
Solution:	It would take the bee 80 s or 1 minute and 20 seconds to fly 20 m.

Measuring speed

☆ ACTIVITY 7.2

You will need:

- stopwatch
- space outside to run

Go outside and measure a distance of 20 m. Mark both ends of this distance with some objects.

Use a stopwatch to determine how long it takes you to run the 20 m.

a **Calculate** your average running speed.

b **Evaluate** whether you think you were running at a constant speed for the whole 20 m.

c **Explain** how you measured the 20 m distance and **discuss** whether you think this was an accurate measurement or not.

7.2 LEARNING CHECK

1 **Convert** the following speed measurements to the units stated.

  a 23 km/h to m/s

  b 10 m/s to km/h.

2 **Calculate** the speed of a dog that runs at a constant speed and covers 100 m in 30 s.

3 **Calculate** the time it would take for a snail to crawl along the edge of a 50 m pool if its speed was 0.0072 km/h.

4 Speed cameras are used on roads to detect drivers who drive above the speed limit. The cameras take time-stamped photos of cars as they move over markings on the road that are a fixed distance apart. Based on what you know about measuring and calculating speed, **explain** how the speed camera is programmed to be able to determine how fast the driver is going.

Shutterstock.com/PhotoIris2021

▲ **FIGURE 7.2.3** Snails move very slowly.

iStock.com/JoeClemson

▲ **FIGURE 7.2.4** Speed cameras take photos of cars moving across road markings.

7.3 Using graphs to determine speed

BY THE END OF THIS MODULE, YOU WILL BE ABLE TO:

✓ interpret distance–time graphs and use them to calculate average speed.

Interactive resource
Label: Speed-time graphs

Interpreting distance–time graphs

Information about the speed of a body can be gained by interpreting a distance–time graph. Figure 7.3.1 shows the distance–time graph and speed–time graph for a car's 1-minute journey.

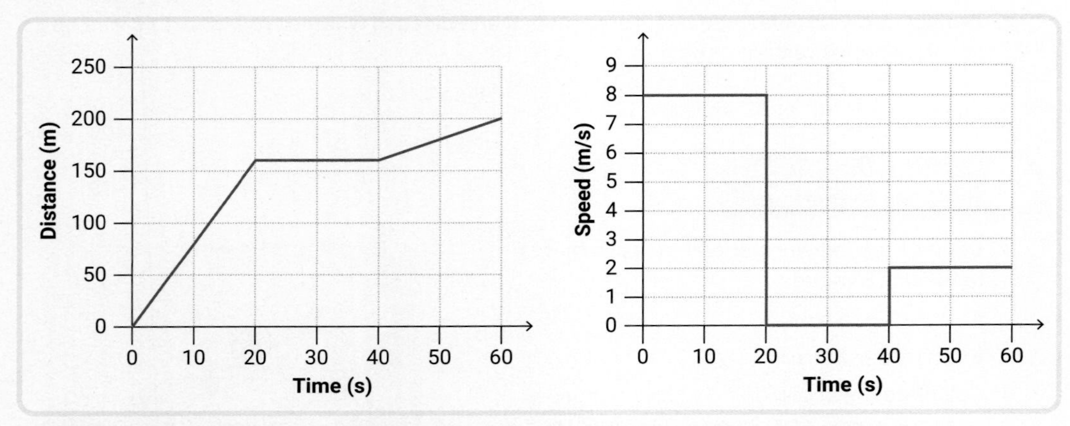

▲ **FIGURE 7.3.1** **(a) A distance–time graph and (b) a speed–time graph for a car's journey**

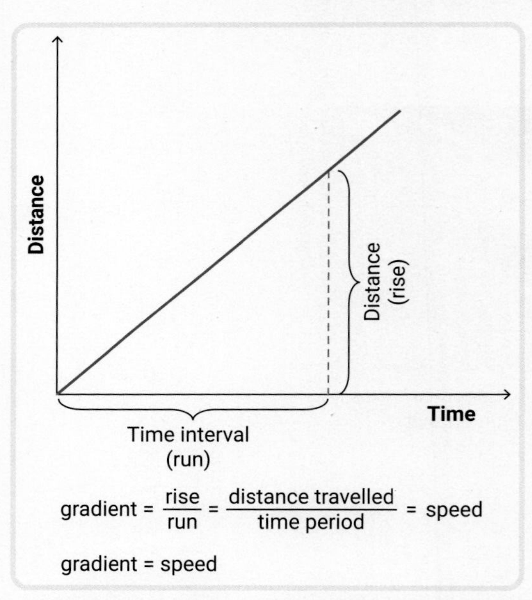

▲ **FIGURE 7.3.2** **The gradient of a distance–time graph is the speed.**

The speed–time graph shows that the car travels at a constant speed of 8 m/s for the first 20 s of the journey. This can be seen on the distance–time graph, as the distance is increasing at a constant rate from 0 s to 20 s. When the car is not moving, the distance–time graph shows a horizontal line from 25 s to 40 s. On the speed–time graph, this is shown by a reading of 0 m/s during the same time.

For the last 20 s of the journey, the car has a constant speed of 2 m/s, as shown on the speed–time graph. On the distance–time graph, the distance increases, but the slope is less steep than it was for the first 20 s. Since the car is travelling much more slowly, it covers less distance in the same period of time. This results in a less steep line and a smaller **gradient**. The gradient is the slope of a linear section of the distance–time graph and can be used to calculate the speed. Figure 7.3.2 shows that the gradient of a distance–time graph is equal to the speed of the body.

gradient
a measure of the slope of a straight line on a graph

WORKED EXAMPLE 7.3.1

Use the distance-time graph Figure 7.3.1 (a) to determine the speed of a car throughout its journey.

THINKING PROCESS	WORKING
Step 1: Determine the car's speed from (0, 0) to (20, 160).	$\text{speed} = \text{gradient} = \frac{\text{rise}}{\text{run}}$ $= \frac{160-0}{20-0} = \frac{160}{20} = 8\text{ m/s}$
Step 2: Determine the car's speed from (20, 160) to (40, 160).	$\text{speed} = \text{gradient} = \frac{\text{rise}}{\text{run}}$ $= \frac{160-160}{40-20} = \frac{0}{20} = 0\text{ m/s}$
Step 3: Determine the car's speed from (40, 160) to (60, 200).	$\text{speed} = \text{gradient} = \frac{\text{rise}}{\text{run}}$ $= \frac{200-160}{60-40} = \frac{40}{20} = 2\text{ m/s}$
Solution:	The car is travelling at a speed of 8 m/s for 20 s, is stationary for 20 s and then travels at a speed of 2 m/s for the final 20 s.

Average speed

Sometimes the speed of a moving body will be changing and the distance–time graph does not have a constant gradient. Figure 7.3.3 shows a distance–time graph for a body that is constantly getting faster across the period shown.

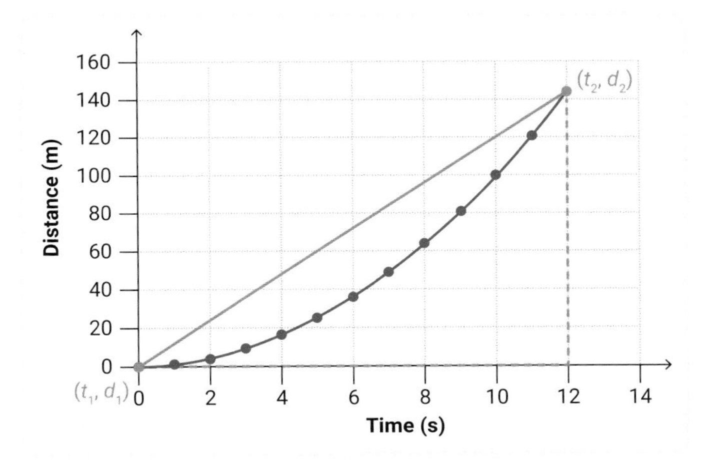

▲ **FIGURE 7.3.3** The average speed over a certain period can be determined for a body that is moving with changing speed.

The speed and, therefore, the gradient are increasing every second. We can determine the average speed of the body over the period by using the same formula for speed when it is constant.

$$\text{Average speed} = \frac{\text{total distance covered}}{\text{time period}}$$

$$v_{av} = \frac{d_2 - d_1}{t_2 - t_1}$$

where d_2 is the distance (m) covered at the end of the time period, d_1 is the distance (m) covered at the start of the time period, t_2 is the end time of the time period, and t_1 is the start time of the time period.

WORKED EXAMPLE 7.3.2

Consider Figure 7.3.4 and determine the average speed of the moving body in the third second (between $t = 2$ s and $t = 3$ s).

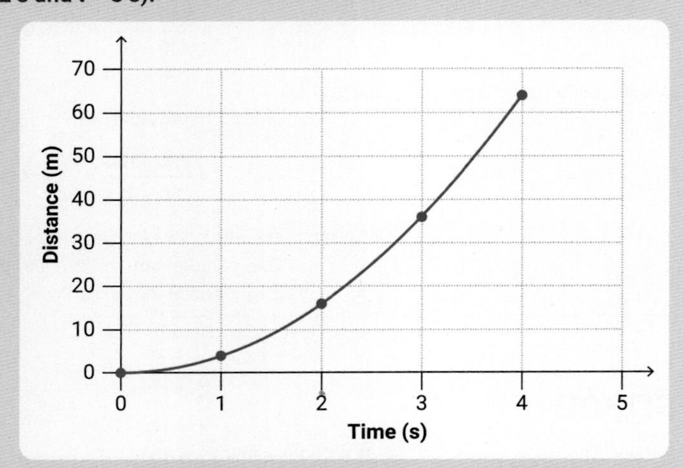

▲ **FIGURE 7.3.4** A distance–time graph for a body with changing speed (accelerating)

THINKING PROCESS	WORKING
Step 1: Identify known and unknown variables.	$v_{av} = ?$ $t_1 = 2$ s $t_2 = 3$ s $d_1 = 16$ m $d_2 = 36$ m

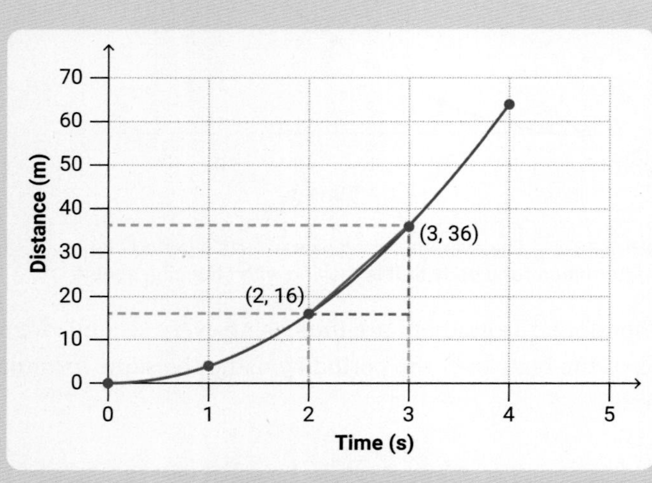

▲ **FIGURE 7.3.5** Using two points on a graph to find the average speed

Step 2: Identify appropriate formula.	$v_{av} = \frac{d_2 - d_1}{t_2 - t_1}$
Step 3: Substitute known variables into the formula.	$v_{av} = \frac{36 - 16}{3 - 2}$
Step 4: Rearrange using algebra (if necessary) and solve.	$v_{av} = \frac{20}{1} = 20\frac{m}{s}$
Solution:	The average speed of the body in the third second is 20 m/s.

7.3 LEARNING CHECK

Consider the distance–time graph shown in Figure 7.3.6.

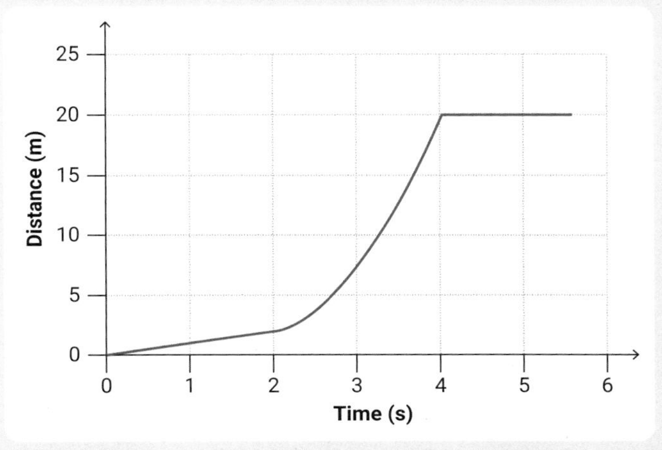

▲ **FIGURE 7.3.6** A distance–time graph for a moving body

1. **Identify** the total distance covered by the body.
2. **Identify** when the body was stationary.
3. **Identify** when the body was travelling with a constant speed.
4. **Identify** when the body was travelling with a changing speed.
5. **Determine** the speed of the body in the first 2 seconds of its motion.
6. **Determine** the average speed of the body between 3 s and 4 s.

7.4 Acceleration

BY THE END OF THIS MODULE, YOU WILL BE ABLE TO:

- ✓ understand the measurement and calculation of acceleration
- ✓ construct and interpret speed–time graphs to represent the motion of moving bodies
- ✓ interpret distance–time graphs and speed–time graphs in terms of acceleration and use speed–time graphs to calculate acceleration.

Video activity
Speed, velocity and acceleration

Extra science investigation
Using ticker timers

GET THINKING

This module is all about acceleration. You might have already heard this word or similar ones such as accelerate or decelerate. Make a list of 10 words that you think have some connection to the term 'acceleration'.

Defining acceleration

acceleration
the rate of change of speed

When a parked car starts and then travels down the road, the speed doesn't change from 0 km/h to 60 km/h (16.7 m/s) instantaneously. The driver of the car presses their foot on the accelerator pedal to increase the speed of the car gradually up to the cruising speed. In this period of increasing speed, the car is accelerating or experiencing **acceleration** (see Figure 7.4.1).

Shutterstock.com/1000 Words

▲ **FIGURE 7.4.1** Cars experience acceleration when their speed is changing.

Acceleration is a measure of how fast the speed of an object changes with time. It is represented with a lowercase *a*, and the units for acceleration are metres per second squared (m/s^2) or metres per second per second ($m\,s^{-1}\,s^{-1}$). The car, for example, might accelerate at a rate of $3\,m\,s^{-1}\,s^{-1}$ where the speed increases by 3 m/s every second.

When starting from rest, after 5 seconds the car will be travelling at 15 m/s. It will have increased its speed from 0 to 3, to 6, then to 12 m/s and finally to 15 m/s.

Acceleration also refers to motion where the speed is decreasing. If the car were to slow from 15 to 0 m/s over 5 seconds, the acceleration would be $-3\ m/s^2$ because the speed decreases by 3 m/s every second. Negative acceleration, where a body slows, can be referred to as **deceleration**.

deceleration
negative acceleration; when speed is decreasing over time

Another example of a body experiencing acceleration is a rocket. When a rocket takes off, it uses very strong **thrust** forces to accelerate it at a rate of $90\ m/s^2$ (see Figure 7.4.2). After 3 seconds, the rocket is already travelling at a speed of 270 m/s (about 972 km/h).

thrust
a force that makes an object move in the opposite direction as a result of expelling mass or fuel

After finishing a race, a runner will decelerate from their top speed to stationary (see Figure 7.4.3). The fastest runner in the world has an acceleration of $-2\ m/s^2$ when slowing from a top speed of 10 to 0 m/s in 5 s after crossing the finish line. That is, if they don't do a victory lap, of course!

NASA/Joel Kowsky

▲ **FIGURE 7.4.2** A rocket taking off is an example of body experiencing acceleration.

Alamy Stock Photo/PA Images

▲ **FIGURE 7.4.3** Runners quickly decelerate at the end of a sprint.

Measuring and calculating acceleration

In the same way that speed can be calculated by the change in distance over time, acceleration can be calculated by the change in speed over time.

$$\text{Acceleration} = \frac{\text{final speed} - \text{initial speed}}{\text{time}}$$

$$a = \frac{v - u}{t}$$

where a is acceleration (m/s^2), v is the final speed of the body (m/s), u is the initial speed of the body (m/s), and t is the time.

WORKED EXAMPLE 7.4.1

Calculate the acceleration of a car that's speed changes from 14 to 8 m/s over a 4 s period.

THINKING PROCESS	WORKING
Step 1: Identify known and unknown variables.	$a = ?$ $u = 14$ m/s $v = 8$ m/s $t = 4$ s
Step 2: Identify appropriate formula.	$a = \frac{v - u}{t}$
Step 3: Substitute known variables into the formula.	$a = \frac{8 - 14}{4}$
Step 4: Rearrange using algebra (if necessary) and solve.	$a = \frac{-6}{4}$ $= -1.5 \frac{\text{m}}{\text{s}^2}$
Solution:	The car's acceleration is -1.5 m/s^2 or is a deceleration of 1.5 m/s^2.

Speed–time graphs allow us to see how speed is changing with time and can help us calculate the acceleration of moving bodies. Figure 7.4.4 shows a speed–time graph for a body that's speed is increasing at a constant rate. The rate at which this speed is changing is the acceleration and can be found by the gradient of the speed–time graph.

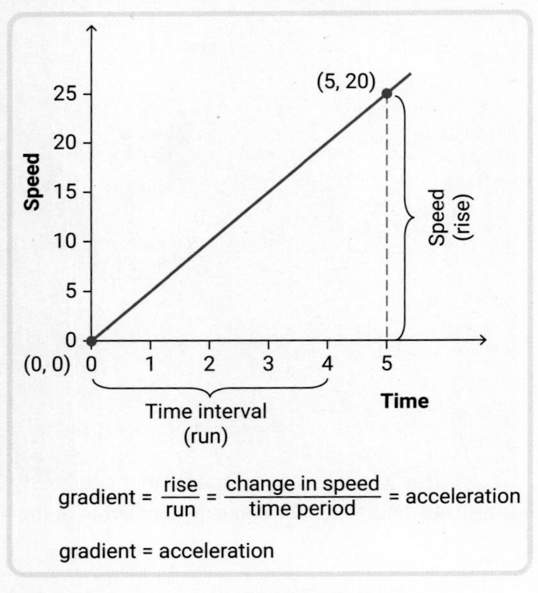

▲ **FIGURE 7.4.4** The gradient of a speed–time graph is the acceleration.

WORKED EXAMPLE 7.4.2

Use the speed–time graph in Figure 7.4.4 to determine the acceleration of the moving body.

THINKING PROCESS	WORKING
Step 1: Find the two points on the graph.	The points are (0, 0) and (5, 20). $t_1 = 0\ \text{s}$ $u = 5\ \text{m/s}$ $t_2 = 5\ \text{s}$ $v = 20\ \text{m/s}$
Step 2: Identify appropriate formula.	$\text{acceleration} = \text{gradient} = \frac{\text{rise}}{\text{run}} = \frac{v-u}{t_2-t_1}$
Step 3: Substitute known variables into the formula.	$a = \frac{20-0}{5-0} = \frac{20}{5} = 4\frac{\text{m}}{\text{s}^2}$
Solution:	The body is accelerating at a rate of $4\ \text{m/s}^2$.

Distance–time graphs can indicate when a body is accelerating or decelerating based on how the speed (gradient) is changing. For example, Figure 7.4.5 shows a body that has a constant speed for the first 2 seconds before slowing for 2 seconds, evident by the slope gradually decreasing during the period of 2 s to 4 s.

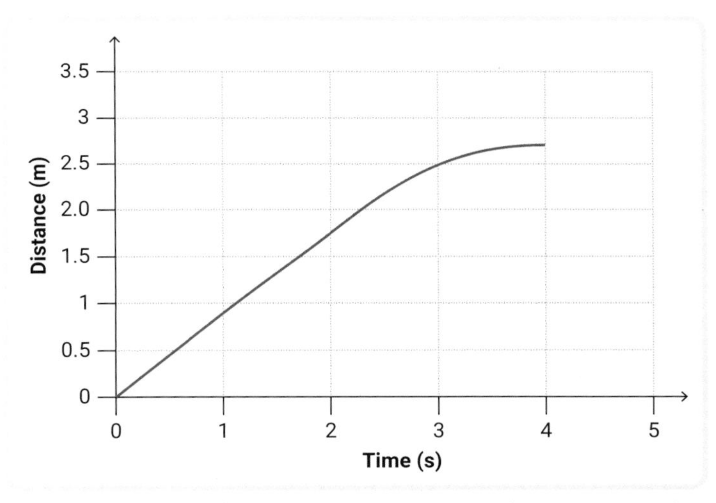

▲ **FIGURE 7.4.5** Distance–time graphs show acceleration by how the gradient is changing.

A summary of the general shape of distance–time graphs and speed–time graphs for different types of acceleration is shown in Table 7.4.1.

▼ **TABLE 7.4.1** The general shape of motion graphs for different accelerations

Acceleration	Speed–time graph shape	Distance–time graph shape
No acceleration	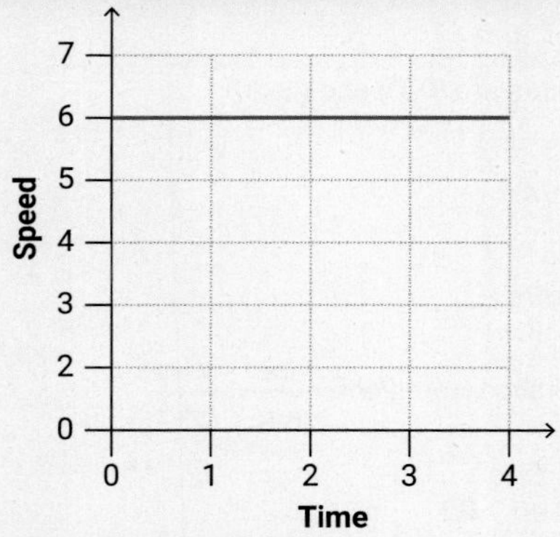 The speed is not changing.	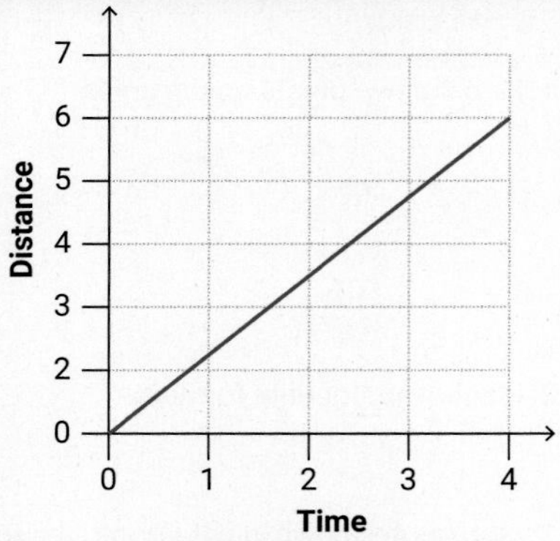 The distance increases at a constant rate (constant gradient).
Constant positive acceleration	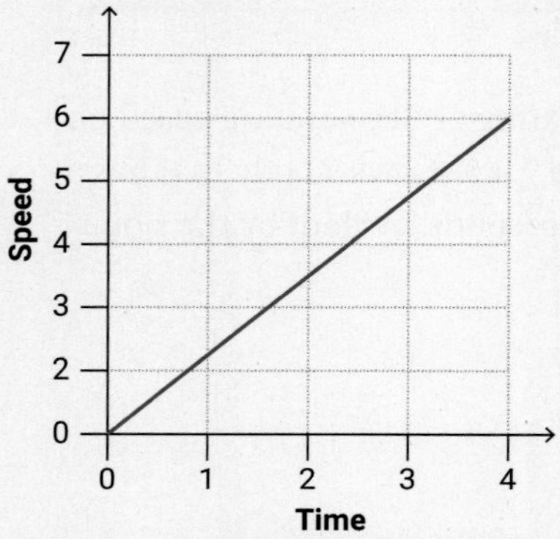 The speed increases at a constant rate (constant gradient).	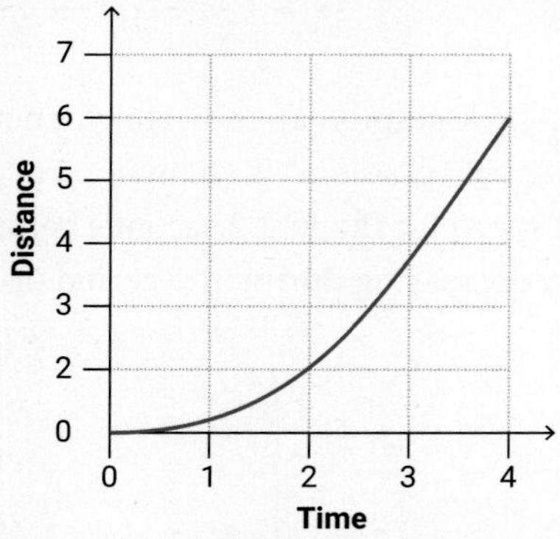 The distance increases at an increasing rate (gradient is increasing).
Constant negative acceleration (deceleration)	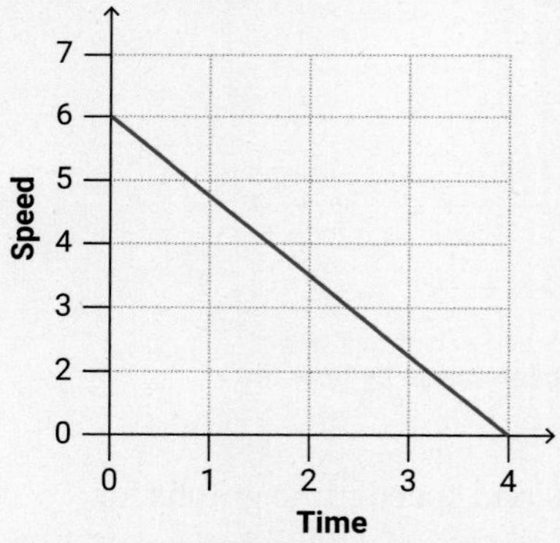 The speed decreases at a constant rate (constant gradient).	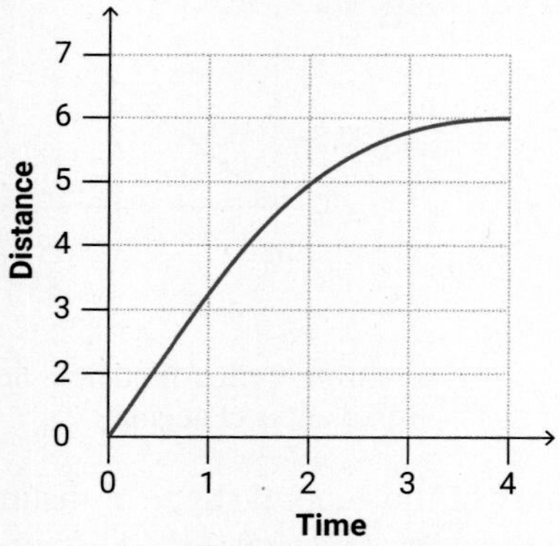 The distance increases at a decreasing rate (gradient is decreasing).

7.4 LEARNING CHECK

1 **Calculate** the acceleration of a bike that starts from rest and is travelling at 9 m/s after 6 seconds.

2 **Calculate** the time it would take for a car to accelerate from 5 to 15 m/s with an acceleration of 2 m/s^2.

3 **Match** the following descriptions to one of the distance–time graphs A–C and one of the speed–time graphs D–E.

   a Pat the dog sat still for a few seconds and then raced off, accelerating at a constant rate.

   b Grandpa walked at a constant speed for a few seconds, stopped for a rest and then continued at a slower speed than before.

   c Jess rode a bike, starting from rest, and accelerated to a constant speed before cruising at this speed.

▲ **FIGURE 7.4.6** A bike accelerates by pedalling with an applied force.

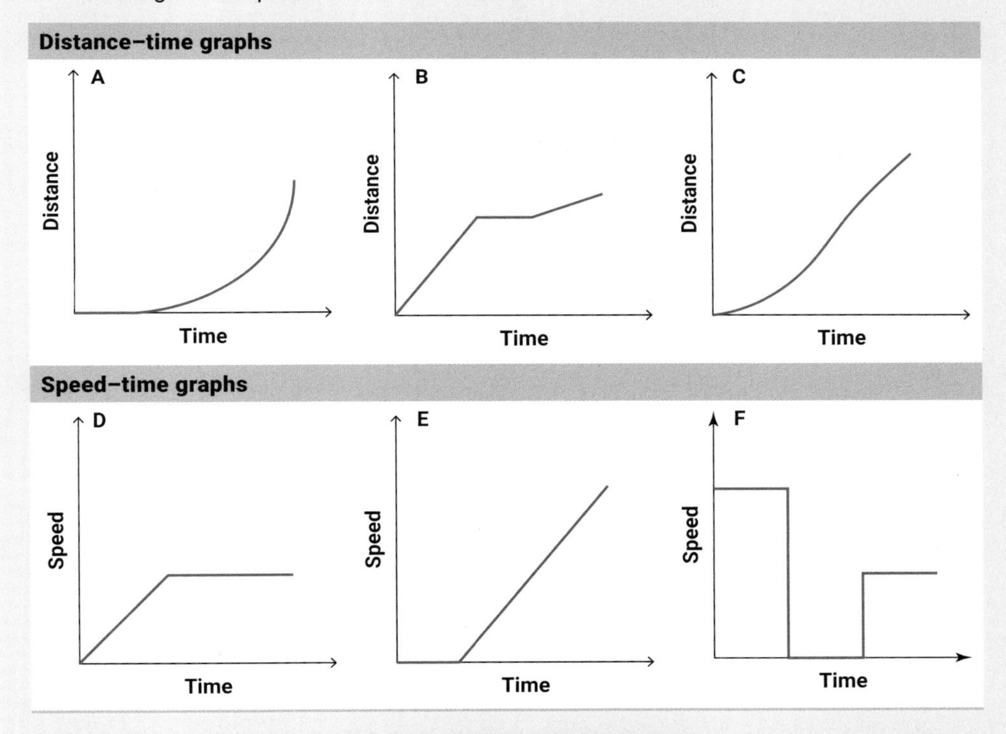

4 **Construct** a speed–time graph for the following situation.

   An aeroplane starts from rest and accelerates at a rate of 0.25 m/s^2 for 900 s before cruising at this final speed for 500 s. The aeroplane then decelerates for 200 s until it is at a speed of 180 m/s.

5 Return to your response to the 'Get thinking' activity. For each of the 10 words in your list, **describe** how this word is connected to the idea of acceleration. If you now think a word isn't connected to acceleration after completing the module, **explain** why.

7.5 Newton's laws of motion

BY THE END OF THIS MODULE, YOU WILL BE ABLE TO:

- ✓ recall Newton's laws of motion
- ✓ use Newton's laws of motion to explain the relationship between forces and motion and make predictions about the motion of bodies when experiencing different forces.

GET THINKING

Skim the pages in this module. Don't read! Just let your eyes move over the pages as though you are taking a photograph. What do you think this module is about?

Video activity
Newton's laws of motion

Interactive resource
Simulation: Forces and motion

Forces and free body diagrams

A **force** is a push, a pull or a twist acting on a body. Force is represented by F and is measured in newtons (N). One newton is approximately equal to the force required to hold an apple (see Figure 7.5.1).

force
a push, a pull or a twist

▲ **FIGURE 7.5.1** 1 N is approximately equal to the force required to hold an apple.

Forces can have any of the following effects on a body:

- make a body start moving
- make a body stop moving
- change the speed of a body (speed up/slow down)
- change the direction of a body
- change the shape of a body.

The first four effects will affect the body's motion.

free body diagram
a diagram that shows the forces acting on a single body as arrows

magnitude
size or extent

Forces are often represented on **free body diagrams**. A free body diagram shows the forces acting on a single body. In a formal free body diagram, the body is represented by a rectangle. However, sometimes the diagram is simplified to use a diagram of the body being examined. Arrows represent forces and point in the direction that the force is acting. Force arrows are different lengths depending on the **magnitude** (size) of the force. The arrow always starts in the centre of the body, regardless of where on the body they are being applied (see Figure 7.5.2).

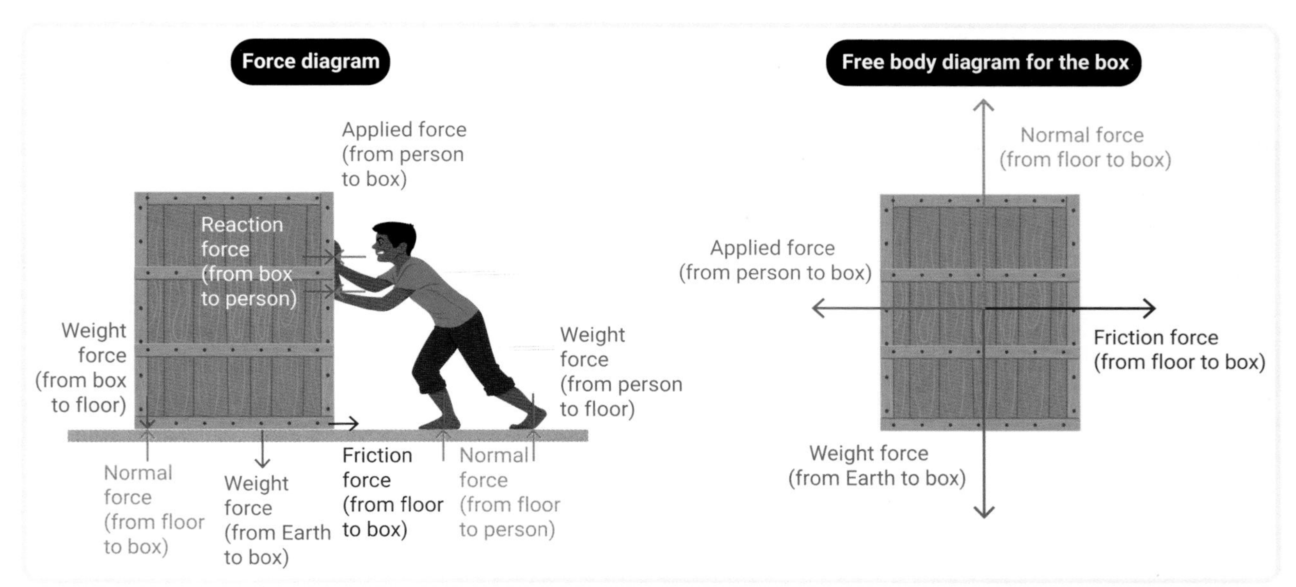

▲ **FIGURE 7.5.2** A free body diagram for a box when it is being pushed only includes the forces acting on the box.

Newton's three laws of motion

Sir Isaac Newton (1643–1727) was an English physicist and is considered one of the world's most influential scientists. While studying motion and trying to make sense of how and why things move, he developed three physical rules or 'laws' by which all his observations could be explained (see Figure 7.5.3). These laws are accepted universally as appropriate to explain and predict motion of most bodies on Earth. The three laws are:

▲ **FIGURE 7.5.3** Stories often suggest that Isaac Newton had some of his big ideas about gravity and motion while sitting under an apple tree and watching the apples fall to the ground.

1 a body will remain at rest or at a constant speed unless acted on by unbalanced forces
2 the acceleration of a body is directly proportional to the unbalanced force applied to it and inversely proportional to the body's mass
3 for every force acting, there is an equal and opposite force reacting.

When learning about these laws for the first time, it is best to look at the third law first, followed by the first law and then the second law. Newton's second law will be covered in the next module.

▲ **FIGURE 7.5.4** Reaction forces when kicking a wall

Newton's third law

Newton's third law states that for every acting force, there is an equal and opposite reaction force. This means that all forces come in pairs, often called action–reaction pairs. When one body applies a force on another body, the body applying the force is known as the **agent** and the body receiving the force is known as the **receiver**. The agent applies a force on the receiver and the receiver applies an equal and opposite reaction force back on the agent.

Consider a person who kicks a wall. The person applies a force to the wall, which is not enough to move the building. The wall applies an equal force back on the person's foot and this force pushes the person away from the wall. The harder the person kicks the wall, the more force is pushed back on them and the further they are pushed away from the wall (see Figure 7.5.4).

agent
a body applying a force on another body

receiver
the body receiving a force from another body

Some other examples of Newton's third law in action are described in Table 7.5.1.

▼ **TABLE 7.5.1** Examples of Newton's third law

Example	Agent	Receiver	Outcome of forces
A foot kicking a ball Shutterstock.com/alphaspirit.it	Foot	Ball	The foot applies a force to the ball and the ball applies the same force back on the foot. The foot feels the force from the ball pushing back on it.
A person diving from a diving block Shutterstock.com/wavebreakmedia	Person	Ground	The person applies a force on the block. The block applies the same force back on the person, pushing them into the air.

Table 7.5.2 describes some common reaction forces that are harder to imagine or see in action.

▼ **TABLE 7.5.2** The types of reaction forces

Reaction force	Description	Example	
Tension	Tension is a reaction force that is exerted by a rope, string or wire on the bodies attached to either end when a force is applied to the rope string or wire.	A crane lifting a heavy mass Shutterstock.com/demamiel62	The weight applies a force on the crane cable and the crane cable applies a tension force, pulling the weight back up.
Upthrust	Upthrust is a reaction force that a liquid or gas exerts on an object floating in it.	A small boat floating in water Shutterstock.com/Jim Vallee	The boat applies a force on the water and the water applies an upthrust force back onto the boat.
Normal	Normal force is a reaction force that a surface exerts on another body when the body applies a force to the surface.	A book sitting on a table iStock.com/Toru Kimura	The book applies a force on the table and the table applies a normal force back onto the book.

Newton's first law

Newton's first law is often known as the law of **inertia**. Inertia is the concept that bodies will continue to stay stationary (at rest) or moving at a constant speed unless acted on by unbalanced forces. When a car is travelling at a constant speed and suddenly stops, your body in the car continues at the same speed because of inertia. The force that made the car stop was applied to the car and not to you. Your seatbelt then applies a force to your body to stop it from continuing at a constant speed through the windscreen.

inertia
a property of a body that resists changes to its motion

For there to be changes in the speed of a body, the body must experience acceleration, which is caused by unbalanced forces acting on the body. Table 7.5.3 shows the different effects balanced and unbalanced forces have on motion.

▼ **TABLE 7.5.3** The different effects of balanced and unbalanced forces on bodies

Forces	Acceleration	Possible effects on motion
Balanced forces	No acceleration	• Body remains stationary • Body remains at constant velocity
Unbalanced forces	Acceleration	• Body starts moving • Body speeds up • Body stops • Body slows down • Body changes direction

Figure 7.5.5 shows the different scenarios for balanced and unbalanced horizontal forces acting on a car.

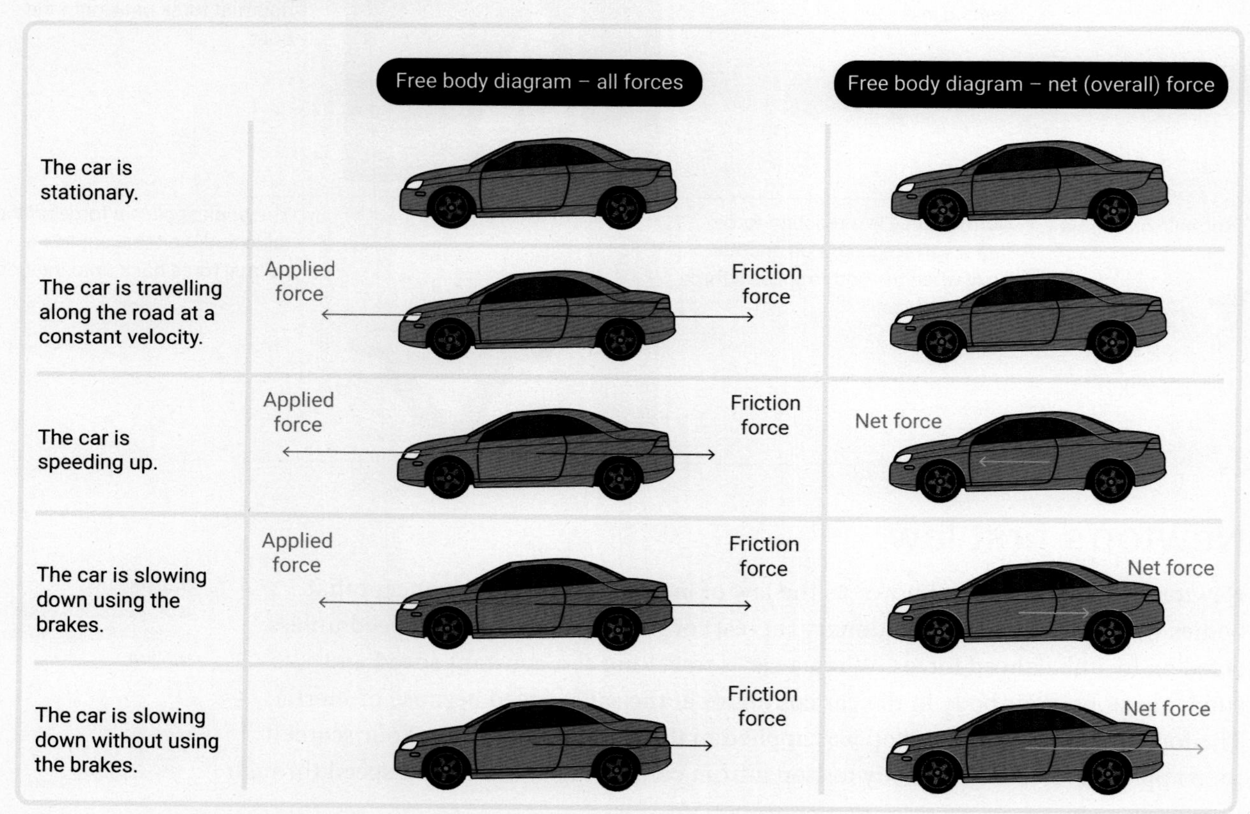

▲ **FIGURE 7.5.5** Different scenarios for balanced and unbalanced horizontal forces acting on a car

7.5 LEARNING CHECK

1 **Define**:
   a force.
   b newton.
   c receiver.
   d normal force.
2 **Construct** a free body diagram for a bike that is speeding up and experiencing friction.
3 When there is only a small amount of tomato sauce left in a bottle, you can shake the bottle forwards very fast and stop quickly to make the sauce more accessible to the opening of the bottle. **Explain** how this is an example of inertia.

iStock.com/Image Source

▲ **FIGURE 7.5.6** Shaking a bottle to pour out sauce is an example of inertia.

4 **Classify** each of the free body diagrams shown in Figure 7.5.7 as showing balanced or unbalanced forces.

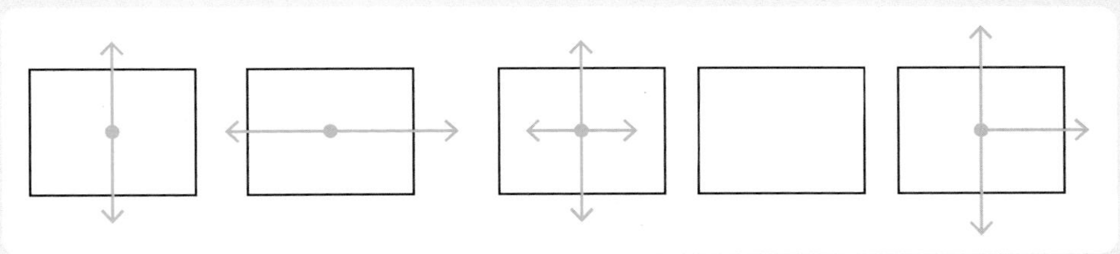

▲ **FIGURE 7.5.7** Examples of free body diagrams

Newton's second law

BY THE END OF THIS MODULE, YOU WILL BE ABLE TO:

- ✓ recall Newton's second law in conceptual and quantitative contexts
- ✓ use the mathematical relationship for Newton's second law to calculate force, mass and acceleration for moving bodies.

GET THINKING

In Module 7.5, we introduced Newton's second law as: 'The acceleration of a body is directly proportional to the unbalanced force applied to it and inversely proportional to the body's mass'. Talk to a partner or write what you think this means in general and what it means for two bodies of the same mass experiencing different forces.

Video activity
Newton's second law of motion

Extra science investigation
Investigating Newton's second and third laws of motion

Newton's second law

Newton's second law states that the acceleration of a body is directly proportional to the unbalanced force applied to it and inversely proportional to the body's mass. In mathematical terms, this law directly translates to the formula:

$$F_{net} = ma$$

where F_{net} is the overall force experienced by a body (N), m is the mass of the body (kg) and a is the acceleration of the body (m/s^2).

The equation shows us that when greater forces are applied to the same body, the acceleration increases. It also shows us that greater masses require greater forces to accelerate them at the same rate. When there is no overall force acting on a body, there is no acceleration and the speed remains constant. This situation is an application of both Newton's first and second laws. This formula can be used in problems where we use mathematical equations to make predictions about the motion of bodies experiencing forces.

WORKED EXAMPLE 7.6.1

Determine the overall force required to accelerate a 150 g apple at a rate of 0.9 m/s^2

THINKING PROCESS	WORKING
Step 1: Draw a free body diagram of the body whose motion is being examined.	a = 0.9 m/s^2 → F_{net} = ? m = 0.150 kg ▲ **FIGURE 7.6.1** A free body diagram of a 150 g apple
Step 2: Identify known and unknown variables.	$m = 150\text{ g} = 0.150\text{ kg}$ $a = 0.9\dfrac{m}{s^2}$
Step 3: Identify appropriate relationship(s).	$F_{net} = m \times a$
Step 4: Substitute known variables into the formula.	$F_{net} = 0.150 \times 0.9$
Step 5: Rearrange using algebra (if necessary) and solve.	$F_{net} = 0.125$ N
Solution:	The overall force required to accelerate a 150 g apple at a rate of 0.9 m/s^2 is 0.125 N.

WORKED EXAMPLE 7.6.2

A 60 kg skier experiences a gravitational force of 164 N pulling them down a slope and a friction force in the opposite direction of 50 N. Determine the acceleration of the skier down the slope.

THINKING PROCESS	WORKING
Step 1: Draw a free body diagram of the body whose motion is being examined.	▲ **FIGURE 7.6.2** A free body diagram for a skier travelling down a slope
Step 2: Identify known and unknown variables.	$m = 65$ kg $F_{net} = 164\text{ N} - 50\text{ N} = 114\text{ N}$
Step 3: Identify appropriate relationship(s).	$F_{net} = m \times a$
Step 4: Substitute known variables into the formula.	$114 = 65 \times a$
Step 5: Rearrange using algebra (if necessary) and solve.	$a = \frac{114}{65}$ $= 1.75\text{ m/s}^2$
Solution:	The acceleration of the skier down the slope is 1.75 m/s^2.

7.6 LEARNING CHECK

1. **Recall** Newton's second law and identify what each symbol represents and the SI units for each.
2. **Calculate** the force required from a bow to accelerate an arrow of mass 0.120 kg at a rate of 5 m/s^2.
3. **Determine** the mass of a block that accelerates at a rate of 0.12 m/s^2 when a force of 16 N is applied to it.
4. A model rocket has a mass of 1.1 kg and experiences a thrust force of 2.5 N from a chemical reaction. **Determine** the acceleration of the rocket.
5. Two siblings are fighting over a remote control that has a mass of 250 g. Sibling A pulls with a force of 25 N on the remote while sibling B pulls in the opposite direction with a force of 10 N.
   a. **Determine** which sibling will end up with the remote control closest to them.
   b. **Determine** the acceleration of the remote control towards that sibling.

7.7 Acceleration due to gravity

BY THE END OF THIS MODULE, YOU WILL BE ABLE TO:

- ✓ recognise that acceleration due to gravity is constant on Earth's surface and recall its value
- ✓ solve problems for falling and rising objects using acceleration due to gravity.

Video activity
Objects falling in a vacuum

Interactive resource
Simulation: Gravity force lab

GET THINKING

Imagine dropping a brick and a feather off a ledge and timing how long it takes for them to reach the ground. Why do the objects fall? Can you predict how long it would take for them to reach the ground? Which object falls faster? If one falls faster than the other, why? Write a brief answer to each question and we will come back to these at the end of the module.

Gravity – a two-way force

gravity
the force of attraction between Earth and objects within its gravitational field

gravitational field
the region where the pull of Earth's gravity is experienced

When we first learn about **gravity**, we usually talk about huge objects like the Sun or Earth as having gravitational forces. We learn that huge and heavy bodies have an attractive force that pulls objects towards them if those objects are within their **gravitational field**. A gravitational field is the three-dimensional space around a body where its gravitational force can be experienced. We are within Earth's gravitational field and are pulled towards Earth by Earth's gravity. Earth is within the Sun's gravitational field and so is pulled towards the Sun and held in orbit by the Sun's gravity.

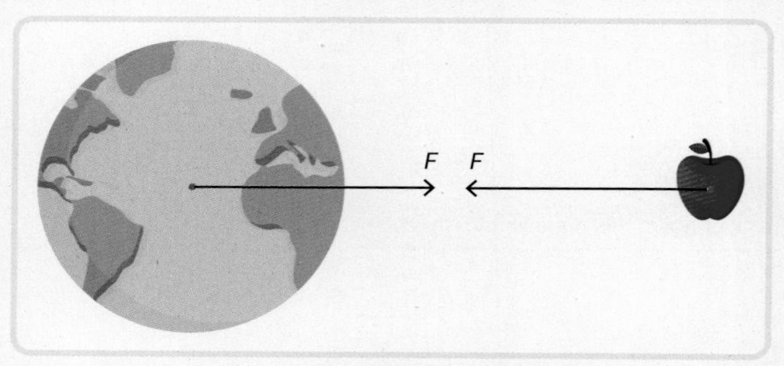

▲ **FIGURE 7.7.1** Gravity is a force that attracts two bodies towards one another. The bodies accelerate differently depending on their mass.

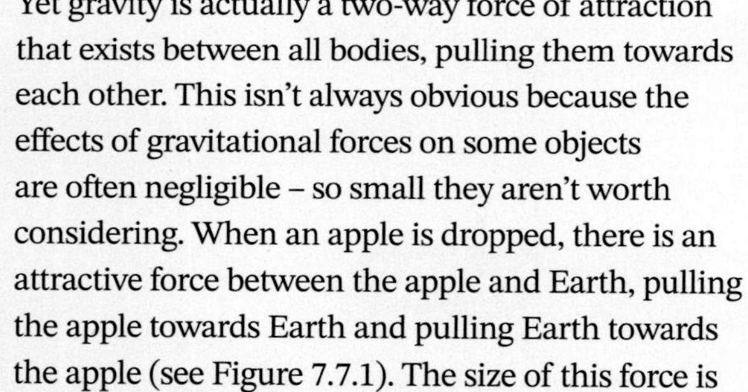

Yet gravity is actually a two-way force of attraction that exists between all bodies, pulling them towards each other. This isn't always obvious because the effects of gravitational forces on some objects are often negligible – so small they aren't worth considering. When an apple is dropped, there is an attractive force between the apple and Earth, pulling the apple towards Earth and pulling Earth towards the apple (see Figure 7.7.1). The size of this force is big enough to move and accelerate the apple because it does not have much mass, but the force is so insignificant that Earth would only move towards the apple by an immeasurably tiny amount.

The gravitational force pulling a body that is close to Earth's surface towards Earth is determined by the mass of Earth and the mass of the body. Heavier bodies are attracted more strongly towards Earth and lighter bodies experience weaker gravitational attraction towards Earth.

Acceleration due to gravity

Galileo Galilei was an Italian physicist in the 1600s who studied gravity and its effects on the motion of objects. In a famous experiment where Galileo supposedly dropped cannonballs from the top of the Leaning Tower of Pisa, he discovered that no matter the mass, objects always accelerated towards Earth's surface at the same rate (see Figure 7.7.2). In other words, the acceleration of bodies due to gravity is constant.

However, this is not our everyday experience. We see that a leaf floats slowly to the ground from a tree while an apple drops very quickly. This difference is caused by other forces acting on the bodies as they fall. Forces such as wind or air resistance may blow the leaf back upwards, while the apple doesn't experience the same effect of these resistance forces. When the only force acting on each object is Earth's gravity, both bodies will accelerate at the same rate towards Earth's surface. If you were to drop a feather and a hammer in a **vacuum**, you would see them fall at the same speed.

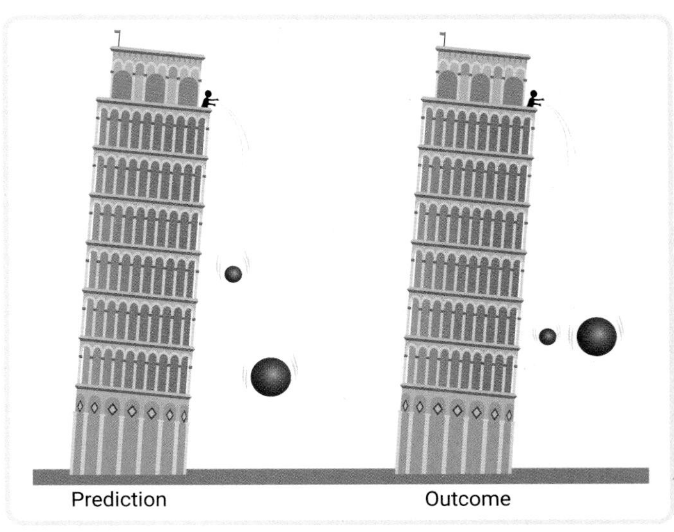

▲ **FIGURE 7.7.2** Galileo's famous Leaning Tower of Pisa experiment suggested that all objects fall at the same rate due to gravity's pull.

vacuum
a space empty of all matter

Experiments can be done to determine exactly what this rate of acceleration is. It can vary slightly, though the accepted value for acceleration due to gravity on Earth's surface is $9.8\,m/s^2$. This value is often symbolised as g.

With this value, predictions can be made about falling and rising objects. When bodies are falling and the only force acting is gravity's pull, the acceleration of the body is $9.8\,m/s^2$. When a body has been projected upwards from Earth's surface and the only force acting is gravity's pull, the body slows at the same rate and hence the acceleration is $-9.8\,m/s^2$.

WORKED EXAMPLE 7.7.1

A coin is dropped from the top of a building and falls for 5 seconds. The only force acting on the coin is gravity. Determine the final speed of the coin.

THINKING PROCESS	WORKING
Step 1: Identify known and unknown variables.	$a = g = 9.8\ m/s^2$ $u = 0\frac{m}{s}$ We know this because the coin was dropped. $t = 5\ s$ $v = ?$
Step 2: Identify appropriate relationship(s).	$a = \frac{v - u}{t}$
Step 3: Substitute known variables into the formula.	$9.8 = \frac{v - 0}{5}$
Step 4: Rearrange using algebra (if necessary) and solve.	$9.8 \times 5 = \frac{v}{5} \times 5$ $v = 49\ m/s$
Solution:	The coin would be travelling at a speed of 49 m/s after 5 seconds.

WORKED EXAMPLE 7.7.2

A netball is thrown directly upwards with a speed of 8 m/s. Determine how long it travels upwards until it stops (and the speed of the netball is 0 m/s) before falling back towards Earth. Assume that gravity is the only force acting on the netball.

THINKING PROCESS	WORKING
Step 1: Identify known and unknown variables.	$a = -g = -9.8\ \text{m/s}^2$ $u = 8\frac{m}{s}$ $v = 0\frac{m}{s}$ $t = ?$
Step 2: Identify appropriate relationship(s).	$a = \frac{v - u}{t}$
Step 3: Substitute known variables into the formula.	$-9.8 = \frac{0 - 8}{t}$
Step 4: Rearrange using algebra (if necessary) and solve.	$-9.8 \times t = \frac{-8}{t} \times t$ $-9.8 \times t = -8$ $\frac{-9.8 \times t}{-9.8} = \frac{-8}{-9.8}$ $t = 0.82\ \text{s}$
Solution:	The netball will stop and start to fall back towards Earth after 0.82 seconds.

Weight force

weight force
the force of gravity pulling an object towards Earth

Newton's second law of motion allows us to quantify the force of gravity that pulls a body towards Earth's surface. For a specific body, the force of gravity pulling it down is often known as the **weight force** or force of weight. The weight force for any object can be calculated using a specialised case of Newton's second law:

$$F_w = mg$$

where F_w is the weight force (N), m is the mass of the body (kg) and g is the acceleration due to gravity and equal to $9.8\ \text{m/s}^2$.

WORKED EXAMPLE 7.7.3

A 30 kg dog sits on a couch. Calculate the weight force that is exerted on the couch by the dog due to gravity's pull.

THINKING PROCESS	WORKING
Step 1: Identify known and unknown variables.	$m = 30$ kg $g = 9.8$ m/s^2 $F_w = ?$
Step 2: Identify appropriate relationship(s).	$F_w = mg$
Step 3: Substitute known variables into the formula.	$F_w = 30 \times 9.8$
Step 4: Rearrange using algebra (if necessary) and solve.	$F_w = 294$ N
Solution:	The dog exerts a force of 294 N on the couch due to gravity's pull.

✩ ACTIVITY

Find two objects that have different masses and won't be damaged if dropped from a 2 m height. With a partner, time how long it takes for each mass to reach the ground when dropped from a 2 m height. **Discuss** similarities and differences in the time, using the ideas learned in this module about acceleration due to gravity.

 7.7 LEARNING CHECK

1. A ball is thrown into the air at an initial speed of 20 m/s. **Calculate** how long it will travel upward before the speed has halved.
2. A scale that measures weight actually measures the weight force exerted on the scale and converts it to a mass value based on the acceleration due to gravity on Earth. If the weight force recorded was 539 N, **determine** the mass of the object on the scale.
3. If a 60 kg person used a scale from Earth on Mars, where the acceleration due to gravity is 3.72 m/s^2, **determine** the:
   a. weight force recorded by the scale.
   b. mass presented by the scale.
4. A 120 kg provisions package dropped from a plane to some trapped hikers takes 6 seconds to reach the ground.
   a. **Calculate** the speed of the package when it hits the ground.
   b. Use your answer to part **a** to **explain** why these kinds of packages require parachutes to aid their landing.

7.8 Applying knowledge of motion

BY THE END OF THIS MODULE, YOU WILL BE ABLE TO:

- ✓ identify examples of how Newton's laws of motion are applied to improve the safety of cars and explain how they work
- ✓ identify examples of Newton's laws of motion in performance-enhancement strategies for sports and explain how they work.

GET THINKING

Newton's laws of motion apply to things we observe and do every day. Look at the subheadings in this module. How do you think Newton's laws could apply to car safety and sports performance?

Video activity
Science of golf

Extra science investigations
Motion in traffic
Stopping distance

Objects in motion are everywhere. Motion is a vital component of many parts of life, including transport, construction, sport and almost every industry. Newton's laws of motion are applied in all these settings to improve the safety and performance of many processes and operations. Some examples of their application in motor vehicle safety and sport performance are discussed in this module.

Motor vehicle safety

In 1925, there were 700 road deaths in Australia, which was about 230 deaths per 100 000 registered vehicles. In 2020, in Australia there were 1095 road deaths. However, factoring in the increased popularity of transport by car in the 21st century, this was equivalent to about five deaths per 100 000 registered vehicles. This dramatic reduction in road toll in the last 100 years has been a result of advances in car safety and design. Some of the most significant safety features in modern cars include seatbelts, airbags and headrests.

▲ **FIGURE 7.8.1** The operation of seatbelts is an example of the application of Newton's first law of motion to stop the motion of bodies during an accident.

Seat belts

When a car stops suddenly during a front-on collision, an unrestrained driver or passenger will continue moving. Newton's first law of motion explains that this is due to the inertia of the occupants. Without a seatbelt, the driver could hit the steering wheel or any of the car's occupants could pass through the windscreen. A seatbelt can apply an unbalanced force to everyone in the car, decelerating their motion and keeping them in their seats (see Figure 7.8.1).

Airbags

Seatbelts do a great job of keeping people in their seats during a collision, but the restraint is directed mostly to the lower part of the body. The upper body can sometimes continue to move forwards due to inertia. For passengers in the back seat, this is not as big a problem because the seats in front of them are usually far away and are a softer barrier. In the front of a car, people's heads can collide with the dashboard, which can cause head injuries. Supplemental restraint systems in modern cars are activated when the car suffers an impact to release airbags automatically. Airbags apply a force to the upper body to decelerate the head and shoulders instead of allowing them to continue at the speed they were travelling before the collision (see Figure 7.8.2).

Alamy Stock Photo/Patti McConville

▲ **FIGURE 7.8.2** Airbags provide an unbalanced force to the upper body of front seat car passengers to decelerate their bodies after a car stops suddenly.

Headrests

Different dangers exist when a car experiences a rear-end collision, since the car is being propelled forward quickly instead of suddenly stopping. The seat of the car applies a force onto most of a passenger's body, moving their body forward with the rest of the car. In old cars, there were often no headrests (see Figure 7.8.3). This meant that when the body was propelled forward with the car, the passengers' heads remained at rest since they hadn't experienced the same applied force from the car seat. The fast movement of the body forward without the head moving with it can cause neck injuries known as whiplash, or spinal injuries. Headrests exist in modern cars to apply the same force to the head as the body, so that the passenger's body and head move forward as a whole.

Alamy Stock Photo/ClassicStock

▲ **FIGURE 7.8.3** Old cars did not have headrests as a safety feature to avoid whiplash injuries.

▲ **FIGURE 7.8.4** Swinging a golf club through the stroke, including the follow-through

Using Newton's laws in sport

The sporting world is another area where Newton's laws of motion are applied. Understanding how things move is the key to enhancing performance in most sports.

Golf

A golfer swings a club to hit a small golf ball out onto the fairway. Part of a golfer's technique is to have a follow-through swing and to accelerate the ball to a high speed so that it can reach long distances. When the golf club makes contact with the ball, the club exerts a force on the golf ball and, in agreement with Newton's first law, the unbalanced forces cause the ball to accelerate. According to Newton's second law, the smaller the mass, the greater the acceleration of the ball for the same applied force. This is why golf balls are light (less than 46 g). Inertia also allows the golf club to follow through its swing, as the forces applied to the club are not great enough to change its motion.

According to Newton's third law, when the club applies a force on the ball, the ball applies an equal and opposite reaction force back on the club. If this force were to cause a significant acceleration for the club in the opposite direction to the swing, this would interfere with the technique of the swing and the club would not be able to follow through, as shown in Figure 7.8.4. Since a golf club is heavier than the ball, the effect of the force on the club is reduced. Newton's second law shows us that when the mass is greater, the acceleration effect of an applied force is much smaller. The same sized force that causes the ball to accelerate out onto the golf course has a deceleration effect of about 10% on the club because of their difference in mass, and the club slows down only slightly during the swing.

net force
the overall force acting on a body; measured in newtons (N)

Cycling

Inertia plays a major role in the performance of a cyclist. A cyclist applies force through pedalling to reach a certain speed, and the inertia of the bicycle and cyclist allows them to move along with this speed until the friction between the tyres and the road slows them down. One of the other resistance forces that cause the cyclist to exert more force to reach higher speeds is air resistance. Air resistance lowers the **net force** forwards for the cyclist and reduces the acceleration. Each of the forces affecting the motion of a cyclist is demonstrated in Figure 7.8.5.

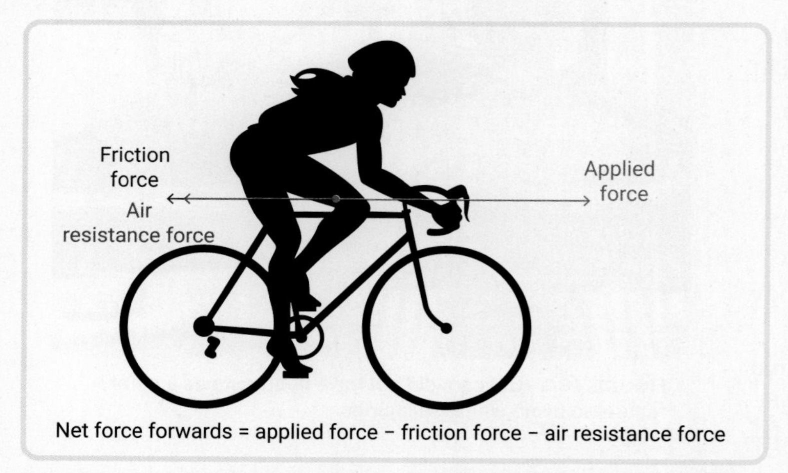

▲ **FIGURE 7.8.5** A summary of the forces affecting the motion of a cyclist

9780170463041

▲ **FIGURE 7.8.6** Cyclists employ many strategies to reduce air resistance and enhance their speed and performance.

Professional cyclists use many strategies to decrease air resistance (see Figure 7.8.6). Some of these strategies include:

- carefully engineered aerodynamic bikes that have a specialised shape to decrease air resistance
- aerodynamic helmets
- skin-tight elastane clothing
- hunched compact riding positions
- shaving their legs to minimise the air resistance caused by their leg hairs.

☆ ACTIVITY

In pairs, take turns completing the following. Run at a jog and when you reach a certain point, bring yourself to a stop. Get your partner to time how long it took you to stop using a stopwatch. Now sprint to the same point and time how long it takes to stop.

1 **Compare** the time taken to slow to a stop from the two different speeds.
2 **Explain** this difference using Newton's laws of motion.

7.8 LEARNING CHECK

1 **Explain** why cars need both seatbelts and airbags in the front seat to protect occupants in front-on collisions.
2 **Justify** the use of skin-tight clothing for a racing cyclist rather than baggy clothing.

7.9 First Nations Australians' spear-throwers

FIRST NATIONS SCIENCE CONTEXTS

IN THIS MODULE, YOU WILL:

✓ investigate the effect of First Nations Australians' spear-throwers on the speed and impact force of an object.

First Nations Australians' spear construction

First Nations Australians use throwing spears for many purposes, including hunting and fishing. Throwing spears are often made with several different materials. The shaft of a spear is usually made of wood for strength and flexibility. The tip is constructed from a material that can withstand the impact force required for the intended target. Prior to colonisation, the Gadigal People of the Eora Nation (Sydney region, New South Wales) constructed fishing spears using stingray barbs as tips. Multiple stingray barbs were attached to the spear shaft in an orientation that ensured the tip could not be dislodged from the flesh of the fish on impact (see Figure 7.9.1).

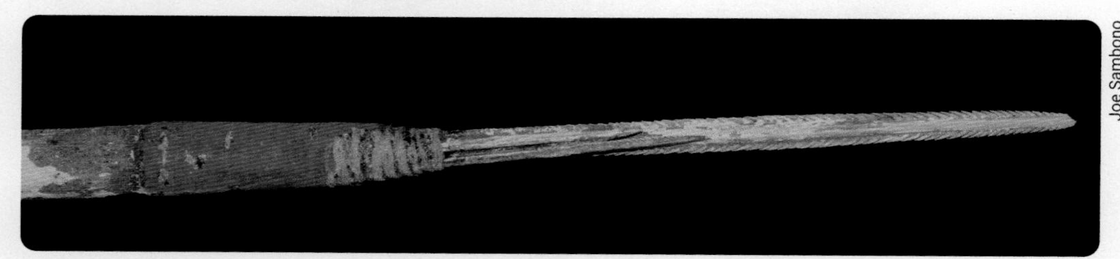

Joe Sambono

▲ **FIGURE 7.9.1** Stingray barbs used as spear tips.

First Nations Australians' use of spear-throwers

Spears can be launched by hand or by using a spear-thrower. A typical spear-thrower consists of a shaft with a small peg that holds the base of the spear, and a handgrip to help launch the spear (see Figure 7.9.2).

Joe Sambono

▲ **FIGURE 7.9.2** A peg on a spear-thrower, used to hold the base of a spear

The spear is placed onto the spear-thrower and is launched with a throwing motion (see Figure 7.9.3).

▲ **FIGURE 7.9.3** How a spear is placed on a spear-thrower

The development of spear-throwers demonstrates First Nations Australians' knowledge of the relationship between mass, force, velocity and impact force. A spear-thrower is designed to increase the speed of a spear and subsequently the impact force. It is a tool that acts as a speed multiplier on a projectile.

The kinetic energy of a moving object is equal to half of its mass multiplied by the square of its velocity. When a spear is launched using a spear-thrower, the kinetic energy is greater than a spear launched by arm. Therefore, the impact force of the spear is greater due to the increased kinetic energy.

Different combinations of spear and spear-thrower affect speed, energy and impact force. Heavier spears launched with a spear-thrower have a lower velocity than lighter spears.

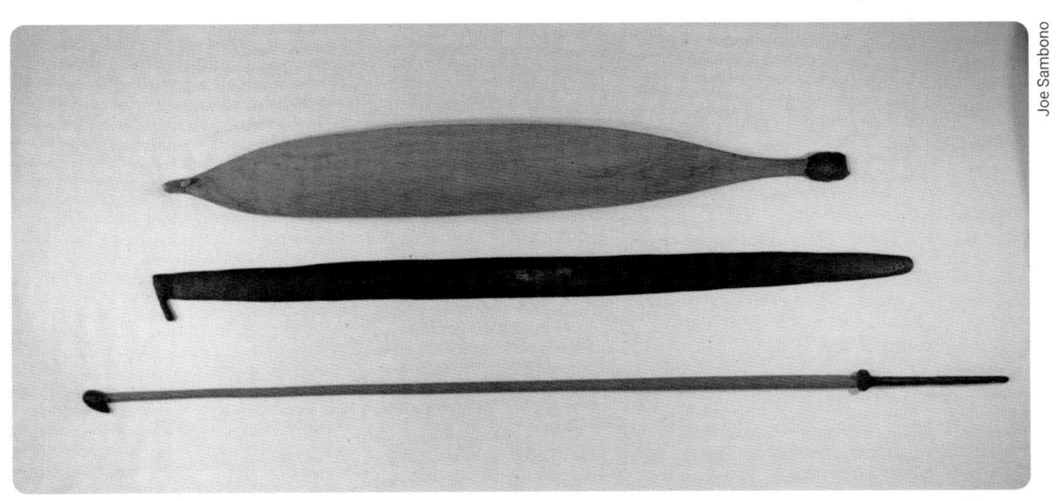

▲ **FIGURE 7.9.4** The weight and shape of the spear- thrower and spear affect speed, energy and the force of impact.

☆ ACTIVITY

Investigating the relationship between mass and impact force

It is not culturally respectful to use a traditional spear-thrower without consulting Elders or Knowledge Holders. It is also not safe to throw spears or sticks. In this activity, you will use a ball thrower to launch a projectile. Like a spear-thrower, a ball thrower acts as a speed multiplier for a projectile.

Materials

- balls with different masses
- ball thrower
- velocity speed gun (or timer)
- measuring tape
- scales

Make balls with different masses by drilling a small hole in a tennis ball and filling it with different amounts of dry sand. Seal the hole with glue. Balls made this way will ensure the hold on the ball in the thrower remains consistent. Alternatively, balls of different weights (e.g. plastic balls, tennis balls or cricket balls) can be used, although they may not fit in the ball thrower. They may also introduce additional variables to the investigation.

Method

1. Throw a tennis ball with your arm.
2. Use a speed sensor to record the speed of the ball. If a speed sensor is not available, estimate the speed by measuring the distance travelled and time from launch to impact.
3. Measure the distance the ball is thrown. You will need to do several trials. Record your data in a table.
4. Throw a tennis ball using the ball thrower.
5. Measure or estimate the speed and record the distance the ball travels. Record your data in the table.
6. Investigate the effect of the mass of the ball by repeating steps 1–4 with balls of different masses. Record your data in the table.

Evaluation

1. What did you observe?
2. How does the lever (ball thrower) affect the speed of the projectile?
3. What effect did the mass of the projectile have on the speed?
4. Consider the impact force of each object. The formula for impact force usually includes deformation distance; that is, how much an object is deformed on collision, such as how much a car is dented in a collision. In this case, the deformation distance of the ball and the ground is negligible and can be ignored. Hence, impact force can be estimated as a function of the kinetic energy of the object (half of the mass multiplied by the square of its velocity).
   a. Estimate the impact force for each ball.
   b. What is the likely effect of the mass of the projectile on impact force?
5. How did First Nations Australians development of spear-throwers overcome the limitations of being able to hit targets at long distances and with light projectiles?

SCIENCE AS A HUMAN ENDEAVOUR

7.10 Crash test dummies for vehicle safety

BY THE END OF THIS MODULE, YOU WILL BE ABLE TO:

✓ evaluate the significance of using different car crash dummies for vehicle safety.

Crash testing motor vehicles

Each year in Australia and New Zealand, about 40 000 adults and children die or are seriously injured in vehicle collisions. Although this number is large, it would be significantly higher if it were not for the safety features that have been mandated for motor vehicles in these countries since the 1960s. Before then, most cars had no built-in protection for vehicle occupants and lacked today's standard safety features such as seatbelts, airbags and sophisticated anti-lock braking systems. These days, safety ratings for new cars are a major selling point. The Australasian New Car Assessment Program (ANCAP) performs safety tests and publishes independent safety ratings for new cars.

Video activity
Female crash test dummies

Extra science investigation
Crumple zones

ANCAP safety tests include destructive crash tests in a lab to simulate common types of road incidents and to assess how these crashes would affect the driver and passengers within different cars. Crash tests use sophisticated crash test dummies that collect impact data for potential human injuries during collisions. The forces experienced by the dummies when travelling at different speeds and decelerating at different rates can be studied to make recommendations about car safety features.

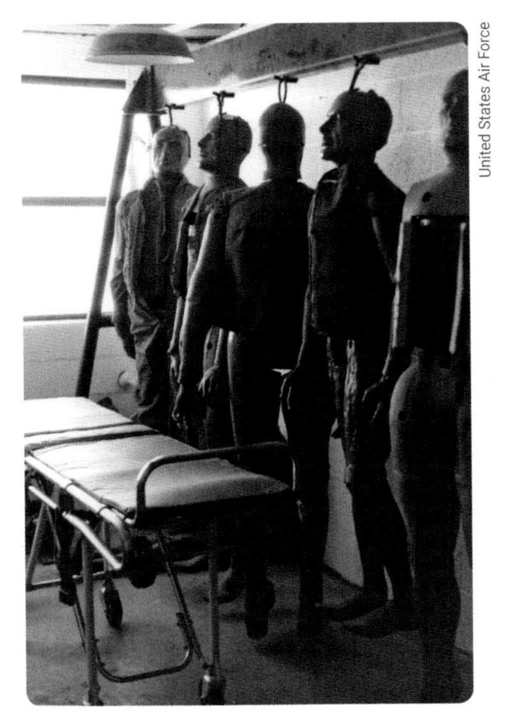

United States Air Force

The development of the crash test dummy

In the 1940s, researchers and automobile engineers used human cadavers in the crash tests. Obviously, there were some ethical issues associated with this practice. This led to the United States Air Force creating Sierra Sam, the first crash test dummy (see Figure 7.10.1). Sam was modelled on the average-sized American male.

◀ **FIGURE 7.10.1** The first crash test dummy, Sierra Sam, was developed in the 1940s for the United States Air Force.

▲ **FIGURE 7.10.2** ANCAP's Crash Test Dummy Family, including the Hybrid III models

A more advanced crash test dummy, Hybrid I, was developed by General Motors in the 1970s. Their third-generation dummy, Hybrid III, is the most commonly used dummy in the world today. It is made in a range of dummy sizes to reflect different sized bodies (see Figure 7.10.2).

Today's crash test dummies are extremely sophisticated, costing between $500 000 and $1 million because of the complex technology and sensors used within them. Unfortunately, in many countries it is not mandated that safety testing uses both male and female dummies, with many only using male dummies. There are many dangerous implications of this, such as seatbelts and airbags designed for men not offering the same protection for women and children, who tend to be smaller than the average man.

There are still questions posed about how well a handful of dummies can accurately predict the impacts of a crash on very diverse body sizes and shapes. Many research institutes are developing virtual testing systems, which many consider to be the future of crash testing. In these tests, computer-generated models of humans with varying size, skeletons and muscle composition can be used to make more accurate predictions about crash impacts for all drivers and passengers. This data can then be used to design and develop modern safety features for road vehicles.

7.10 LEARNING CHECK

1. **Define** 'crash test dummy'.
2. **Suggest** why the first crash test dummy and those that are commonly used today are based on the male body.
3. **Evaluate** the importance of crash test dummies to society.
4. **Compare** the following scenarios for crash testing to evaluate the advantages and disadvantages of each.
   a. Using only male crash test dummies
   b. Using male and female Hybrid III crash test dummies
   c. Using a virtual crash test system for varied body types and compositions

SCIENCE INVESTIGATIONS

7.11 Analysing linear data

SCIENCE SKILLS IN FOCUS

IN THIS MODULE, YOU WILL FOCUS ON LEARNING AND IMPROVING THESE SKILLS:

- analysing proportional data using two techniques to identify relationships and determine constant values
- selecting appropriate mathematical relationships and graphical analysis to organise and process data to identify trends and data anomalies or errors
- investigating the relationship between speed, time and acceleration by collecting and processing data in tables and graphs.

Using graphs to analyse linear data

Scientists often use collected experimental data to identify the relationship between variables and to determine or verify physical quantities. In this module, we will learn about how this can be done for variables that should be directly proportional to one another. Directly proportional means that, when one variable is multiplied, the other variable multiplies by the same amount, so there is a linear relationship between the two variables (see Figure 7.11.1). For example, if you walk for three times as long at a constant speed, the distance travelled is three times as far.

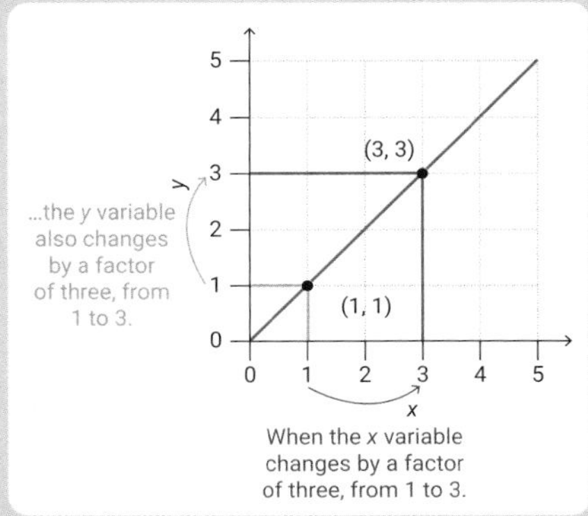

FIGURE 7.11.1 Directly proportional data

To investigate this, we will use some data from an experiment that investigated Newton's second law. In the experiment, a ball of unknown mass was accelerated at four different rates and the force required to accelerate the ball was measured. The relationship should theoretically be linear because force is directly proportional to acceleration when mass is constant (double the acceleration requires double the force). We know this theoretical relationship as Newton's second law:

$$F = ma$$

The results of the experiment are presented in Table 7.11.1.

TABLE 7.11.1 The experimental results for the force required to accelerate a ball at different rates

Acceleration (m/s^2), independent variable (x)	Force (N), dependent variable (y)
0	0
0.4	0.81
0.6	1.14
0.8	1.60
1.0	2.00

Non-graphical analysis methods

Using the data from Table 7.11.1, we can start to identify trends by describing how the dependent variable (force) changes as the independent variable (acceleration) increases. However, it is difficult to determine a mathematical relationship just by looking at the numbers. Since we know the theoretical relationship should be Newton's second law, we could apply this mathematical formula to determine the mass of the ball. To identify the trend accurately using the formula $F = ma$, we need to calculate the mass for each data point and then calculate the average mass.

Table 7.11.2 demonstrates how to use all data to find an average with the formula $F = ma$.

▼ **TABLE 7.11.2** Non-graphical analysis methods to determine the mass of the ball

Acceleration (m/s^2), independent variable	Force (N), dependent variable	Mass (kg)
0	0	–
0.4	0.84	2.1
0.6	1.14	1.8
0.8	1.60	2.0
1.0	2.00	2.0

$$m_{av} = \frac{2.1+1.8+2.0+2.0}{4} = 1.98 \text{ kg}$$

This average mass value would be more reliable than if we only used one point to determine a trend, as the average is less likely to be affected by error. The more data points we have, the more we can minimise the impact of random errors. However, this method is still limited, as it considers all data points equally when determining the mass.

▶ Graphical analysis method

If a graphical approach is used, a linear relationship can be easily identified for the experimental data by the shape of a graph. Figure 7.11.2 shows a scatter plot of the data with the dependent variable (force) on the *y*-axis and the independent variable (acceleration) on the *x*-axis.

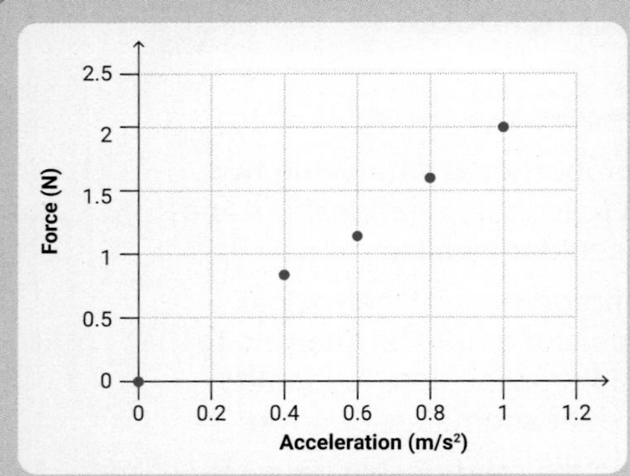

▲ **FIGURE 7.11.2** Graphing force against acceleration

Problem data points can be identified easily when they don't fit the general trend. This means we can disregard them when fitting a linear trendline. Fitting a trendline can be done by hand or using graphing software. The trendline shows us the experimental relationship between the data. If we know the theoretical relationship, we can align it with the trendline to interpret and give meaning to the gradient and *y*-intercept of the trendline. Figure 7.11.3 shows that a linear trendline has been fitted to the data so that it goes through the most points possible. The gradient and *y*-intercept for the straight line can be determined by hand or using graphical analysis software. We can then interpret the data's experimental relationship by comparing it to the theoretical relationship.

Using this method, the mass of the ball can be found since it is equal to the gradient of the straight line, which is 1.99 kg (see Figure 7.11.3).

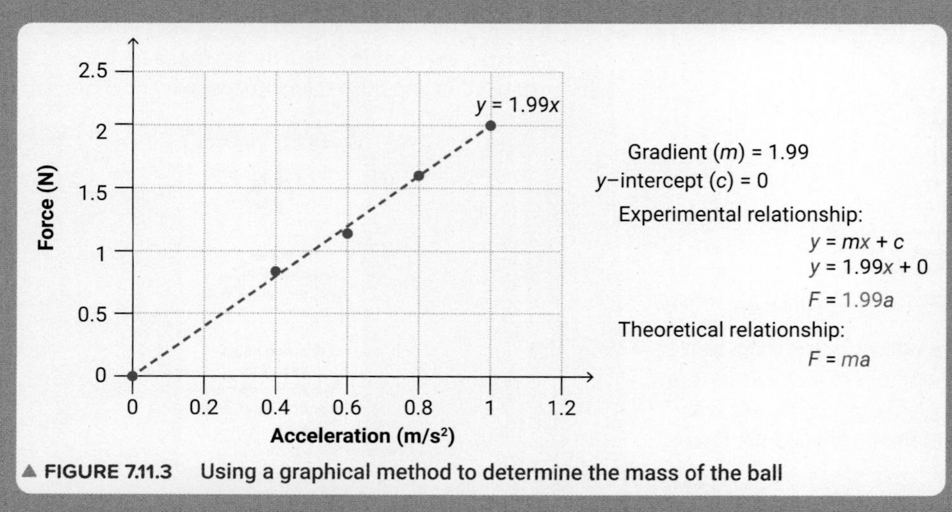

▲ **FIGURE 7.11.3** Using a graphical method to determine the mass of the ball

INVESTIGATION 1: INVESTIGATING THE ACCELERATION, SPEED AND TIME RELATIONSHIP USING A MOTION SIMULATION

In this activity, you will apply the methods from the Science Skills in Focus to investigate the acceleration, speed and time relationship:

$$a = \frac{v - u}{t}$$

During the investigation, a constant force will be applied to a body, and Newton's second law tells us that a constant force on a constant mass will result in constant acceleration. The body will start from rest with an initial speed of 0 m/s. Therefore, this relationship can be rearranged as:

$$v = at$$

This form of the equation shows us that it is, theoretically, a linear relationship, with the speed (v) being the dependent variable (y), acceleration (a) being the constant variable (m) and time (t) being the independent variable (x).

AIM

To investigate the relationship between time and speed for a body experiencing a constant net force, and apply mathematical and graphical analyses to interpret the raw data

MATERIALS

- ☑ PhET simulation 'Forces and Motion: Basics'
  https://phet.colorado.edu/en/simulations/forces-and-motion-basics
- ☑ spreadsheet software
- ☑ stopwatch

METHOD

1 Construct an appropriate results table for the relationship you are investigating.

2 Open the simulation and choose the 'Motion' module from the start-up page.

3 Pause the simulation using the pause (II) button to set up for the experiment.

4 Use the following settings for your experiment.
   - **a** Uncheck 'Force'.
   - **b** Check 'Values'.
   - **c** Leave 'Masses' unchecked.
   - **d** Check 'Speed'.
   - **e** Set the 'Applied Force' to 30 newtons.
   - **f** Use the box only as the mass on the skateboard.

5 Start the timer at the same time as you press the Play button on the simulation.

6 Record the speed of the box every 5 seconds for 30 seconds in your results table.

RESULTS

Present the results table appropriately, including headings and units.

ANALYSIS AND EVALUATION

1 Apply a graphical method to determine the acceleration of the body using a linear trendline. Use the spreadsheet software to generate the graph and add a trendline with an equation.

2 Based on the graph produced, justify why a linear trendline is appropriate for this data.

3 Identify if there were any anomalous data points based on the graph produced.

4 Identify possible limitations or random errors that may have occurred when collecting data in this investigation.

CONCLUSION

1 Draw a conclusion about the relationship between speed and time for a body starting from rest and experiencing a constant net force.

2 Draw a conclusion about the acceleration of the box used in your simulation when a net force of 30 N was applied.

3 Use Newton's second law to calculate the mass of the block.

4 On the simulation, turn on 'Masses' to see the mass of the box you used. Compare this mass with your experimentally determined mass. Discuss reasons for the difference.

INVESTIGATION 2: INVESTIGATING NEWTON'S SECOND LAW AND MEASURING ACCELERATION DUE TO GRAVITY

In this activity, you will apply the methods from the Science Skills in Focus to investigate Newton's second law in the context of falling objects. Remember from Module 7.6 that Newton's second law is applied to the force produced by weight in the following way:

$$F = ma$$

$$F_w = mg$$

This equation shows a theoretically linear relationship with weight force (F_w) as the dependent variable (y), acceleration due to gravity (g) as constant (m as in gradient) and mass (m) as the independent variable (x).

AIM

To investigate the relationship between the mass of an object and its weight force and apply mathematical and graphical analyses to interpret raw data and experimentally measure Earth's acceleration due to gravity

MATERIALS

- ☑ spring balance
- ☑ 5 different hanging masses
- ☑ spreadsheet software

METHOD

1. Construct an appropriate results table for the relationship you are investigating.
2. Ensure the spring balance is calibrated and that when there is no mass hanging from the balance it reads 0 N.
3. Attach the lightest mass to the spring balance and record the force shown on the spring balance.
4. Increase the mass and record the force shown on the spring balance.
5. Repeat step 4 another three or four times so that there is a minimum of five data points recorded.

RESULTS

Present the results table appropriately, including headings and units.

ANALYSIS AND EVALUATION

1. Apply a graphical method to determine the acceleration due to gravity using a linear trendline. Use the spreadsheet software to generate the graph and add a trendline with an equation.
2. Based on the graph produced, justify why a linear trendline is appropriate for the data.
3. Based on the graph produced, identify if there were any anomalous data points.
4. Identify possible limitations or random errors that may have occurred when collecting data in this investigation.

CONCLUSION

1. Draw a conclusion about the relationship between mass and weight force for a hanging body on Earth's surface.
2. Draw a conclusion about the acceleration due to gravity.
3. Compare the experimental value for g with the theoretical value of 9.8 m/s^2. You might include a percentage error calculation using the formula:

$$\text{Percentage error} = \left(\frac{|\text{theoretical value} - \text{experimental value}|}{\text{theoretical value}}\right) \times 100\%$$

9780170463041

7 REVIEW

REMEMBERING

1 **State** the symbol used to represent each of the following measurements and the SI units they are measured in.
   a Distance
   b Speed
   c Acceleration
   d Force
   e Acceleration due to gravity
   f Mass
   g Time

2 **State** the formula used to calculate the average speed of a moving body.

3 **Define** 'acceleration' in words and using a mathematical formula.

4 **Recall** Newton's three laws of motion.

5 **Copy and complete** the following sentence: According to Newton's second law, the net force applied to a body is proportional to its ____ and its ____.

6 **Recall** the form of Newton's second law that is specifically used for the force of gravity pulling bodies towards Earth's surface. Identify each variable and its SI units.

UNDERSTANDING

7 Imagine your friend had never heard the word 'distance' before. How would you describe to them what it means?

8 **Describe** the concept of inertia, giving an example in your response.

9 **Explain** what a negative acceleration value means for a car travelling on a road.

10 **Explain** why the 'normal force' is considered a type of reaction force and is an example of Newton's third law.

11 **Copy and complete** the following sentence: According to Newton's second law, for a body experiencing a constant net force, if the mass increases the acceleration will ____.

12 **Explain** why the gravitational attraction force between Earth and an elephant is different from the gravitational attraction force between Earth and a puppy.

APPLYING

13 If it takes a snail 2 minutes to travel 10 cm, **determine** how many metres the snail can travel in 1 hour.

14 **Give an example** of Newton's third law in action.

15 A dog sitting on a skateboard is pushed by a small child, who applies a force of 45.5 N. The frictional force that resists the forward motion of the skateboard is 12.6 N. If the dog weighs 15 kg and the skateboard is accelerating at a rate of 1.6 m/s^2, **determine** the mass of the skateboard.

16 For a 120 g apple that falls from a very tall tree, **determine** the:
   a weight force pulling the apple towards the ground.
   b speed of the apple when it hits the ground if it is falling for 2.4 seconds.

17 **Explain** the need for head rests in cars as a safety feature.

ANALYSING

18 What knowledge of motion do First Nations Australians utilise when developing and using spear-throwers?

19 The distance–time graph below shows the journey of a teenager's afternoon jog.
   a **Identify** the distance of the entire jog.
   b **Identify** how long the teenager was out on their jog.
   c **Identify** when and where the teenager took a break.
   d **Determine** the speed of the teenager across the whole journey.

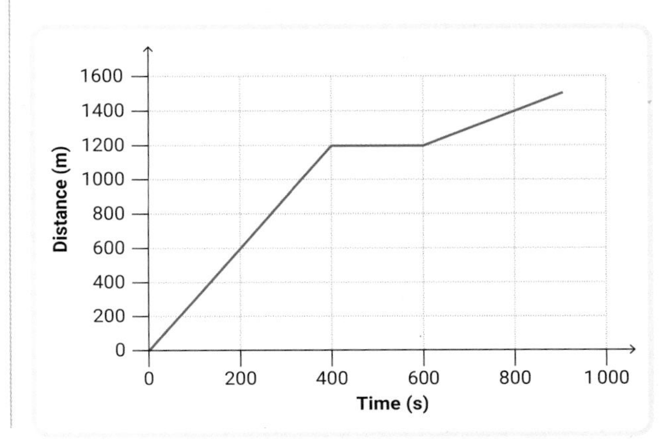

20 A cyclist is riding down a hill. Their speed is recorded on the speed–time graph shown below.

- **a** **Identify** the maximum speed of the cyclist.
- **b** **Identify** the period in which the cyclist is travelling at a constant speed.
- **c** **Identify** the period when the cyclist is accelerating.
- **d** At what time did the cyclist most likely start using the brakes? **Justify** your answer.
- **e** **Determine** the acceleration of the bicycle in the first 25 seconds.

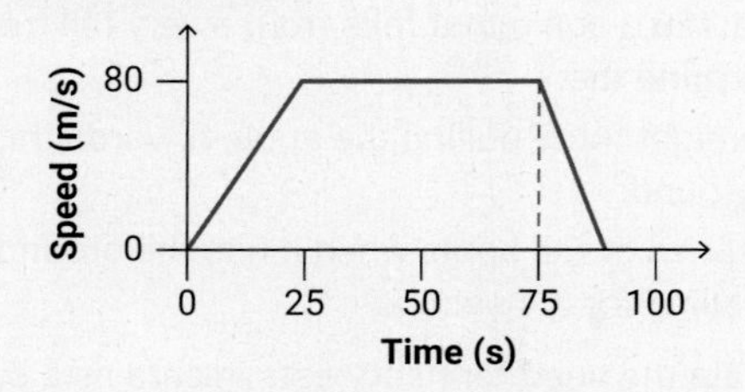

21 Two dogs, Spot and Max, were each sitting on a 2 kg skateboard being pushed by their owner, and their speed after five seconds was recorded. Use the information in the following table to **determine** the mass of each dog.

Dog	Force applied by owner (N)	Speed after 5 s (m/s)
Spot	20	5
Max	16	2.5

22 **Copy and complete** the following table for a body that has been thrown into the air, assuming there is no air resistance.

Time (t)	Speed (m/s)	Acceleration (m/s²)
0	60	–
2	40.4	
4		
	11	

EVALUATING

23 When a hazard such as a pedestrian appears on the road, the driver of a car should react by putting their foot on the brake to decelerate the car to a stop before they reach the point of the hazard (the pedestrian). **Evaluate** the importance of having a fast reaction time when driving a car on suburban roads.

24 When you stand on scales, the scales actually measure the weight force your body exerts on them and converts this force to a mass using Newton's second law and the acceleration due to gravity ($g = 9.8\ \text{m/s}^2$), before displaying a mass value on the screen. Imagine standing on a set of scales on the Moon or another planet, and **discuss** the implications this may have on your weight, as shown by the scales.

25 During a triathlon, a triathlete swims through water, runs along the road and cycles on a road. **List** five tips for a triathlete that they could employ before and during the race to help enhance their performance. **Justify** your recommendations using Newton's laws of motion.

CREATING

26 **Create** an infographic of a person driving a car; include annotations that describe different ways Newton's laws of motion are relevant to driving. Include at least one application of each law.

27 **Create** a promotional poster for a piece of cycling equipment that is marketed for professional cyclists with the selling point that it will improve their performance. Use what you have learned from this chapter, particularly Module 7.7, to complete this.

BIG SCIENCE CHALLENGE PROJECT

1 Connect what you've learned

In this chapter, you have learned about many ideas within the topic of motion and many applications for these concepts. Create a mind map to show how these main ideas are connected, annotating the connections with explanations of their relationship.

2 Check your thinking

When introducing the Mars rovers at the start of this chapter, we asked the following questions.

- How did they all get there?
- How far is it between Earth and Mars?
- How long did it take to fly there?
- How fast did they travel?
- How were these speeds achieved?

Perform some research to find answers to these questions. Write a short paragraph about what you think the biggest challenges were for NASA and the China National Space Administration's astrophysicists in designing the rockets and rovers and planning for their journey to Mars.

3 Make an action plan

Imagine you are working with the team at NASA that planned the landing of the Perseverance rover on Mars. Using the topics from this chapter, make a list of 10 key questions you and your team would need to ask and answer to ensure that the rover landed safely on the surface of Mars. Explain why each of your questions is important for this project.

4 Communicate

Create an infographic titled 'Planning for Perseverance's Landing' using a photo or a sketch similar to the one on this page. Use the information from your action plan as the labels and annotations.

NASA/JPL-Caltech

The Perseverance Rover landing on Mars

8 Psychology

8.1 Introduction to psychology (p. 313)

Psychology is the scientific study of human thoughts, emotions and behaviours.

8.2 History of psychology (p. 316)

Psychological theories and practices change over time as new evidence is gathered and existing theories are challenged or supported through ongoing research.

8.3 The nervous system (p. 320)

The nervous system is made up of two parts: the central nervous system and the peripheral nervous system.

8.4 The brain (p. 324)

The brain is responsible for controlling all the functions of the body; interpreting all internal and external information; and controlling memory, speech, emotions, thoughts and behaviours.

8.5 Brain research (p. 327)

Psychology's scientific understandings are based on research that has been repeated over time, using systematic methods to collect and analyse data.

8.6 Consciousness (p. 330)

Consciousness is our level of awareness of the internal and external stimuli that are taking place at any one time.

8.7 Sensation and perception (p. 333)

All the information we get from the world around us enters through our senses. Perception occurs once the process of sensation is complete.

8.8 Nature versus nurture (p. 335)

Psychologists study the effect of genes and environmental influences on social, emotional, physical and cognitive development.

Assessments

- End-of-chapter test
- Portfolio assessment: Research investigation

Interactive resources

- Label: Scientific method in psychology **(8.1)**; The lobes of the brain **(8.4)**; Hierarchy of awareness **(8.6)**
- Simulation: Neurons **(8.3)**

Videos

- Video activities: What is psychology? **(8.1)**; Phineas Gage **(8.2)**; Sigmund Freud **(8.2)**; The nervous system **(8.3)**; Neurons as cells **(8.3)**; Introduction to the brain **(8.4)**; The history of the brain **(8.5)**; Monkey business **(8.6)**; Optical illusions **(8.7)**; Nature v. nurture **(8.8)**

Nelson MindTap

To access these resources and many more, visit: **cengage.com.au/nelsonmindtap**

8.1 Introduction to psychology

BY THE END OF THIS MODULE, YOU WILL BE ABLE TO:

- ✓ define 'psychology'
- ✓ compare psychology, psychiatry and social work
- ✓ identify the role of research in psychology
- ✓ describe the scientific method
- ✓ explain how the scientific method strengthens or weakens a theory.

GET THINKING

You may have seen an online personality test such as those that use your name or your date of birth, or even compare you to a type of animal, to create a personality profile.

Before you read any further, write a few sentences explaining whether you think these are good examples of psychology. There are no right or wrong answers at this point. Compare your ideas with one other person.

Video activity
What is psychology?

Interactive resource
Label: Scientific method in psychology

Psychology, psychiatry and social work

Psychology is the **scientific** study of human thoughts, emotions and behaviours (see Figure 8.1.1).

psychology
the scientific study of human thoughts, emotions and behaviours

scientific
based on systematic measurement and analysis of observable evidence

Shutterstock.com/FGC

▲ **FIGURE 8.1.1** Psychology is the scientific study of human thoughts, emotions and behaviours.

Psychologists study individuals and groups to better understand how people, communities and societies function. Psychologists work with individuals and groups to help them find better ways of managing their lives so that they can function effectively in society. Psychologists must study at university for 4–10 years and be accredited by the relevant psychology board.

psychologist
a qualified professional who works with human thoughts, emotions and behaviours in a variety of ways

Psychiatry and social work are similar professions that also focus on helping individuals and groups improve their lives. **Psychiatrists** are medical doctors who specialise in diagnosing and treating **mental illness** and can prescribe medication. They have specialised in psychiatry after completing their medical degree.

psychiatrist
a medical doctor who specialises in diagnosing and treating mental illness

mental illness
any health condition that affects a person's thoughts, mood and behaviour, and causes distress or dysfunction in daily life

Shutterstock.com/VH-studio

▲ FIGURE 8.1.2 Psychologists, psychiatrists and social workers provide support to a range of people to help them make their lives better.

Social workers are allied health professionals who help people to improve their wellbeing and support them in crises to connect to services and their communities. They may work with people with mental illness, but they also support people whose life circumstances prevent them from participating fully in society. For example, they assist people who experience poverty, people with intellectual or physical disabilities, people who have suffered from violence or other forms of abuse, and migrants who may need support.

social worker
an allied health professional who helps people improve their wellbeing and supports them in crises by connecting them to services and communities

The three professions often work together or refer people to each other. Some people may have a care plan that includes support from all three professions.

Psychology is a science

Psychology follows the scientific method to develop and test theories that can be applied to **populations**. In psychology, the term 'population' refers to the target group who are being studied. Instead of referring to everyone, a population to a psychologist consists of a particular group, such as children aged five, families living in rural areas, athletes with anxiety or teenagers who use TikTok.

population
the target group that is being studied

Psychology's scientific method involves systematically investigating a **theory**, idea or question by collecting evidence, recording data, analysing data and evaluating outcomes (see Figure 8.1.3). Once the method has been applied and evaluated, it is reported and **peer-reviewed** (checked for scientific accuracy by other psychologists) before being published.

theory
a framework for explaining different scientific concepts and systems

peer-reviewed
checked for scientific accuracy by other scientists in the same field

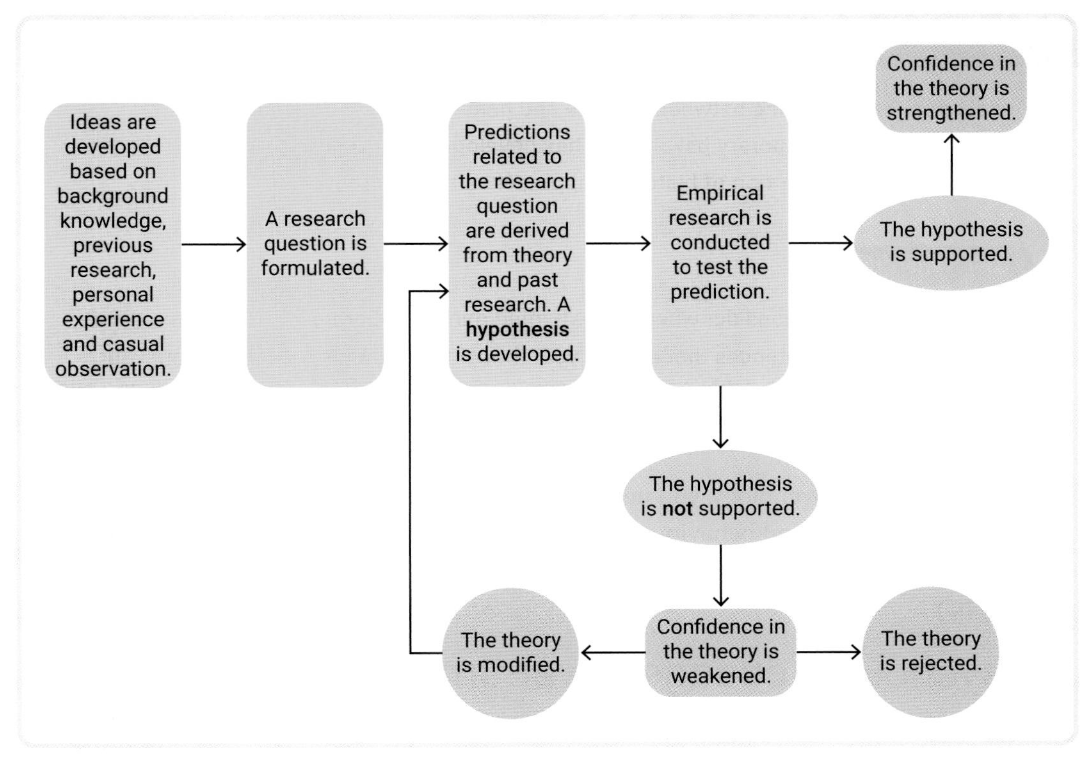

▲ **FIGURE 8.1.3** The scientific method in psychology

Psychological theories explain concepts and understandings that apply to human thoughts, emotions and behaviours. Theories develop and change as the scientific method is applied. A theory becomes 'stronger' each time its hypothesis is supported by the data from an investigation or research project.

Fields of psychology

Since its origins, psychology has expanded into many different **fields** where the research and theories may be specialised. Examples include cognitive psychology, social psychology, educational psychology, developmental psychology, personality psychology, community psychology, biological psychology and psychotherapy.

fields
specific areas of study within psychology

8.1 LEARNING CHECK

1 **Define** 'psychology'.
2 **Describe** two differences between a psychologist and a psychiatrist.
3 **Explain** why psychology is described as a science.
4 **Create** a diagram that shows the steps in the scientific method a psychologist would follow to investigate memory improvement for school students.

Extension

5 **Describe** one field of psychology of interest to you. Include details about the area(s) of focus, qualifications required and who would benefit from working with this type of psychologist. Gather your information from at least three different sources and include a reference list.

8.2 History of psychology

BY THE END OF THIS MODULE, YOU WILL BE ABLE TO:

✓ describe the difference between science and pseudoscience
✓ explain how contemporary psychology has developed over time
✓ describe the contribution of historical approaches to modern psychology.

Video activities
Phineas Gage
Sigmund Freud

GET THINKING

Take 30 seconds to consider whether the human mind is separate from the human body. Write down three sentences that explain your thoughts about this topic. Compare your answer with that of one other person and discuss your ideas. Review your answer and add at least two more sentences.

When you study psychology, you need to understand that the ways we explain human thoughts, emotions and behaviours are constantly changing. This happens as new evidence is gathered, and existing theories are challenged or supported through ongoing research.

Psychology changes over time

Although theories about human nature and the mind have existed since ancient times, psychology was only identified as a separate scientific field in the late 1800s.

Originally, most of the research in the field of psychology was focused on mental illness and **criminology**. People with mental illnesses were locked in asylums or placed in prisons, away from the everyday world. Therapy was limited and often cruel.

criminology
the scientific study of crime and criminal behaviour

Psychology's scientific approach has changed attitudes and behaviours towards the treatment of people with mental illnesses. Psychologists now have a better understanding of how to help people improve their lives. Research continues to change the way we think about mental illness and mental wellbeing.

▲ **FIGURE 8.2.1** René Descartes (1596–1650)

From philosophy to science

Psychology today is considered a science, but this has not always been the case.

Before the 1800s, the human mind and its function were widely discussed by both philosophers and physicians. There was much debate about whether the human mind and body were separate or worked together. Limited methods were available to collect evidence about the human brain and this made it difficult to investigate scientifically.

Philosopher René Descartes (see Figure 8.2.1), asserted that 'I think, therefore I am'. This summarised the idea that personality or identity are not contained in the body but in the mind, which is separate.

However, early physicians proposed that all thoughts and emotions were linked to the physical body. By curing the physical illness, they believed that the mental illness would also be cured.

An example from history – phrenology

In the late 1700s, Franz Gall, a physician from Vienna, proposed that the shape of the human skull could predict the way individuals behave and that it indicated their abilities and tendencies. He suggested that the size and location of indentations and bumps on the skull had arisen in the process of brain growth and were linked to specific characteristics. For example, a large bump on either side of the head near the eyes could mean that a person was mathematically talented. Gall's theory became known as phrenology (see Figure 8.2.2).

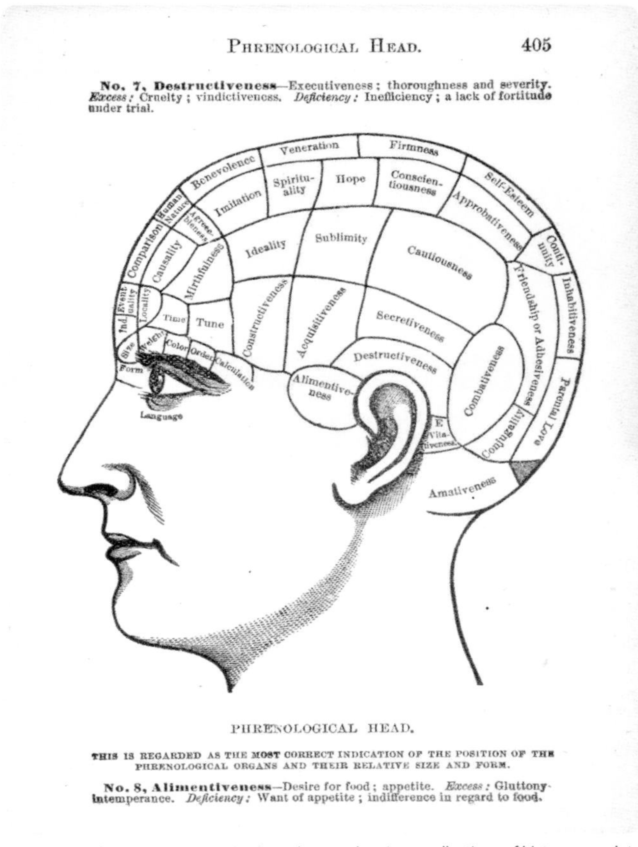

Wellcome Collection/Forty years in phrenology: embracing recollections of history, anecdote, and experience/[Nelson Sizer]

▲ **FIGURE 8.2.2** A phrenologist's map

Johann Gaspar Spurzheim, Gall's student and research assistant, developed the theory further to suggest that, by exercising the brain, changes could be made to improve a person's life.

Phrenology was popular in the late 1800s, but it was always controversial. Even though it was based on measuring the head and its features, there was no scientific evidence linking these findings to human behaviour, thoughts and emotions, and it was soon discounted as **pseudoscience**.

Although phrenology is no longer accepted in psychology, it is a useful example of how psychological thinking and theories have developed over time. The idea of linking the shape, size and location of parts of the brain with human behaviour and personality was one step towards the development of modern neuroscience.

pseudoscience
an idea or practice that claims to be scientific but does not use the scientific method

The origins of modern psychology

Until the development of modern brain imaging technology, the only way to see the brain was to dissect dead bodies or to observe and examine living patients. As it is not ethical to cause harm to living creatures, our understanding of the human brain was often developed through **case studies**. These involve a detailed examination of specific individuals or groups when experiments are not possible due to ethical or practical reasons.

case study
a detailed examination of specific individuals and/or groups

▲ **FIGURE 8.2.3** A drawing showing how the metal rod passed through Gage's skull and brain.

Case study: Phineas Gage

In the United States in 1848, a railway worker named Phineas Gage was using explosives and had a metal rod shot through his skull (see Figure 8.2.3). Surprisingly, Gage did not die of his injuries, even though the rod damaged parts of his brain.

In addition to the physical effects of the injury, such as the loss of sight in one eye, the doctors noticed that Gage suffered significant changes to his personality that were difficult to explain. As a result of Gage's injuries, psychologists began to **hypothesise** that different parts of the brain have different functions. This concept is referred to as **localisation**.

hypothesise
to make a testable scientific prediction

localisation
the idea that different parts of the brain have different functions

psychoanalyst
a qualified professional who treats patients using psychoanalysis

Gage is a famous case study used by psychologists, but many other case studies have also contributed to our understanding of the human brain.

The theory of psychoanalysis – Sigmund Freud

▲ **FIGURE 8.2.4** Sigmund Freud (1856–1939)

During the late 1800s, Sigmund Freud (see Figure 8.2.4) began to develop a theory of personality formation. He proposed it was a person's past experiences that had the most impact on their development.

He also proposed that humans are influenced by the unconscious (parts of the mind that we are not aware of) which can influence us to behave in particular ways. He believed that mental illnesses could be treated through self-exploration and guided therapy with an expert **psychoanalyst**.

Much of Freud's theory was controversial at the time because it lacked evidence, but it challenged traditional approaches to mental illness. It led to more research into the ways that understanding personal history, thoughts and actions can be used to improve mental health. Freud is credited with the initial use of 'talking therapy', whereby a patient expresses their mental state by speaking with a psychologist, who provides alternative ways of thinking that may lead to a healthier approach to problems in life.

9780170463041

Further developments in psychology

Psychologists like Freud did not work in isolation. At the time he was developing psychoanalysis, other theories were also developing and emerging. In contrast to Freud's theory, behaviourists proposed that personality developed solely as a result of direct experience with our environment (physical and social). According to this theory, humans create associations between thoughts and feelings that motivate them to behave in a particular way.

Although behaviourism and psychoanalysis were originally in direct opposition, aspects of these theories were incorporated into new approaches over time. One example is cognitive behaviour therapy, an evidence-based form of talking therapy currently used widely in Australia. It draws on the talking therapy approach originally developed by Freud and utilises understandings from behaviourism regarding how to create new ways of thinking. Clients meet regularly with a psychologist and work together to change unhelpful or damaging thoughts so that the person can function more effectively in their life.

LEARNING CHECK

1 **Define** 'phrenology'.
2 **Distinguish** between science and pseudoscience.
3 **Discuss** why Phineas Gage was an important case study in psychology.
4 **Describe** Freud's theory of personality formation.

Extension

5 Select one historical theory of human behaviour, emotions and/or thought developed more than 100 years ago and research it further. Examples of historical theories that could be used for this task include those of Sheldon (somatotypes), Galen (humours), Freud (psychoanalysis), Freeman (lobotomy), James (functionalism), Watson (behaviourism), Thorndike (reinforcement) and Broca/Wernicke (localisation).
   a **Explain** the key understandings and methods that apply to this approach.
   b **Describe** how this approach was or was not considered scientific.
   c **Explain** how the historical approach may have contributed to the development of one field of psychology.

8.3 The nervous system

BY THE END OF THIS MODULE, YOU WILL BE ABLE TO:

- ✓ describe the structure of the nervous system
- ✓ explain the function of different divisions of the nervous system
- ✓ identify, draw and label the key structures of the neuron
- ✓ distinguish the functions of sensory neurons, motor neurons and interneurons
- ✓ create a labelled diagram that shows the process of neurotransmission.

Video activities
Neurons as cells
The nervous system

Interactive resource
Simulation: Neurons

unconscious
completely lacking awareness

central nervous system (CNS)
the part of the nervous system comprising the brain and spinal cord

peripheral nervous system (PNS)
the part of the nervous system that carries messages to and from the CNS

GET THINKING

As you read the text in this box, quickly list everything that you are doing right now. Keep your list handy – you will add information to your list later in the chapter.

At any given time, you have many conscious and **unconscious** processes occurring that let you function in daily life. This includes all bodily functions such as breathing, mental processes such as reading and motor functions such as writing. All of this is possible because of the nervous system.

The parts of the nervous system

The nervous system is made up of two parts: the **central nervous system (CNS)** and the **peripheral nervous system (PNS)**, as shown in Figure 8.3.1.

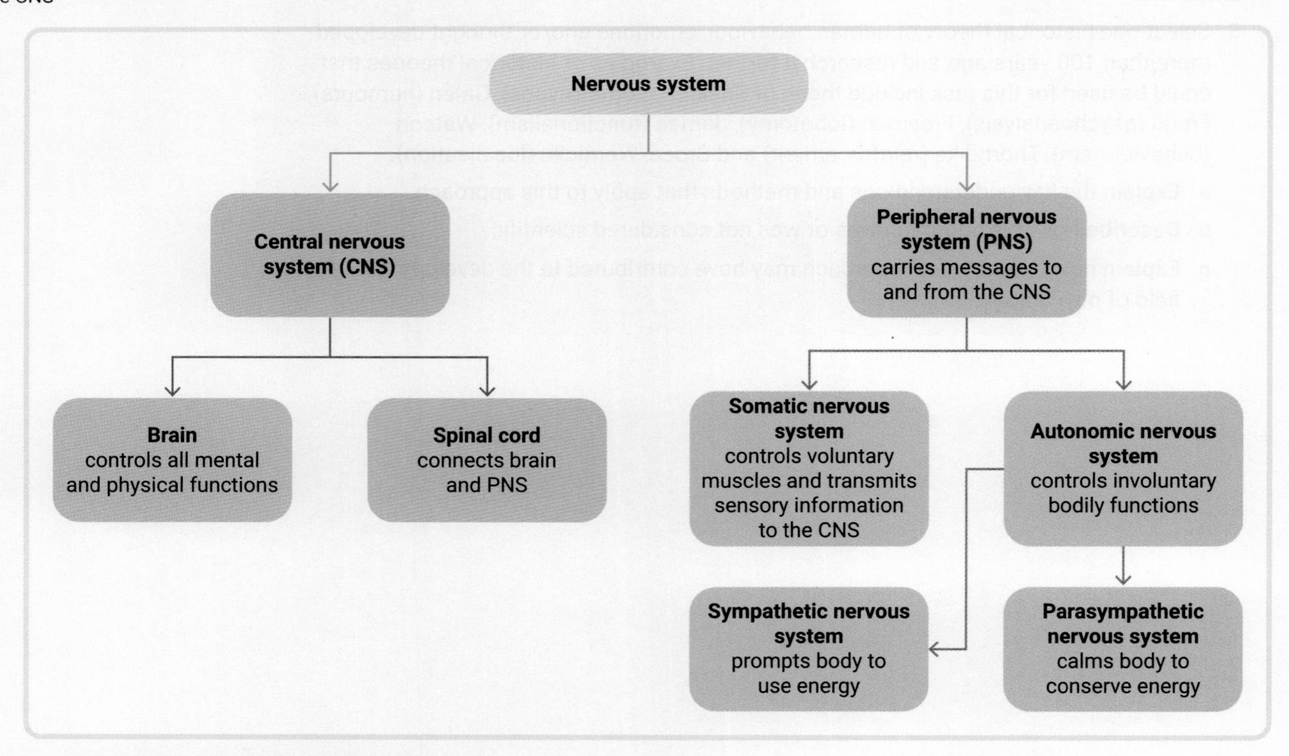

▲ **FIGURE 8.3.1** The nervous system

The CNS consists of the brain and spinal cord. It is the control centre for all our mental and physical processes.

The PNS consists of the **somatic nervous system** and the **autonomic nervous system**. The autonomic nervous system is divided further into the **sympathetic nervous system** and the **parasympathetic nervous system**.

The PNS consists of all the nerves outside the brain and spinal cord. It is responsible for communicating information to and from the body's organs, muscles and glands to the CNS. It relays sensory information to the CNS, which then sends a motor message back via the PNS, so the body is able to respond.

Shutterstock.com/Grigvovan

▲ **FIGURE 8.3.2** Moving your hand away quickly away from hot water is an example of the CNS and PNS in action.

For example, when you put your hand in the shower or under a tap to check the water temperature, the sensory information is received by the PNS and sent to the CNS, which interprets the temperature and relays the required motor response to the PNS. This is why you pull your hand away quickly if the water is too hot.

The somatic nervous system oversees voluntary muscle movements such as walking, running and throwing a ball. The autonomic nervous system controls involuntary body functions such as heart rate, digestion and hormone regulation.

The sympathetic nervous system is the part of the nervous system that triggers the fight–flight–freeze response. Have you ever been on a roller coaster or got nervous before a test? You might have felt your heart beat a bit faster, the hair on the back of your neck stand on end and your breathing rate increase. This is your sympathetic nervous system in action. It gives you the energy you need to stay and defend (fight), run away from the situation (flight) or not react at all (freeze). The parasympathetic nervous system calms you down and returns your body to its normal state once the threat has subsided.

somatic nervous system
the part of the PNS that controls voluntary muscles

autonomic nervous system
the part of the PNS that controls involuntary body functions

sympathetic nervous system
the part of the autonomic nervous system that is activated during a fight–flight–freeze situation

parasympathetic nervous system
the part of the autonomic nervous system that calms the body and restores it to its normal state following the activation of a sympathetic nervous system response

Shutterstock.com/Jacob Lund

▲ **FIGURE 8.3.3** On a roller coaster you can feel your sympathetic nervous system in action.

neuron
a specialised type of cell in the nervous system that transmits a signal between the CNS and PNS

sensory neuron
neurons that transmit electrical impulses from the sensory receptors to the CNS

motor neuron
a neuron that transmits movement instructions from the CNS to the muscles

interneuron
a neuron that conveys impulses from one neuron to another

dendrite
the branching network at the end of a neuron that receives information from other neurons

axon
a long, thin fibre that carries electrical impulses

myelin sheath
a protective coat around an axon that increases the speed of nerve impulses

Neurons

Neurons are the basic units of the nervous system and are involved in all our behaviours and bodily functions. They receive and send electrochemical signals between the CNS and PNS.

There are three main types of neurons:

- **sensory neurons**, which transmit signals in the form of electrical impulses from the sensory receptors to the CNS
- **motor neurons**, which transmit movement requests from the CNS to the muscles in the body
- **interneurons**, which connect neurons to each other in the CNS. They make up most of the neurons in the brain and spinal cord.

The different types of neurons exist in a variety of shapes and sizes, but they all have **dendrites**, a nucleus, a cell body (soma), an **axon** and a **myelin sheath** (see Figure 8.3.4).

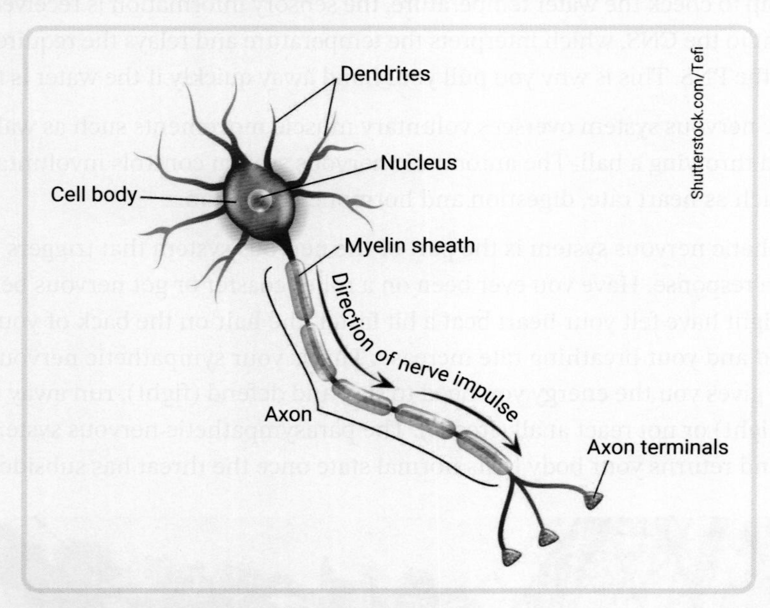

▲ **FIGURE 8.3.4** The main parts of a neuron

Neurotransmission

Signals in the form of electrical impulses pass between neurons and between neurons and muscles or other parts of the body. This is called neurotransmission.

neurotransmitter
a chemical messenger

The dendrites receive a chemical signal in the form of a **neurotransmitter**. They convert this to an electrical impulse that passes through the cell body to the axon. The electrical impulse travels down the axon and reaches the axon terminal. Here it is converted back to a neurotransmitter.

Neurons do not actually touch each other. There is a small gap between them called a **synapse** that the neurotransmitter travels across to be collected by the dendrites of the receiving neuron, as shown in Figure 8.3.5.

synapse
the small gap between the axon terminals of one neuron and the dendrites of the next

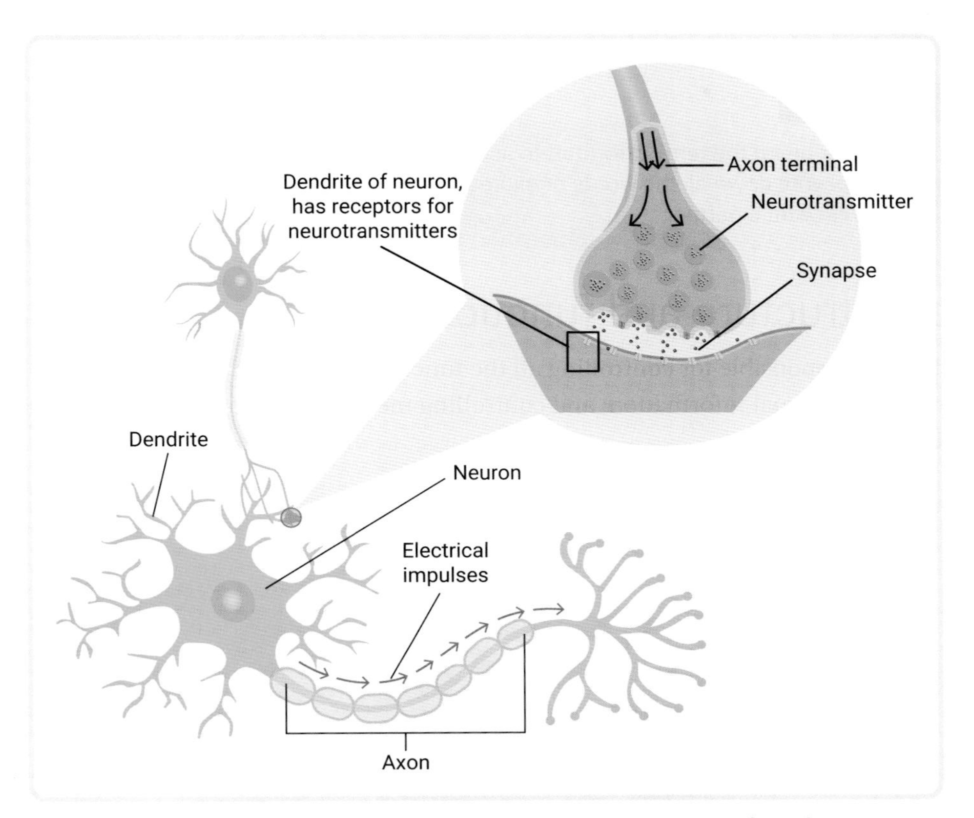

▲ **FIGURE 8.3.5** Neurotransmitters pass messages across a synapse from the axon terminal of one neuron to the receptors on the dendrite of the next neuron.

When the same group of neurons communicate often, the connection between them strengthens and they form neural pathways. If the communication is repeated often enough, the myelin sheath thickens, allowing messages to transmit faster. This is how even complex behaviours, like riding a bike or writing, become automatic.

Neural pathways are not fixed. As we change a behaviour, new pathways are formed. If we repeat the new behaviour, this strengthens the new pathway. This explains why practice is the key to change in our lives.

8.3 LEARNING CHECK

1 **Draw** a diagram of the nervous system that clearly labels the function of each part.
2 **Explain** the role of the dendrite, soma, axon and axon terminal.
3 **Describe** the process of neural transmission, using an example from real life to illustrate this process.

8.4 The brain

BY THE END OF THIS MODULE, YOU WILL BE ABLE TO:

✓ explain the function of the four lobes of the brain and describe what happens if they are damaged.

Video activity
Introduction to the brain

Interactive resource
Label: The lobes of the brain

GET THINKING

Have you ever heard the term 'grey matter' used to describe a part of the brain? Take 2 minutes to write down all the words or expressions you know that relate to the brain. Share your list with a partner.

Brain structure and function

The brain is responsible for controlling all the functions of the body; interpreting all internal and external information; and controlling memory, speech, emotions, thoughts and behaviours.

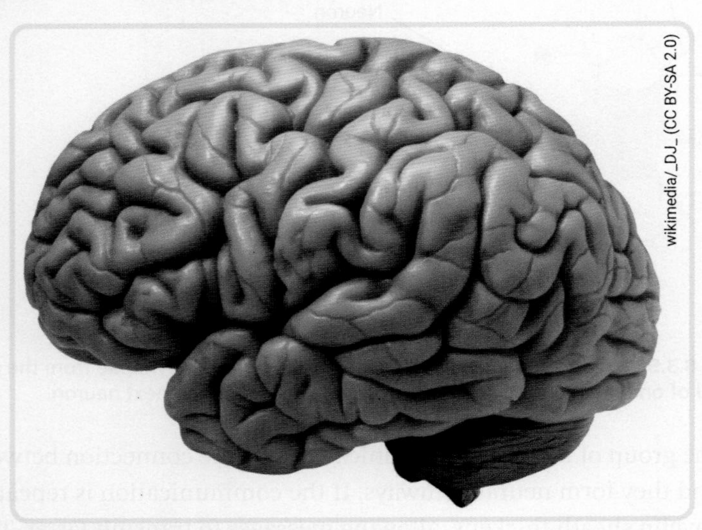

▲ **FIGURE 8.4.1** The outer part of brain

cerebral cortex
the outer layer of the brain, consisting of two hemispheres

hemispheres
the two halves of the cerebral cortex, connected by the corpus callosum

corpus callosum
the thick band of nerve fibres that connects the two hemispheres of the brain and allows messages to be passed between them

The outer part of the brain that we are most familiar with is called the **cerebral cortex**. It has a wrinkled appearance with grooves (sulci) and bumps (gyri). The cerebral cortex is involved in higher-order thinking and information processing and directs complex behaviours such as problem-solving and planning.

The cerebral cortex is divided into two **hemispheres**: the left and the right (see Figure 8.4.2a). The left hemisphere is involved in verbal skills such as language, reading and writing. It controls the right-hand side of the body. The right hemisphere is involved in non-verbal skills, such as spatial awareness and drawing. It controls the left-hand side of the body. The two hemispheres are connected via the **corpus callosum** (see Figure 8.4.2b). This allows the two hemispheres to communicate so that the left and right sides of the brain work together.

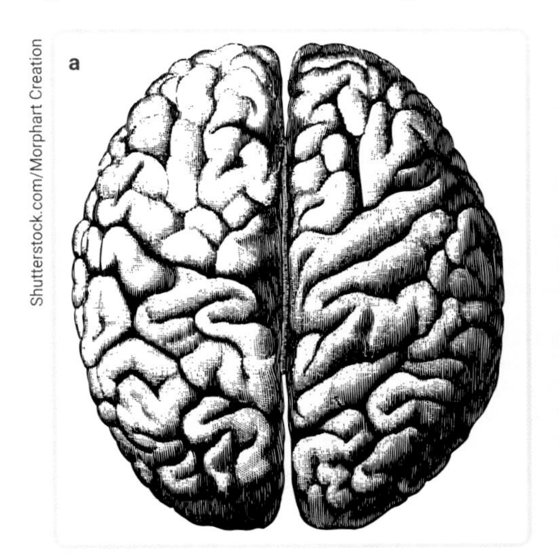

Shutterstock.com/Morphart Creation

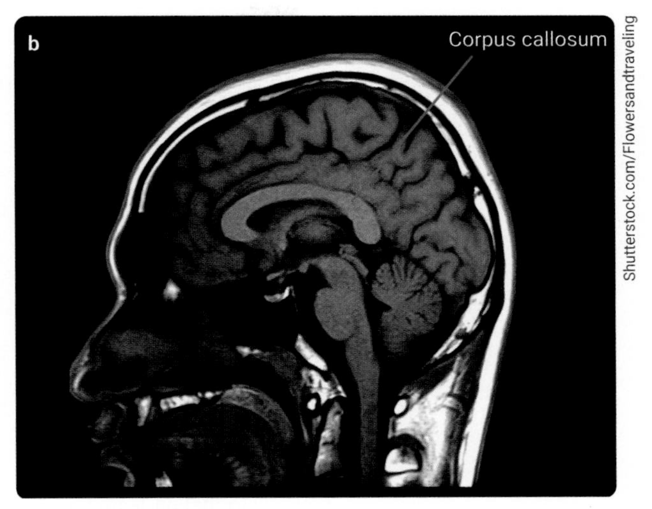

Shutterstock.com/Flowersandtraveling

▲ **FIGURE 8.4.2** **(a)** The cerebral cortex consists of a left hemisphere and a right hemisphere. **(b)** The corpus callosum connects the two hemispheres of the cerebral cortex.

Each hemisphere is divided into four lobes: the **frontal lobe**, the **temporal lobe**, the **occipital lobe** and the **parietal lobe** (see Figure 8.4.3). Each is associated with specific processes, but for us to be able to function the lobes must all work together.

frontal lobe
the front part of the cerebral cortex, associated with voluntary movement and higher-order mental abilities such as planning and thinking

temporal lobes
the lobes on each side of the cerebral cortex, associated with auditory information and speech

occipital lobe
the back part of the cerebral cortex, associated with vision

parietal lobe
the lobe on the top of the cerebral cortex, associated with body sensations such as touch and temperature and spatial awareness

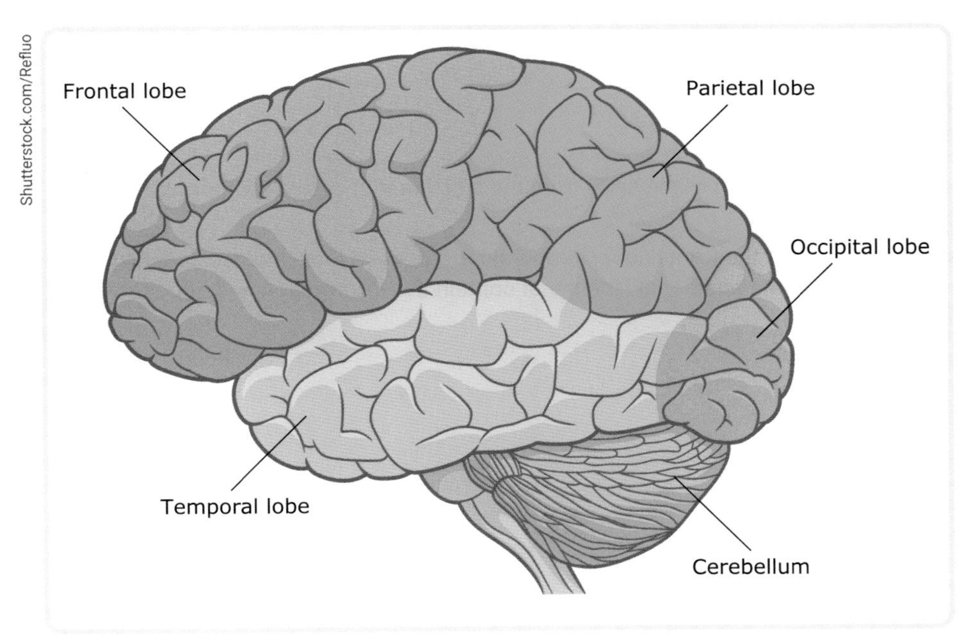

Shutterstock.com/Refluo

▲ **FIGURE 8.4.3** The four lobes of the brain

Brain disease or injuries can affect people in many ways, depending on which lobe is damaged (see Table 8.4.1).

▼ **TABLE 8.4.1** The function of each lobe and the potential effects of damage

Lobe	Function	Effect of damage to lobe
Frontal	Involved in higher mental processes such as planning, thinking and decision-making	• Difficulty concentrating • Difficulty processing information and answering questions • Poor balance • Difficulty problem-solving • Difficulties with organisation
Parietal	Concerned with body sensations such as touch and spatial awareness	• Loss of sensation in a part of the body • Confusion between left and right • Disorientation • Inability to recognise one side of the body as belonging to them (spatial neglect) • Inability to perceive temperature/touch
Temporal	Associated with hearing, memory formation, learning and processing of emotions	• Difficulty learning and retaining new information • Emotional disturbance • Difficulty interpreting sounds • Difficulty with facial recognition • Problems with understanding spoken words
Occipital	Responsible for processing visual information	• Difficulty locating objects • Inability to recognise words (word blindness) • Difficulty identifying colours • Hallucinations • Problems with recognising drawn objects

8.4 LEARNING CHECK

1 **Create** your own drawing of the brain from the side (lateral) view. **Label** each of the four lobes.

2 **List** the functions of the frontal lobe, temporal lobe, parietal lobe and occipital lobe. You may choose to create additional labels that contain this information and add them to your drawing.

8.5 Brain research

BY THE END OF THIS MODULE, YOU WILL BE ABLE TO:

✓ describe three brain scanning techniques.

GET THINKING

As we have learned in the previous modules, the human brain is very complex. This is why brain research is such an important and diverse field of science. Brain mapping is one example of a brain research technique. Research brain mapping and write down five facts to add to a whole-class mind map.

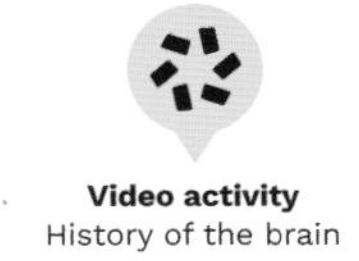

Video activity
History of the brain

Psychology's scientific understandings are based on research that has been repeated over time, using systematic methods to collect and analyse data. All research is now ethically bound to ensure that participants are not harmed and that they are fully informed about their roles.

Brain research is conducted in multiple ways, including the use of neuroimaging technology, participant questionnaires, observations and the measurement of bodily responses to stimuli.

Split-brain research

Much of what is known about the hemispheres of the brain originated from Roger Sperry's numerous experiments on animals and humans who had had their corpus callosum cut. Sperry was a renowned American neuropsychologist and Nobel laureate. He developed the experiments during the 1960s to test if the brain's different functions were specific to the two hemispheres. In one study in 1968, Sperry gathered 11 participants whose corpus callosum had been cut in an attempt to treat severe epilepsy. This was the **experimental group**. The treatment was classified as an experiment and the performance of the 11 participants on various tasks was compared with that of people whose corpus callosum was intact (the **control group**). The **independent variable** was whether the person had their corpus callosum intact or not and the **dependent variable** was the participants' performance on the tasks.

experimental group the group that is exposed to the independent variable

control group the group that is not exposed to the independent variable

independent variable what you are changing

dependent variable what you are measuring

Method

One eye of each participant was blindfolded. They were asked to focus on the middle of a screen where a black dot acted as a fixation point to separate the left- and right-hand sides. Words were then projected on either the left or the right side of the fixation point, and Sperry asked participants to tell him what they saw.

Findings

When participants saw the word with their right eye, it was processed by the left hemisphere of the brain, and they were able to say what they saw. This is because the left hemisphere is responsible for language. When participants saw the word with their left eye, it was processed by the right hemisphere, and they could not identify the word (see Figure 8.5.1).

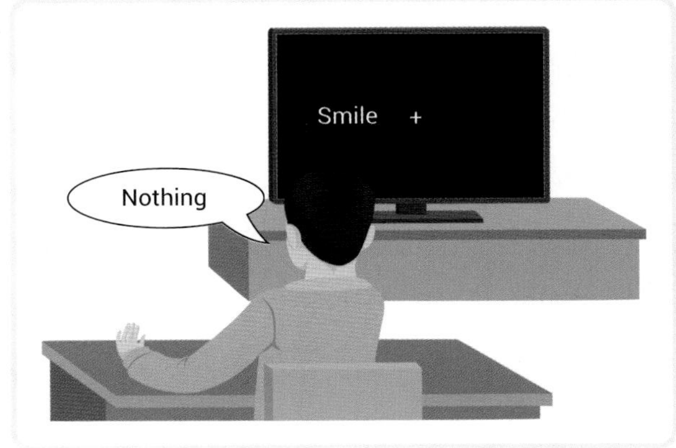

▲ **FIGURE 8.5.1** Sperry used split-brain experiments with participants who had their corpus callosum cut.

Sperry and his team conducted other experiments using different methods and it was concluded that not only do each of the hemispheres have specialised functions, but they also have their own memories and experiences.

Case study: the effect of damage to the brain of an AFL player

A young football player was diagnosed with a brain injury after losing some of his brain function, the symptoms of which included memory loss, some hearing difficulties and an inability to concentrate. Doctors linked his symptoms to his history of blows to the head during his football career. Certain aspects of the sport present a high risk for head injuries (see Figure 8.5.2).

AAP Images/JOE CASTRO

▲ **FIGURE 8.5.2** Many players of Australian Rules and other football codes have suffered brain injuries caused by blows to the head.

A concussion is usually diagnosed after a blow to the head. It is limited to a minor amount of bruising, some swelling and some damage to the brain tissue. In this case, the footballer suffered many concussions during his playing career from when he was a child to when he retired at 28 years old. Multiple episodes of concussion are more likely to lead to long-term consequences such as dementia, Alzheimer's disease and Parkinson's disease, and so the footballer decided to end his career to avoid further damage.

This case illustrates that, as our understanding of brain functions improves, a diagnosis can be made earlier, which may mean more effective interventions that can improve the quality of life for people with an acquired brain injury. Changes to the AFL's rules, such as the introduction of concussion protocols that enforce breaks from play and training for concussed players, also result from this improved understanding.

Techniques to investigate the brain

Before the development of neuroimaging technology, scientists had to wait until a person had died before they could examine their brain. Today, there are a variety of techniques that enable researchers and medical professionals to pinpoint specific areas of the brain. This assists them to identify the location and extent of the damage caused by disease or injury.

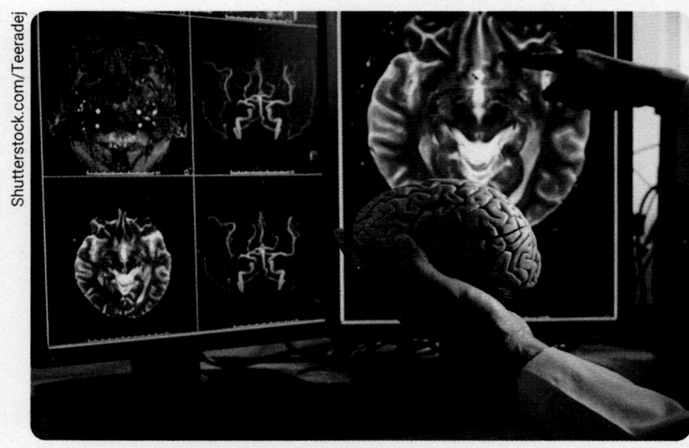

Shutterstock.com/Teeradej

▲ **FIGURE 8.5.3** There are many types of neuroimaging techniques that take scans of the brain.

By observing and measuring the brain, it has been possible to identify which parts of the brain are linked to specific functions and even how the different parts of the brain interact. As the technology continues to improve, researchers are learning more about the complex interactions within the brain and are beginning to explain that, while each lobe may have areas of specialisation, they work together at different times and in different situations.

Some commonly used brain scan techniques are listed in Table 8.5.1.

▼ **TABLE 8.5.1** Different brain scan techniques

Technique	Type	Method	Strengths and limitations
Computed axial tomography (more commonly known as a CAT scan, or CT scan) Alamy Stock Photo/Science Photo Library	Still images	• A type of X-ray scans through 180° and takes a measurement every 1° • A cross-sectional picture of the brain is produced • Used to detect tumours, strokes and Alzheimer's disease	• The images are clearer than a standard X-ray • Have been used in research to identify abnormalities in brain structure related to mental health issues • The images show only the brain structure • The image quality is not as good as magnetic resonance imaging
Magnetic resonance imaging (MRI) iStock.com/Springsky	Still images	• Uses a magnetic field to produce a 3D image of the area of the body that was scanned, especially soft tissue • Images resembling 2D slices can be produced to identify tumours, strokes, cancer and traumatic brain injuries	• The 3D image is highly detailed • The magnetic field is harmless to the patient • The scanner can be claustrophobic • If the patient has anything magnetic in them (e.g. a pacemaker), an MRI cannot be used
Functional magnetic resonance imaging (fMRI) Alamy Stock Photo/Cultura Creative Ltd	Dynamic scan	• Examines the brain's activity by measuring blood oxygen levels and blood flow. Higher levels/flow indicates a more active brain area. The different colours in the scan represent different levels of activity • Used to identify epilepsy, strokes, Alzheimer's disease and traumatic brain injuries	• Produces a colour image of the brain, which provides more information about brain activity • More expensive than other scans

8.5 LEARNING CHECK

1 **Explain** what Roger Sperry's experiment showed us about the brain.

2 **Compare and contrast** an MRI and an fMRI.

Extension

3 **Investigate** the link between head injuries in sport and dementia or Parkinson's disease. Prepare a short report that outlines the effect of multiple head injuries and identify ways that these can be prevented. Ensure that you provide a reference list.

8.6 Consciousness

BY THE END OF THIS MODULE, YOU WILL BE ABLE TO:

- ✓ define 'consciousness'
- ✓ describe the different levels of consciousness
- ✓ describe how consciousness can be measured.

consciousness
awareness of internal and external stimuli

selective attention
a state of total awareness, focusing on one event

divided attention
awareness of more than one thing at a time

daydreaming
focusing on internal thoughts and processes rather than external ones

meditating
a practice in which an individual achieves a state of physical and mental relaxation that they are aware of and control by themselves

GET THINKING

List the first five things you are aware of right now. Then make a new list identifying one thing you are aware of using each of your five basic senses.

Did you notice any difference between your first and second lists? Write at least two sentences explaining why there were or were not differences.

Compare your lists with at least one other person's. Did you both have the same lists? Why or why not?

Is it possible to measure awareness? Justify your response by explaining why or why not.

What is consciousness?

Consciousness in psychology refers to our level of awareness of the internal and external stimuli that are taking place at any one time. We are constantly bombarded with information, both internal (inside our bodies) and external (in the environment around us) and it is not possible to be conscious of everything simultaneously.

People choose what they are going to pay attention to and, therefore, what it is they are going to be aware of. Right now, you are choosing to read this sentence. At the same time, you may have chosen to ignore the sound of another person talking and you are not aware of your breathing. However, did you find that, when you read this, you were suddenly aware of these things?

Our level of awareness determines the amount and type of information we are able to take in and remember. This is called the hierarchy of awareness (see Figure 8.6.1).

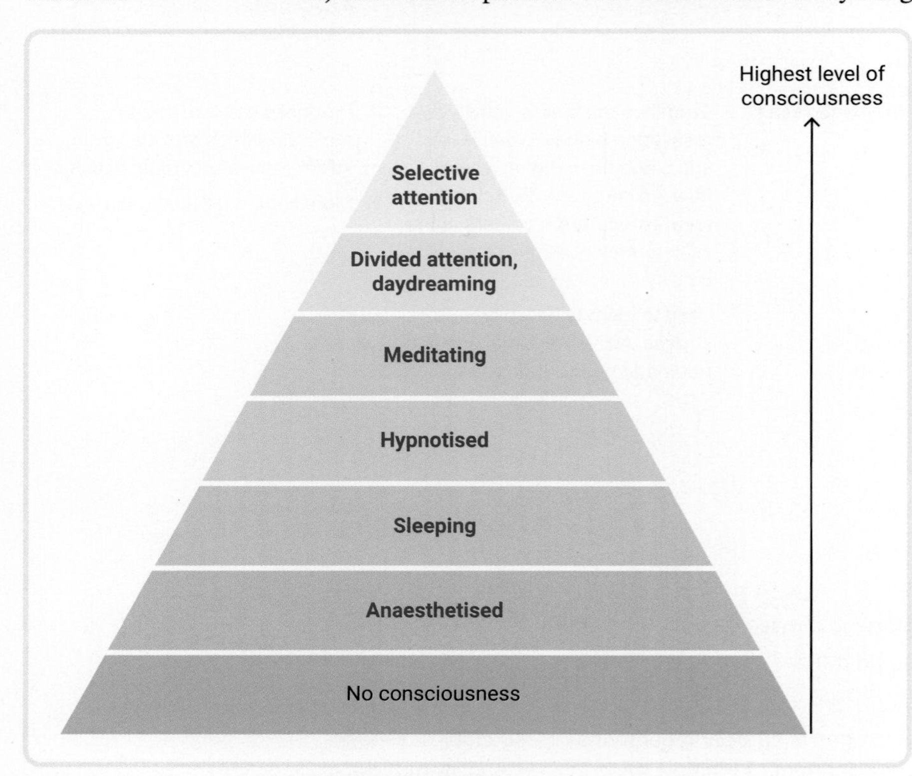

▲ **FIGURE 8.6.1** The hierarchy of awareness

Types of attention

Selective attention is when a person is completely aware of what they are focusing on, to the exclusion of all other information. This explains why a person listening to their favourite music does not hear their parent or carer telling them it's time for homework.

In 1999, researchers Christopher Chabris and Daniel Simons described an investigation now known as 'The invisible gorilla' wherein students were asked to watch a video of a basketball game and count the number of times one team caught the ball. The participants were so focused on counting the catches that they did not notice someone in a gorilla costume walking through the middle of the game (see Figure 8.6.2).

▲ **FIGURE 8.6.2** How did students watching the video not see the gorilla?

Studying while listening to music or driving while having a conversation with someone in the passenger seat are examples of divided attention. When our attention is divided, our level of awareness decreases. This is one reason why it is illegal to use mobile phones while driving.

Measuring states of consciousness

Physiological measures are most commonly used to measure states of consciousness. They are more accurate than **self-reporting measures** where a person has to report their own results, such as completing a survey, which the individual can influence and manipulate.

An electroencephalograph (EEG) measures the electrical activity in the brain, using electrodes attached to the scalp. The activity is recorded in the form of brain waves (see Figure 8.6.3).

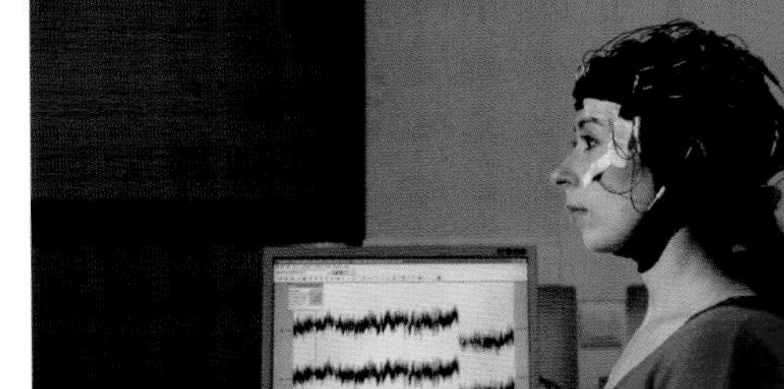

▲ **FIGURE 8.6.3** An EEG recording electrical activity in the brain. An EEG reading shows this as brain waves.

hypnotised
an artificial sleep-like state that induces a state of relaxation

sleeping
a partial or full lack of awareness, dependent on the type of sleep (e.g. light or deep sleep)

anaesthetised
a state where an individual has no pain sensation as a result of being given an anaesthetic

physiological measures
methods that record and measure a physical response

self-reporting measures
methods that record data that has been directly reported by the participant in a study or trial

Video activity
Monkey business

Interactive resource
Label: Hierarchy of awareness

The different patterns of the waves can be used to diagnose different issues such as epilepsy, brain tumours and sleep issues. Different wave patterns represent different states of consciousness. For example, when we are awake and alert, the brain waves are fast (high frequency) and small (low amplitude). These are known as beta waves (see Figure 8.6.4a). When we are in non-rapid eye movement sleep, we show slow (low frequency) and big (high amplitude) brain waves, known as delta waves (see Figure 8.6.4b).

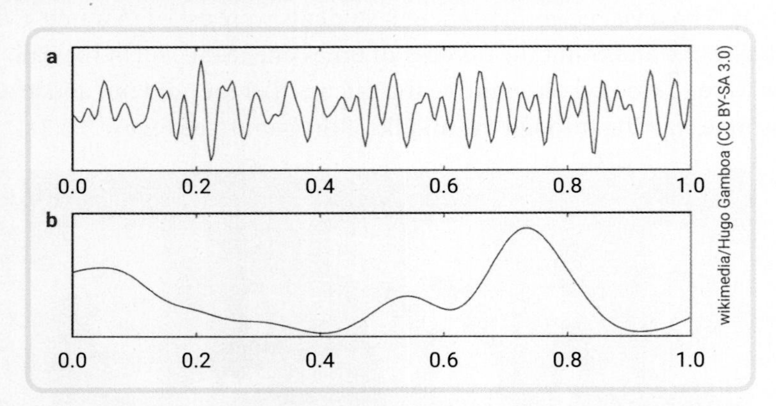

▲ **FIGURE 8.6.4** (a) Beta waves from an EEG; (b) delta waves from an EEG

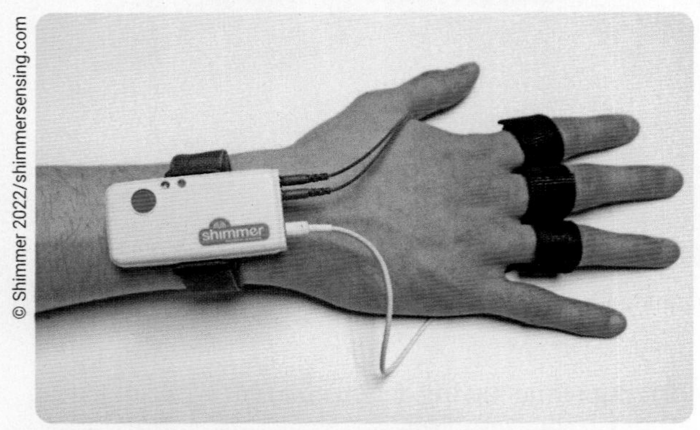

▲ **FIGURE 8.6.5** A galvanic skin response monitor attached to a human hand

The galvanic skin response measures the electrical conductivity of the skin. When we are in different states of consciousness, our sweat glands produce more or less sweat. This either increases or decreases the skin's electrical conductivity, which can be measured using a special monitor (see Figure 8.6.5).

8.6 LEARNING CHECK

1 **Define** 'consciousness'.
2 **Explain** the difference between selective attention and divided attention, using examples.
3 **State** two physiological methods used to measure consciousness.

8.7 Sensation and perception

BY THE END OF THIS MODULE, YOU WILL BE ABLE TO:

- ✓ explain the process of perception
- ✓ explain how visual illusions occur.

GET THINKING

Our five basic senses inform how we perceive the world around us. Think about the last time you smelled or heard something pleasant. Describe the smell or sound. When and where did you smell or hear it? List all the reactions to this smell or sound that you can remember. For example, did it remind you of a person, a place or an event? How do you feel when you think of this smell or sound?

Video activity
Optical illusions

From sensation to perception

All the information we get from the world around us enters through our senses. There are three steps in the process of **sensation: reception, transduction** and **transmission**. Once these processes have been completed, **perception** occurs. The three parts to perception are **selection, organisation** and **interpretation**.

Perception is about expectancy and takes place in the occipital lobe. It is affected by our previous experiences, conditions of the specific situation that we are in and other factors such as culture.

Visual illusions

Although illusions can be based on any of the senses, **visual illusions** have attracted a great deal of research and popular interest. Visual illusions are all around us. They can occur naturally or can be human made. A visual illusion occurs when we misinterpret visual information due to disease or misperception. An example is the Storseisundet Bridge in Norway, which looks very dangerous but is actually quite safe (see Figure 8.7.1).

Shutterstock.com/Manuel Compaan

▲ **FIGURE 8.7.1** This road in Norway looks very dangerous but it leads to the Storseisundet Bridge, which curves downward and is safe.

sensation
the first stage in the process of perception; in this stage, important elements of sensory stimuli are selected for further processing

reception
the detection of a stimulus by a sense organ, such as the ear receiving sound waves

transduction
the conversion of a stimulus into electrochemical energy, which is a form of energy that can be processed by the brain

transmission
the transfer of electrochemical energy to the brain to be processed

perception
the process of interpreting sensory information

selection
the retention of important features of the stimulus with discarding of unimportant features

organisation
the grouping together of selected features of the stimulus to form a whole

interpretation
the meaning given by the brain to the whole

visual illusion
a mismatch between the reality we are looking at and our perception of it

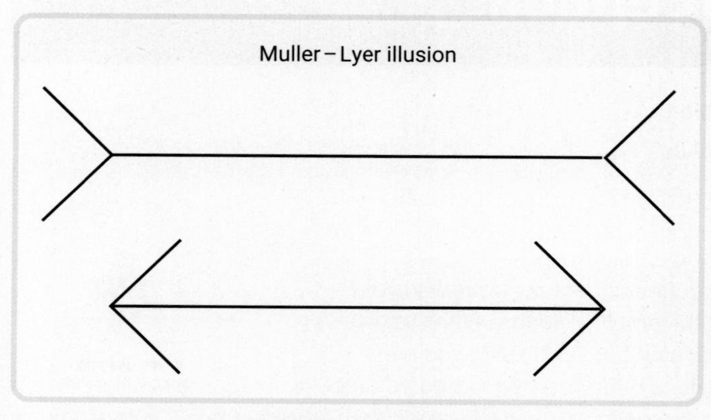

▲ **FIGURE 8.7.2** The Muller–Lyer Illusion; which line is longer?

The Muller–Lyer illusion is a famous illusion that shows two lines equal in length with inward or outward pointing tips (see Figure 8.7.2).

The illusion gives the impression that the lines are different lengths. It is thought that the eyes are drawn past the line with feather tails (the outward tips) to make the line look longer. For the line with the inward tips (arrow heads), it is thought that the eyes are drawn within the lines to give the perception that the line is shorter.

Illusions are important because they inform psychologists about how humans perceive the world and how the processes of sensation and perception take place. They show that the brain can make errors. These errors occur because our brain is always processing so much information that we create something called a **perceptual expectancy** – that is, we see what we expect to see. The illusion is created when our brain interprets what we see based on our previous experiences, rather than seeing what is actually there.

perceptual expectancy
seeing what we expect to see

8.7 LEARNING CHECK

1 **Create** a flow chart to illustrate how a loud bang would move from sensation to perception.

2 **Explain** how visual illusions occur.

3 **Describe** how a perceptual expectancy might affect the way we perceive a visual image.

Extension

4 Find three different visual illusions. For each, briefly **explain** what the eyes are doing, what the brain is doing and how the information is being processed.

5 Scenarios are used in psychology to illustrate the concepts being studied. Read the scenario below and answer the questions that follow.

A person is walking along a track in a park. A magpie swoops and hits the person's head. The next time they walk along a track, a leaf blows into their head. The person thinks they see a magpie (even though there is not one there).

a **Explain** the role of sensation and perception in this scenario.

b **Explain** the role of perceptual expectancy in the person's reaction to the leaf hitting their head.

8.8 Nature versus nurture

BY THE END OF THIS MODULE, YOU WILL BE ABLE TO:

- ✓ describe the difference between nature and nurture
- ✓ explain why studying identical twins provides information useful to the nature versus nurture debate.

GET THINKING

Are we born with a fixed level of intelligence? Spend 30 seconds forming an opinion, then write three sentences in response to this question.

Compare your answer with that of at least one other person and discuss the similarities and differences between your answers. Can you explain why you think this?

Video activity
Nature v. nurture

Like some other scientists, psychologists are interested in whether human behaviour is based on **nature** or **nurture**. While medical sciences discuss this mostly in the context of disease and medical conditions, psychologists focus on how these influence social, emotional, physical and cognitive development.

nature
inherited through our genes

nurture
environmental influences

identical twins
twins who have the same genetic sequence because they were formed from a single fertilised egg

fraternal twins
twins who are genetically similar but not identical because they were formed from two different fertilised eggs

Twin studies

Twins can be either **identical twins** or **fraternal twins** (see Figure 8.8.1). Since they are often raised in the same environment, psychologists can study the similarities and differences between twins and identify characteristics that are influenced by both genetics and environment.

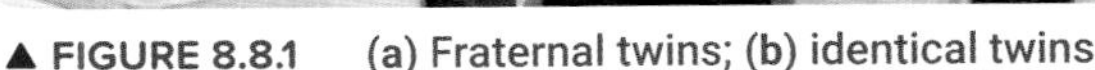

▲ **FIGURE 8.8.1** (a) Fraternal twins; (b) identical twins

Even though identical twins are genetically the same, they often develop different likes, dislikes, skills and understandings. However, there are many examples of identical twins raised apart who have striking similarities.

The 'Jim' identical twins

longitudinal study research that takes place over an extended time (i.e. many years)

The 'Jim twins' are identical twins who were separated early in life and reared apart. They became important participants in the Minnesota Twin Study, a **longitudinal study** that investigated the genetic and environmental impacts on child development in identical twins who were raised apart (see Figure 8.8.2).

Alamy Stock Photo/Ira Berger

▲ **FIGURE 8.8.2** The 'Jim twins' – identical twins who were reunited aged 39

Jim Lewis was adopted when he was 4 weeks old. Years later, when they were 39, he was reunited with his identical twin brother – Jim Springer.

The twins discovered some amazing similarities in their lives.

- Both twins were married to women named Betty and divorced from women named Linda.
- One had a son called James Alan, while the other had a son named James Allan.
- Both named their pet dog 'Toy'.
- Both had worked or studied in law-enforcement.
- Each did well in mathematics but struggled with spelling.
- Each did carpentry and mechanical drawing.
- Both twins suffered similar health issues, such as both having tension headaches at 18, and having similar weight gain at the same time.

At first glance, the Jim twins' story seems almost impossible to believe. However, the Minnesota Twin Study is a peer-reviewed study (meaning it was approved by other scientists) that has provided much valid and reliable data for ongoing research. The study illustrates how properly constructed scientific investigations can produce interesting, surprising and useful information.

The Minnesota study found that identical twins raised together or apart were far more similar on many psychological tests than non-identical twins and siblings raised in the same family. For example, the Jim twins scored very similar scores on a series of personality and intelligence tests and even had similar handwriting.

Children raised by animals

Some unfortunate children suffer such extreme neglect that they end up being sheltered and cared for by animals. One famous example is Oxana Malaya, a Ukrainian girl who at the age of three was left outside with dogs by her parents. She crawled into the dogs' shelter and lived as a dog until she was rescued in 1991 at the age of eight.

When discovered, Oxana could not walk upright, had no human language except 'yes' and 'no' and had no knowledge of human social interactions. Her case showed that much of what humans learn is obtained from the environment in which we live.

Many developmental psychologists regard walking upright as an important milestone in child development that is preceded by crawling. Oxana's case suggests that, unless the environment provides the correct stimulus, many skills that humans consider 'natural' are, in fact, learned.

Psychologists recognise the importance of both nature and nurture for healthy human development. Socialisation refers to the acquisition of social skills that assist us to participate effectively in society. Ethical considerations mean it is extremely difficult to research the effects of isolation on child social and emotional development. The children raised by animals provide examples of the effects of isolation on humans that cannot be studied in other ways, such as in a laboratory.

8.8 LEARNING CHECK

1 Can identical twins be different sexes? **Explain** your answer.

2 **Explain** what the case of Oxana Malaya shows about the role of nurture in child development.

Extension

3 **Discuss** the contribution of twin studies to our understanding about nature versus nurture.

4 **Research** another example of a child raised by animals and **explain** what this shows about the role of nature and nurture in human development. Examples you could research: Genie Wild Child, Amala and Kamala, Marina Chapman, Prava the Bird Boy.

5 Ask members of your family if they are willing participate in an observational study to find family similarities. Conduct your study by following the steps below.

   a Develop a list of family features (e.g. height, weight, hair colour and eye colour) that you could collect data about.

   b Create a list of five questions that you will ask each person in the family (these could be for interview or as a survey). For example, you might want to find out about their likes/dislikes, talents/abilities, habits, and if they consider themselves extraverted/introverted.

   c Collect your data.

   d Identify the information that is common across family members.

   e Write a description of your findings.

Glossary

absolute dating determining the age of a fossil in years

absolute magnitude the brightness of an astronomical object, as though all objects were 10 parsecs from Earth

absolute zero the lowest possible temperature (0 K or −273.15°C); the point at which there is no particle movement and no energy

absorption spectrum the dark lines characteristic of an element or compound in a continuous spectrum

acceleration the rate of change of speed

acid a chemical that can donate hydrogen ions and undergo chemical reactions with metals and bases

acid rain rain that is acidic due to chemicals dissolved in water in the atmosphere

activation energy the minimum energy particles need to have for a successful collision to occur

activity series a list of metals ranked by their chemical reactivity

agent the body applying a force on another body

allele an alternative form of a gene

allele frequency the measure of how common an allele is in a population

allopatric speciation speciation due to a barrier

allotrope different forms of the same element with different physical and chemical properties

amino acid the building block of proteins

anaesthetised a state where an individual has no pain sensation as a result of being given and anaesthetic

analogous structures structures in different species that are anatomically different but have the same or similar function

anaphase the third phase of cell division, when the chromosomes are pulled apart

aneuploidy having additional or missing chromosomes

anion an ion with a negative charge

anisotropies small variations in temperature in the universe due to cosmic microwave background radiation that are not consistent in all directions

anticodon complementary to the codon; found on the tRNA molecule

apoptosis programmed cell death

apparent magnitude the brightness of an astronomical object, as seen from Earth

artificial selection the process whereby humans breed organisms for desired traits

asteroid a rocky object smaller than a planet that orbits the Sun

astronomical a term to relate ideas or objects to astronomy

astronomical unit (AU) approximately 150 million kilometres, equivalent to the distance between the centre of Earth and the centre of the Sun

atmosphere the gaseous layer surrounding Earth and retained by Earth's gravity

atom a particle of matter made up of protons, neutrons and electrons

atomic number (*Z*) the number of protons in the nucleus; the same for every atom of the same element

autonomic nervous system the part of the PNS that controls voluntary body functions

autosomal the inheritance of genes on the autosome chromosomes

autosomal inheritance inherited genes that are located on the body chromosomes

autosomes all the chromosomes in a cell, except for the sex chromosomes

Avogadro's number the number of particles in one mole of chemical substance; 6.02×10^{23}

axon a long, thin fibre that carries electrical impulses

base a group of chemicals that react to neutralise acids; may include metal hydroxides, metal carbonates or metal oxides

Big Bang theory the generally accepted theory for the formation of the universe

biochemical evidence evidence of evolution based on the fact the same enzymes are found in the cells of most organisms

biodiversity the variety of all living organisms in an ecosystem

bioinformatics the use of technology to collect and analyse biological data, such as the sequence of amino acids or proteins

biosphere all the parts of Earth and the atmosphere that support life, including all living things

biotechnology a branch of science that manipulates living organisms, systems or DNA to produce products that are useful to humans

black dwarf a star at the end of its life cycle that no longer emits light

black hole a cosmic body of extremely intense gravity from which nothing, not even light, can escape

blueshift the shift of spectra from local stars moving towards Earth towards the blue end of the spectrum

body an object, person or thing that is made of matter and has mass

bonding pair a pair of electrons shared between two atoms, where one electron comes from each atom

cancer uncontrolled cell division resulting in a growth or tumour

carbon footprint the amount of carbon dioxide emitted by a person, an organisation or an event

carbon sequestration the process of capturing and storing atmospheric carbon dioxide

carbon sink something that naturally stores large quantities of carbon dioxide

carcinogen an agent that increases the likelihood of developing cancer

carrier a female with a recessive allele on one of her X chromosomes

case study a detailed examination of specific individuals and/or groups

catalyst a substance that increases the rate of a chemical reaction by lowering the activation energy of the reaction without itself being changed

cation an ion with a positive charge

cell cycle the series of events that takes place in a cell as it grows and divides

central nervous system (CNS) the part of the nervous system comprising the brain and spinal cord

centrioles organelles in the cell that produce spindle fibres

centromere the point on a chromosome where the two chromatids are joined

cerebral cortex the outer layer of the brain, consisting of two hemispheres

chemical bonding the joining of atoms by transfer, loss or sharing of electrons to achieve a stable electron configuration

chemical bonds the forces of attraction that hold atoms together in compounds or molecules

chemical equations word equations or balanced equations that are used to represent the substances in chemical reactions

chiasma the point on the chromatids where crossing over occurs

chromatid one half of a duplicated chromosome

chromatin unpackaged DNA found within the nucleus of a non-dividing cell

chromosomal mutation a change in the number of chromosomes or an arm of a chromosome

chromosome a thread-like structure found in the cell, composed of DNA

climate the average weather conditions in a particular region over an extended period

climate change a change in global or regional climate patterns

cluster a group of stars or galaxies

co-dominant when alleles for a particular gene are neither dominant nor recessive and both are expressed in the phenotype but are not blended

codon a sequence of three nucleotides on mRNA that code for a specific amino acid

collision theory the theory that a reaction will only occur if particles come into contact with sufficient energy and at the correct orientation to cause a successful collision

comet an object made of rock and ice that orbits the Sun at regular intervals

common ancestor the ancestor that two or more descendants have in common

comparative anatomy the study of body structures of organisms to show the adaptive changes made from a common ancestor

compound two or more different elements joined by a chemical bond

consciousness awareness of internal and external stimuli

constellation a group of stars that looks like a picture or pattern

continental ice sheet an extensive sheet of permanent ice covering a large area of the land surface

continuous light spectrum a spectrum in which all radiation 'colours' are emitted

control group the group that is not exposed to the independent variable

convection current a flow of materials (such as air, water or molten rock) and energy caused by differences in densities due to temperature differences

COP26 the 26th Conference of the Parties to the United Nations Framework Convention on Climate Change

Coriolis effect the apparent deflection of large masses of air and water due to Earth's rotation on its axis

corpus callosum the thick band of nerve fibres that connects the two hemispheres of the brain and allows messages to be passed between them

corrosion a chemical process that often happens to metals exposed to oxygen

cosmic microwave background radiation remnant radiation from the early stages of the universe in the microwave wavelength; emitted approximately 300 000 years after the Big Bang

cosmologist a scientist who studies the universe

covalent bond the electrostatic force of attraction between a shared pair of electrons and the nuclei of the atoms sharing the electrons

covalent molecule a distinct structure formed when two or more non-metal atoms join through covalent bonding

covalent network a structure formed when non-metal atoms join in a covalently bonded lattice

criminology the scientific study of crime and criminal behaviour

crossing over the exchange of genetic material between non-sister chromatids on homologous chromosomes

cyanobacteria colonies of blue-green algae capable of photosynthesis

cytokinesis the division of the cytoplasm after mitosis

D

dark energy a kind of negative pressure that is thought to be responsible for the accelerating expansion of the universe

dark matter matter that cannot be seen but is known to exist because its effect on objects can be observed

daughter cells the two cells that are produced as a result of cell division

daydreaming focusing on internal thoughts and processes rather than external ones

deceleration negative acceleration; when speed is decreasing over time

decomposition the breaking down of chemical substances into two or more smaller products

decomposition reactions reactions involving the breakdown of compounds into simpler elements or compounds

deep ocean currents the water movement occurring below a depth of 400 m caused by changes in water temperature and salinity

deforestation the removal of large areas of forest to enable the land to be used for other purposes

deletion the removal of a nucleotide from a DNA sequence

delocalised electrons electrons in a metallic lattice that do not belong to any particular atom

dendrite the branching network at the end of a neuron that receives information from other neurons

deoxyribonucleic acid the molecule that determines the characteristics of most living things

deoxyribose sugar one of the components of a nucleotide in DNA

dependent variable what you are measuring

diploid the full complement of DNA, represented as $2n$

displacement the distance and direction that a body's position is relative to the starting point; measured in metres (m) with a direction

displacement reactions reactions involving the replacement of atoms in compounds

distance the length of the pathway taken by a moving body

divided attention awareness of more than one thing at a time

DNA ligase an enzyme that facilitates the repair of the sugar–phosphate backbone; it 'fills the gaps' in the DNA

DNA polymerase the group of enzymes responsible for DNA replication by ensuring nucleotides are added in the complementary sequence to the existing strand of DNA

DNA profile an image used to determine an individual's DNA characteristics

DNA replication the process whereby DNA makes a copy of itself

dominant an inheritance that identifies the dominant allele in a genotype

dominant allele the allele that will be expressed in the phenotype

double displacement reaction a reaction in which atoms of the reactants displace each other from their compounds

double helix the shape of DNA, similar to a twisted ladder

dry climate a climate zone where seasonal evaporation exceeds seasonal precipitation

ductile can be stretched into a wire

Earth system Earth as a complex whole made up of the four interacting spheres: atmosphere, biosphere, hydrosphere and geosphere

elastic potential energy the energy stored in a body that is stretched or compressed out of its natural rest position

electromagnetic spectrum the range of electromagnetic waves sequenced by energy, frequency and wavelength from gamma rays to radio waves

electron a negatively charged particle in the atom; it moves in space around the nucleus

electron configuration the arrangement of electrons in electron shells in an atom

electron dot diagram a representation of the electron arrangement in the valence shells of atoms in covalent molecules

electron shell a level around a nucleus containing electrons of the same energy

electronegativity the ability of an atom to attract shared electrons when forming a chemical bond

electrostatic bond the force of attraction between positive and negative particles

element matter consisting of atoms with the same number of protons in their nucleus

embryology the study of the early stages of development

emission spectrum the pattern of lines formed from the movement of electrons between energy shells

energy the ability to do work measured in joules (J)

energy shells regions of space around a nucleus that contain electrons of the same energy

energy transfer the movement of energy from particle to particle, body to body or space to space without changing the type of energy

energy transformation a movement of energy where energy converts from one form into another during a process or an action

enhanced greenhouse effect strengthening of the natural greenhouse effect due to an increase in greenhouse gases caused by human activities

equation of motion a relationship between motion variables such as displacement, velocity, acceleration and time that can be used for moving bodies

evolution the gradual change in characteristics of a species over many generations resulting in new and different species

excited the state of an atom or electron when it absorbs energy

exon a coding section of DNA in a specific gene sequence

experimental group the group that is exposed to the independent variable

extreme weather event unexpected, unusual, severe or unseasonal weather

fields specific areas of study within psychology

filial first set of offspring from a cross between parents

finite resource a limited resource

first filial generation the first set of offspring from a parent cross

force a push, a pull or a twist

fossil any preserved trace of a once-living organism

fossil record the list of fossil finds, their classification and their age

fraternal twins twins who are genetically similar but not identical because they were formed from two different fertilised eggs

free body diagram a diagram that shows the forces acting on a single body as arrows

frequency the number of repeated cycles per unit of time; expressed in hertz (Hz)

frontal lobe the front part of the cerebral cortex, associated with voluntary movement and higher-order mental abilities such as planning and thinking

galaxy a collection of gas, dust and billions of stars and their solar systems, all held together by gravity

gametes the sex cells of a sexually reproducing organism

gel electrophoresis a technology used to separate DNA by size to produce a DNA profile

gene a section of DNA that codes for a protein or a certain trait

gene mutation a change in the DNA sequence in one or more genes

gene pool the total amount of genetic material available in a population

gene therapy the introduction of functional genes into cells to replace defective or missing genes to treat genetic disorders

generation the time taken for one individual to produce offspring

genetic code the sequence of nitrogen-rich bases in an organism's DNA

genetic disorder disease symptoms produced when a DNA sequence is different from normal

genetic drift the change in allele frequency seen in small populations due to chance events from one generation to the next

genetic engineering the deliberate modification of an organism's DNA

genetically modified organism an organism whose genes have been altered in the laboratory to produce a desired trait

geneticists scientists who study genetics and inheritance

genome an organism's full set of DNA

genotype the combination of alleles for a specific gene

geosphere the rocks, minerals and landforms of the surface and interior of Earth

germline cells the cells that form the ovum and the sperm

germline mutations mutations that occur in sperm or ova

global convective cells three large atmospheric pressure cells that occur in both the northern and southern hemispheres

gonads the sex organs of an organism; where meiosis occurs

gradient a measure of the slope of a straight line on a graph

gravitational field the region where the pull of Earth's gravity is experienced

gravitational potential energy the energy stored in a body due to its position within a gravitational field

gravitational waves ripples in space-time produced when a massive body is accelerated or otherwise disturbed

gravity the force of attraction between Earth and objects within its gravitational field

greenhouse effect a natural process that traps energy within the atmosphere, raising Earth's surface temperature

greenhouse gases gases in the atmosphere that can trap heat and affect global surface temperatures and other aspects of climate

groups the vertical columns on the periodic table

halogens the name for the group 17 elements on the periodic table

haploid having one copy of each chromosome, represented as n

helicase the enzyme that unwinds and unzips the DNA molecule

hemispheres the two halves of the cerebral cortex, connected by the corpus callosum

heterozygous having two different alleles on homologous chromosomes

homologous carrying the same genes for characteristics at the same locations on a chromosome

homologous pairs maternal and paternal chromosomes with genes found at the same location

homologous structures body parts that can be found in a range of organisms that have a similar structure but different functions

homozygous dominant having two of the same dominant alleles on homologous chromosomes

homozygous recessive having two of the same recessive alleles on homologous chromosomes

Hubble's law the law that states that the observed velocity of a receding galaxy is proportional to the distance from the observer

hydrogen bond a type of attraction between molecules

hydrosphere the part of Earth containing all forms of water

hypnotised an artificial sleep-like state that induces a state of relaxation

hypothesise to make a testable scientific prediction

identical twins twins who have the same genetic sequence because they were formed from a single fertilised egg

incomplete dominance when alleles for a gene are neither dominant nor recessive and are expressed as a blend in the phenotype

independent assortment the random lining up of maternal and paternal chromosomes during metaphase I

independent variable what you are changing

index fossil a fossil that can be used to compare the relative age of rock strata from different locations

inertia the property of a body that resists changes to its motion

inflation the initial rapid expansion of space-time just after the Big Bang

insertion the addition of a nucleotide into a DNA sequence

intermolecular forces forces of attraction between covalent molecules

interneuron a neuron that transmits impulses from one neuron to another

interphase the resting phase of the cell cycle

interpretation the meaning given by the brain to the whole

intron a non-coding section of DNA in a specific gene sequence

inversion mutation a chromosomal mutation in which part of a chromosome is reversed end-to-end

ion a charged particle formed when an atom loses or gains valence electrons

ionic bond the force of attraction between positive metal ions and negative non-metal ions

ionic bonding bonding between metal and non-metal atoms involving a transfer of electrons

ionic formula the chemical representation of an ionic substance showing the number and type of atoms present

ionic lattice an organised structure of alternating positive metal ions and negative non-metal ions

ionic salt a chemical containing a metal ion and a non-metal ion

ionisation energy the energy required to remove an electron from the valence shell

isolation a mechanism or barrier to separate breeding populations

isotopes atoms of an element with the same number of protons but a different number of neutrons

karyotype a picture of an organism's complete set of chromosomes

kinetic energy the energy or ability to do work contained in a body that has mass, due to its current state of movement

Köppen climate classification a system used since 1900 that classifies climate zones based on temperature, amount and type of precipitation, and vegetation

land clearing the removal of natural vegetation and habitats, such as forests

law of conservation of energy the fundamental law of physics; when energy is transferred or transformed, the total amount of energy remains the same

law of conservation of mass the total mass of reactants and products in a chemical reaction is equal

light-year an astronomical unit of distance equivalent to the distance that light travels in 1 year; approximately 9.4607×10^{12} km

localisation the idea that different parts of the brain have different functions

lone pair a pair of electrons in the valence shell of the central atom but not involved in bonding

longitudinal study research that takes place over an extended time (i.e. many years)

lustrous shiny when cut or polished

magnitude size or extent

magnitude scale a scale that shows the brightness of an astronomical body; negative numbers are very bright, and positive numbers are less bright

main sequence stars that fuse hydrogen to helium

malleable can be beaten into different shapes

mass number (*A*) the total number of protons and neutrons in the nucleus of an atom

master mix a premixed solution used in PCR techniques to make copies of DNA

mechanical energy the energy associated with being able to cause the movement of a body; includes kinetic energy, gravitational potential energy and elastic potential energy

meditating a practice in which an individual achieves a state of physical and mental relaxation that they are aware of and control by themselves

meiosis cell division producing cells that will specialise into gametes with half the number of chromosomes of the parent cell

mental illness any health condition that affects a person's thoughts, mood and behaviour, and causes distress or dysfunction in daily life

messenger RNA (mRNA) RNA that carries the copy of the DNA code out of the nucleus

metal displacement reaction a chemical reaction where a more reactive metal displaces a less reactive metal from a solution

metal hydroxide an ionic salt containing a metal ion and a hydroxide ion

metallic bond the force of attraction between the metal cations and delocalised electrons

metallic lattice the organised structure formed with rows of metal cations surrounded by delocalised electrons

metaphase the second phase of cell division, when chromosomes line up in the centre of the cell

mitigation reducing the severity of an impact or event

mitosis cell division for the growth, replacement and repair of somatic cells

mode of inheritance the manner in which a genetic trait or disorder is passed from one generation to the next

molar mass the mass in grams of one mole of a chemical substance

mole the quantity of a chemical substance that contains 6.02×10^{23} particles

molecule a group of atoms bonded together

monohybrid cross a cross between two organisms with two alleles at one gene location

motor neuron a neuron that transmits movement instructions from the CNS to the muscles in the body

molecule a group of atoms bonded together

mutagen an agent that increases the likelihood of mutation

mutation a spontaneous and permanent change to a DNA sequence

myelin sheath a protective coat around an axon that increases the speed of nerve impulses

natural selection the process in which an environmental factor acts on a population, resulting in some individuals being more likely to survive and reproduce

nature inherited through our genes

nebula a cloud of cosmic gas and dust

net force the overall force acting on a body; measured in newtons (N)

net zero greenhouse emissions when emissions of greenhouse gas do not exceed the amount of gases absorbed or stored

neuron a specialised type of cell in the nervous system that transmits a signal between the CNS and PNS

neurotransmitter a chemical messenger

neutralisation the chemical reaction between an acid and a base to produce a salt and water

neutron an uncharged particle found in the nucleus of an atom

neutron star a small, super-dense star that has spent all its fuel and is at the end of its burning life

nitrogenous base a base that contains nitrogen: adenine (A); thymine (T); cytosine (C) and guanine (G)

noble gases the name for the group 18 elements on the periodic table; they have 'full' valence shells

non-disjunction the incorrect separation of chromosomes or chromatids at the centromere, resulting in gametes with an unusual chromosome number

nuclear fusion the process of combining two nuclei to make a third nucleus

nucleotide the building block of DNA

nucleus the structure in the centre of an atom containing protons and neutrons

nurture environmental influences

occipital lobe the back part of the cerebral cortex, associated with vision

open system a system in which energy and matter can be exchanged with their surroundings

ore a compound containing a metal and other elements that is mined from the ground and processed to extract the metal

organisation the grouping together of selected features to form a whole

origin a defined point that indicates the starting point or a displacement of zero

parasympathetic nervous system the part of the autonomic nervous system that calms the body and restores it to its normal state following the activation of a sympathetic nervous system response

parietal lobe the lobe on the top of the cerebral cortex, associated with bodily sensations such as touch, temperature and spatial awareness

parsec an astronomical unit of distance equivalent to 3.26 light-years

pedigree chart a diagram showing patterns of inheritance over generations; also called a family tree

peer-reviewed checked for scientific accuracy by other scientists in the same field

perception the process of interpreting sensory information

perceptual expectancy seeing what we expect to see

periodic table a method of arranging elements by increasing atomic number

periods the horizontal rows on the periodic table

peripheral nervous system (PNS) the part of the nervous system that carries messages to and from the CNS

permafrost permanently frozen soil, sediment or rock

pH a measure of the acid or base levels in a solution, measured between 0 and 14

phenotype the observable characteristics of the genotype

phosphate group one of the components of a nucleotide in DNA

photon a form of elementary energy particle

photosynthetic able to photosynthesise, converting carbon dioxide and water into sugar and oxygen

phylogenetic tree a diagram representing lines of evolutionary descent from a common ancestor

physiological measures methods that record and measure a physical response

planetary nebula a ring-shaped nebula formed by an expanding shell of gas around an ageing star

planetoid a small body that is like a planet but does not meet specific criteria, such as minimum diameter

plasmid a small circular fragment of DNA in the cytoplasm of bacteria

polar climate a climate zone with ice, snow and temperatures too low to support most vegetation

polyatomic ion an ion that contains more than one non-metal element and has an overall charge

polymer a large chemical made in a synthesis reaction from repeating, simpler chemicals called monomers

polymerase chain reaction (PCR) a technology used to make many copies of DNA

population a group of individuals of the same species living in the same place at the same time; the target group that is being studied

position the location of a body

potential energy the energy stored in a body that can be released in some way

precipitate a solid formed from certain combinations of positive and negative ions

precipitation liquid or solid water in the form of rain, snow, sleet or hail

products chemical substances that form in chemical reactions

progeny offspring

prophase the first phase of cell division, when chromosomes duplicate and condense

proportion the relative abundance of elements dispersed throughout the universe

protein a long chain of amino acids; essential in all living organisms

proton a particle in the nucleus of an atom with a positive charge

protostar an object in the process of forming a star, before nuclear fusion begins

pseudoscience an idea or practice that claims to be scientific but does not use the scientific method

psychiatrist a medical doctor who specialises in diagnosing and treating mental illness

psychoanalyst a qualified professional who treats patients using psychoanalysis

psychologist a qualified professional who works with human thoughts, emotions and behaviours in a variety of ways

psychology the scientific study of human thoughts, emotions and behaviours

pulsar a highly magnetised rotating neutron star that emits pulses of radiation at regular intervals

purebred the same alleles for a given gene, see homozygous dominant/recessive

quark a type of elementary particle that is the fundamental component of matter

radiation a type of energy generally related to the electromagnetic spectrum

radiometric dating a dating method that measures the decay of radioactive isotopes to determine the age of fossils

random assortment the way chromosomes line up during metaphase I, which results in random combinations of genes

rate of reaction a measurement of how fast a reaction is proceeding

reactants chemical substances that, when added together, react to form products in a chemical reaction

receiver the body receiving a force from another body

reception the detection of a stimulus by a sense organ, such as the ear receiving sound waves

recessive an inheritance that identifies the recessive allele in a genotype

recessive allele the allele that is masked by a dominant allele and is only expressed in the homozygote

recombinant DNA DNA that has been manipulated by combining DNA from other organisms to produce a transgenic gene

red giant a dying star in the final stages of its life cycle

red supergiant similar to a red giant but much bigger; the biggest stars in the universe

redshift the shift of spectra from local stars moving away from Earth towards the red end of the spectrum

relative dating determining if a fossil or rock is older or younger than that of another

relative humidity the amount of moisture that air holds compared with the amount it could hold if saturated at a given temperature

rest the state of being stationary, having a speed of 0 m/s

restriction enzymes special enzymes that cut DNA at specific recognition sites; isolated from bacteria

ribonucleic acid (RNA) a single-stranded nucleic acid

ribose sugar a five-carbon sugar that is a component of RNA nucleotides

ribosomal RNA (rRNA) RNA that forms the ribosomal organelle

RNA polymerase a group of enzymes responsible for the formation of mRNA during transcription

saline containing salt, salty

salinity the concentration of dissolved salt

scalar quantity a measurement that has a magnitude only and no direction

scientific based on systematic measurement and analysis of observable evidence

second filial generation the set of offspring from the first filial parent cross

selection the retention of important features of the stimulus with discarding of unimportant features

selection pressure the effect the selective agent has on the population

selective agent the environmental factor acting on the population

selective attention a state of total awareness, focusing on one event

self-reporting measures methods that record data that has been directly reported by the participant in a study or trial

sensation the first stage in the process of perception; in this stage, important elements of sensory stimuli are selected for further processing

sensory neuron neurons that transmit electrical impulses from the sensory receptors to the CNS

sex chromosomes a pair of chromosomes that determine the sex of an individual

sex-linked the inheritance of characteristics found only on the X and Y (sex) chromosomes

singularity a point where density and matter are infinite

sleeping a partial or full lack of awareness, dependent on the type of sleep (e.g. light or deep sleep)

social worker an allied health professional who helps people to improve wellbeing and supports them in crises by connecting them to services and communities

somatic relating to the cells that make up the body other than the reproductive cells

somatic nervous system the part of the PNS that controls voluntary muscles

space-time the interconnected nature of the dimensions of space and time

speciation a process in which two groups become so genetically different they can no longer breed with each other under natural conditions to produce fertile offspring

species a group of organisms capable of reproducing under natural conditions to produce fertile offspring

spectroscope an instrument used to split light into its component colours

spectrum the different bands of colour that are visible when white light is refracted through a prism, as seen in rainbows

speed a measure of how much distance a body covers per unit time, typically metres per second (m/s)

spindle fibres protein structures that pull apart the chromosomes during cell division

staggered cut a cut from a restriction enzyme that results in 'sticky ends', overhanging unpaired nucleotides

state of consciousness the level of awareness that can be measured

stationary keeping a constant position or not moving in any direction; at rest

storm surge a brief increase in sea levels due to the combined effect of storms, low air pressure, strong winds and tides

stratigraphy comparing strata or layers of rock to determine the relative age of fossils

structural formula a representation of a chemical structure showing covalent bonds as lines

struggle for existence the competition between individuals for required resources such as food, water or space

subatomic particle a particle found inside an atom, such as a proton, a neutron or an electron

substitution the swapping of a nucleotide within a DNA sequence

successful collision a collision that results in products forming

supercluster a large group made up of clusters of galaxies

supernova a massive explosion caused by a massive star suddenly collapsing; new atomic nuclei are formed

surface temperature a measure of the relative hotness or coldness of Earth's surface

survival of the fittest the idea that individuals with the best-suited characteristics will survive, reproduce and pass their traits on to the next generation

sympathetic nervous system the part of the autonomic nervous system that is activated during a fight–flight–freeze situation

sympatric speciation speciation due to reproductive isolation

synapse the small gap between the axon terminals of one neuron and the dendrites of the next

synthesis reactions reactions in which two or more elements and compounds combine to form a more complex substance

telophase the fourth and final phase of cell division, when the nucleus re-forms and the chromosomes unravel

9780170463041

temperate climate a climate zone with moderate temperature and precipitation that exists between the extremes of tropical climates and polar climates

temporal lobes the lobes on either side of the cerebral cortex, associated with auditory information and speech

theory a framework for explaining different scientific concepts and systems

thermal cycling the repeated process of heating and cooling required in PCR

thermal expansion an increase in the volume of materials as they get hotter

thrust a force that makes an object move in the opposite direction as a result of expelling mass or fuel

transcription the process of copying DNA to make a new molecule of mRNA

transduction the conversion of a stimulus into electrochemical energy which is a form of energy that can be processed by the brain

transfer RNA (tRNA) RNA that carries an anticodon and an amino acid

transgenic organism an organism that contains DNA sequences from an unrelated organism that have been artificially introduced

translation the process of building a protein based on mRNA instructions at the ribosome

translocation the result when part of a chromosome detaches and reattaches to a different chromosome

transmission the transfer of electrochemical energy to the brain to be processed

ubiquitous proteins proteins that are found in nearly all organisms and carry out the same function

unconscious completely lacking awareness

upwelling the movement of water from the depths of the ocean to the surface, bringing nutrients and carbon dioxide with it

uracil a nitrogen-rich base found only in RNA; a complement to adenine

vacuum a space empty of all matter

valence electrons electrons in the highest electron shell of an atom

valence shell the highest energy shell of an atom that contains electrons

variation a difference in characteristics due to different genes

vector quantity a measurement that has both a magnitude and a direction

velocity the rate of change of displacement with time, a vector quantity for speed; measured in metres per second (m/s) that includes a direction

vestigial organs organs that are retained in a species despite no longer being functional as they were in the ancestral species

visual illusion a mismatch between the reality we are looking at and our perception of it

weather the conditions in the lower part of the atmosphere at a given time and place

weight force the force of gravity pulling an object towards Earth

white dwarf a dim star that has exhausted its fuel and is at the end of its burning stage

zygote a diploid ($2n$) cell resulting from the joining of two haploid gametes

Additional credits

Chapter 1

- **p. 4 (Figure 1.1.1 (b)):** The Ava Helen and Linus Pauling Papers (MSS Pauling), Oregon State University Special Collections and Archives Research Center, Corvallis, Oregon.
- **p. 7 (Figure 1.2.2):** Shutterstock.com/Olga Danylenko
- **p. 40 (Figure 1.11.2):** The Ava Helen and Linus Pauling Papers (MSS Pauling), Oregon State University Special Collections and Archives Research Center, Corvallis, Oregon.

Chapter 2

- **p. 53 (Figure 2.1.2):** Source: Cornell, B. 2016. Random Assortment. [ONLINE] Available at: http://ib.bioninja.com.au. Accessed October 2022.
- **p. 59 (Figure 2.3.1):** Hug, L., Baker, B., Anantharaman, K. et al. A new view of the tree of life. Nat Microbiol 1, 16048 (2016). https://doi.org/10.1038/nmicrobiol.2016.48 (CC BY 4.0)
- **p. 66 (Figure 2.5.1):** Adapted from https://scholarworks.umass.edu/cgi/viewcontent.cgi?filename=1&article=2328&context=dissertations_2&type=additional. Photos: Horse Shutterstock.com/tashh1601; Donkey Shutterstock.com/photomaster; Mule Shutterstock.com/seksan wangjaisuk
- **p. 74 (Table 2.7.1):** Top row: Alamy Stock Photo/VPC Travel Photo; Second row: Alamy Stock Photo/blickwinkel; Third row: Shutterstock.com/olpo; Bottom row: © 2016, Masao et al. from eLife, New footprints from Laetoli (Tanzania) provide evidence for marked body size variation in early hominins (CC BY 4.0)

Chapter 3

- **p. 101 (Figure 3.0.1):** (b) Shutterstock.com/Sanit Fuangnakhon; (c) iStock.com/sergioph
- **p. 139** Bridge: Shutterstock.com/ChameleonsEye; Bag: Shutterstock.com/Sanit Fuangnakhon; Wire: iStock.com/sergioph

Chapter 4

- **p. 148 (Figure 4.3.2):** Top row, left to right: Shutterstock.com/Karol Kozlowski; Shutterstock.com/Vile_82; Shutterstock.com/woe. Bottom row, left to right: Shutterstock.com/Aksenenko Olga; Shutterstock.com/Lorena Fernandez; Shutterstock.com/Anastasia_Fisechko

Chapter 5

- **p. 218** Shutterstock.com/Designua

Chapter 6

- **p. 254 (Figure 6.9.2):** NASA Earth Observatory image by Lauren Dauphin, using VIIRS data from NASA EOSDIS/LANCE and GIBS/Worldview, and the Suomi National Polarorbiting Partnership.
- **p. 260 (Figure 6.11.4):** NOAA Climate.gov Data:NCEI

Index

C

D

E

I

N

O

P

Q

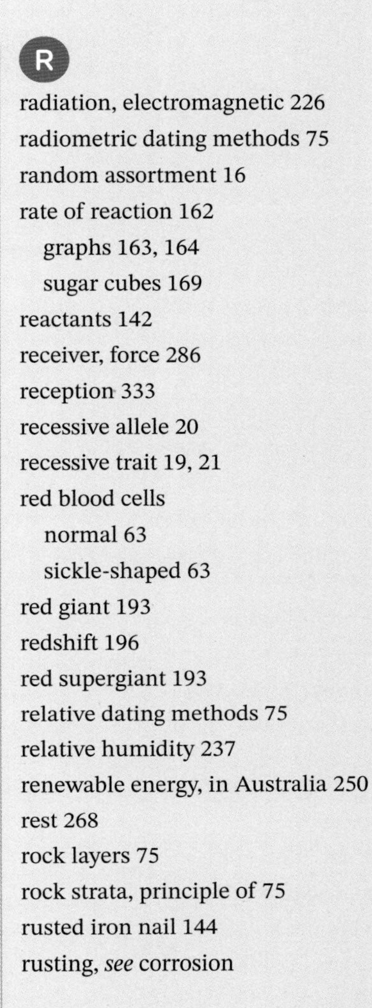
R

S

T

U